Facing
Facts

Facing Facts

Realism in American Thought
and Culture, 1850–1920

David E. Shi

New York   Oxford
OXFORD UNIVERSITY PRESS
1995

Oxford University Press

Oxford   New York   Toronto
Delhi   Bombay   Calcutta   Madras   Karachi
Kuala Lumpur   Singapore   Hong Kong   Tokyo
Nairobi   Dar es Salaam   Cape Town
Melbourne   Auckland

and associated companies in
Berlin   Ibadan

Copyright © 1995 by David E. Shi

Published by Oxford University Press, Inc.,
200 Madison Avenue, New York, New York 10016

Library of Congress Cataloging-in-Publication Data
Shi, David E.
Facing facts : realism in American thought and culture, 1850–1920
/ David E. Shi.
p. cm.   Includes bibliographical references and index.
ISBN 0-19-503892-4
1. United States—Intellectual life—1865–1918.
2. United States—Intellectual life—1783–1865.   3. Realism.
4. United States—Civilization—1865–1918.
5. United States—Civilization—1783–1865.   I. Title.
E169.1.S5558   1994
973—dc20   93-33716

1 3 5 7 9 8 6 4 2

Printed in the United States of America
on acid free paper

For Susan

PREFACE

Some books are written out of a sense of inspired urgency; others reflect new discoveries, theories, or methodologies. This book resulted from a more mundane impulse. In teaching courses dealing with developments in American thought and the arts during the nineteenth century, I grew frustrated by the lack of an interdisciplinary study suitable for undergraduates. I therefore set out to write an interpretive synthesis that would consolidate fresh knowledge from a variety of disciplines. My hope was to satisfy a practical need and also to illuminate some key issues and developments in American cultural history.

Yet in trying to master the wide array of primary sources and the ever-increasing volume of secondary literature about intellectual, social, economic, and cultural developments in the post-Civil War era, I found myself struggling in the midst of limitless material. Too late I came to appreciate the advice of Joe Kett, a close friend and astute social historian at the University of Virginia, who early on discouraged me from undertaking the project. "What you propose staggers me," he warned. "Realism in literature alone is enough to gag the best of us."

By then, however, I had gone too far into the briar patch of realism to turn back. What eventually saved me was the lifeline of scholarly generosity. Two colleagues at Davidson College—Larry Ligo and Tony Abbott—read chapters at an early stage and provided suggestions as well as encouragement. Other friends and scholars from around the country were kind enough to read parts of the manuscript at various stages. They included Casey Blake, Charles Cornwell, Anne Croce, Robert Crunden, George Daniels, Steve Forman, Daniel Horowitz, William E. Leverette, Jr., David Levin, Townsend Ludington, H. Wayne Morgan, Robert Scholnick, Gretchen Schoel, Robert Twombly, and Larzer Ziff.

I owe a special debt to a group of scholarly kindred spirits—James Gilbert, Suellen Hoy, Joe Kett, and Walter Nugent—each of whom read the entire manuscript and sent along detailed assessments. I also learned much from my seminar students at Davidson College. They proved to be discerning—and often witty—critics of this work-in-progress. Sheldon Meyer, my editor at Oxford University Press, pruned my prose without bruising my ego, and Stephanie Sakson displayed a falcon's eye as copy editor. Gerry McCauley, a baseball coach masquerading as a literary agent, took valuable time away from his All-Star team to offer pep talks when I most needed them.

Grants from the Huntington Library, the National Endowment for the Humanities, and Davidson College furnished me with vital interludes of unfettered time to conduct research and write. Staff members at the Huntington Library, the Beinecke Rare Book Library at Yale University, the Alderman Library at the University of Virginia, and the Houghton Library at Harvard University were unfailingly gracious to an outlander. The good folks at the Davidson College Library, as always, afforded me unexcelled service and support. I also want to thank my erstwhile and charming student research assistants at Davidson College—Lisa Culbertson, Jennifer Ely, Jeff Fraley, Adrienne Goins, Drew Henderson, Kimberly Short, Elizabeth Spainhour, Trevor Wade, and Luise Weinrich—as well as Alison Gurney at Furman University.

My deepest gratitude is reserved for my wife Susan. She nurtured this work-in-progress with characteristic enthusiasm and patience. Not only does she remain my most willing reader, but she also helps me remember what really matters in our household. Without her unstinting support, this book would have never been completed, and for this, and for much else, it is dedicated to her.

Greenville, S.C.   D.E.S.
January 1994

CONTENTS

Part V　　An Epoch of Confusion

Facing
Facts

INTRODUCTION

I n 1858 Charles Godfrey Leland, the outspoken editor of *Graham's Magazine*, announced that Americans had developed a rapturous devotion to "literal *facts*." Everywhere, it seemed, people were preoccupied with the pursuit of verifiable knowledge and tangible concerns. Leland predicted that such a "realistic" outlook would transform the nation's intellectual and artistic life. Three years later, the *Christian Examiner* echoed Leland's observations when it reported that Americans were "eager for facts and positive verification." This reverence for concrete facts and visible realities soon coalesced into a self-conscious cultural movement. By 1866 a precocious twenty-three-year-old book reviewer named Henry James could proclaim that the "taste of the age is for realism."[1]

Indeed, a realistic outlook seeped into every corner and crevice of intellectual and artistic life during the second half of the nineteenth century, a period that historians have labeled the Gilded Age, the Age of Enterprise, or the Victorian era. Realists of all sorts—scientists, philosophers, writers, artists, architects, and tastemakers—muscled their way onto center stage of American culture and brusquely pushed aside the genteel timidities, romantic excesses, and transcendental idealism then governing affairs of the mind. By the end of the century, the various reality-seeking systems of discourse and artistic expression seemed triumphant. "Old idols," writer Hamlin Garland claimed in 1894, "are crumbling in literature and painting as well as religion. . . . This is our day."[2]

Garland claimed more for realists than they accomplished, but in important ways it *was* their day, and this book endeavors to tell their story. Intended as much for students and general readers as for scholars, *Facing Facts* offers a synthetic overview of the major events, ideas, and individuals that combined to generate the various types of "realistic" expression and determine their fate in the marketplace of taste between the mid-nineteenth century and the end of

World War I. Of course, regiments of specialists in different disciplines have already explored the "age of realism," and I have greatly benefited from their research and insights. Most previous studies, however, have focused on developments in separate fields—social thought, philosophy, natural science, literature, drama, painting—and few have done justice to the intellectual and social context within which the realistic perspective arose.[3]

This book uses a wider lens to survey the landscape of realism and map its essential features and overlapping boundaries. In the process I have tried to produce an accessible, integrative work that makes connections across the disciplines and places the findings of specialized research in a larger context. Although H. L. Mencken insisted that a "professor must have a theory as a dog must have fleas," I do not offer any startling new thesis about realism as a conceptual or aesthetic category. No overarching interpretation can adequately encompass the diverse motives and methods associated with the realistic impulse. Instead I use the era's preoccupation with "facing facts" as an organizing frame within which to examine the emergence of various realistic sensibilities in the natural and social sciences, philosophy, literature, art, and architecture.

This book addresses several basic questions: When and why did realism appear as a viable new intellectual stance and cultural style? What were its aesthetic principles and social aims? How did the realistic movement reflect emerging patterns of mass production and consumption and shifting gender roles? What did it mean in practice to write realistic novels, produce realistic paintings, and design realistic structures? Why did realists at the end of the century become obsessed with the brutality and sordidness of lower-class life? How did critics and the larger public respond to such initiatives?

These questions assume that readers know what realism meant to Americans during the nineteenth century. But there was—and is—considerable confusion as to the term's precise meaning. For some, realism served as an urgent imperative and a guiding creed; for others it was simply a new point of view or distinctive temperament. Several reputed members of the "realist school" were unaware that they had enrolled. "I didn't even know I was a realist until [critics] wrote and told me," confessed Mary Wilkins Freeman, the writer of pungent stories about New England rural life. People loosely employed the realist label throughout the century, leading one bewildered writer to grouse that there "is much talk of the 'realistic school' nowadays, but seldom any very definite comprehension of what that phrase does or should mean."[4]

Despite its inherent ambiguity and inconsistent application, however, realism as a cultural category gains precision when studied in the fullness and complexity of its historical moment. Those who were labeled scientific realists or positivists during the nineteenth century affirmed the existence of a physical realm independent of the mind, a coherent and accessible world of objective facts capable of being known through observation, understood with the use of reason, and accurately represented in thought, literature, and the arts. As the editor of *Harper's Monthly* announced, "we cheerfully accept reality in whatever way it concerns us — in our life, our philosophy, our science and our literature." Such enthusiasts rarely speculated on the nature of reality; it seemed self-evident. Edward Everett Hale, a Boston clergyman, author, and literary critic, baldly asserted that "truth consists of the facts of life. The realist, then, busies himself with the facts of life."[5]

Such an uncritical equation of the visible and tangible with the true defined the initial realistic impulse. "Realism," proclaimed the art critic Leo Stein, "means the spirit of fact predominant, and the sheer acceptance of reality." In facing the facts of their time, people often assumed that what they *saw* was manifestly what it *seemed to be*. The facts, in other words, spoke for themselves. "There's no denying facts," exclaims a character in Frank Norris's *Vandover and the Brute*. Facts, Norris later remarked, were "as concrete as the lamp post on the corner, as practical as a cable car, as real and homely and workaday and commonplace as a bootjack."[6]

This ingenuous confidence in the ability of people to recognize and represent the world of fact led writers and artists to engage in what the ancient Greeks called *mimesis*, the imitation of contemporary life as seen and experienced. Mimetic art and literature sought to mirror the activities of nature and human society as perceived through the senses. "My one ambition," said Theodore Dreiser, "is to represent my world, to conform to the large, truthful lines of life." Yet the tendency of Dreiser and others to highlight the utilitarian, drab, and even repulsive aspects of life led critics such as Ambrose Bierce to charge that realism involved "the art of depicting nature as it is seen by toads, the charm suffusing a landscape painted by a mole, or a story written by a measuring-worm." To its many advocates, however, realism meant intellectual and artistic honesty rather than romantic exaggeration. It also involved a direct confrontation with life rather than an art-for-art's-sake aestheticism. William Dean Howells, the prolific novelist and influential editor, asserted that realism was "nothing more and nothing less than the truthful treatment of material."[7]

It was not so simple. What some realists offered as "truthful treatments" of American life appeared scandalous or superficial to others. Realists often debated among themselves about how candid they should be in exposing the facts as they saw them. For all of their claims of objectivity, positivists viewed "things out there" through a lens of confining social conventions and moral inhibitions. Considerations of the marketplace, class consciousness, racial and gender prejudices, and deeply embedded standards of morality and decorum often narrowed the borders of the realistic impulse, while idiosyncratic personal tastes ensured considerable variety of expression. By the end of the century, many "realistic" thinkers, writers, and artists had come to recognize the danger of equating reality with objective facts. So-called pragmatic realists stressed that the mind is more than a passive mirror; it engages rather than reflects reality and in the process shapes its meaning in quite subjective ways.

What all realists held in common was a *language* of rebellion against the genteel elite governing American taste. They repeatedly invoked such terms as "sentimental," "romantic," "artificial," "anachronistic," and "effeminate" to express their disdain for the prevailing modes of idealism dominating thought and expression. Threaded together more by such oppositional discourse than by a uniform creed, realists could communicate with and applaud each other, but the very fact that they operated primarily out of a rhetorical rather than a philosophical framework allowed them a marked degree of independence from one another.

Realism, of course, was not simply an American phenomenon. It originated in Europe during the early nineteenth century, and its British, French, Spanish, Scandinavian, and Russian practitioners exerted a continuing influence in the United States. What differentiated American realists from their European peers was the piquant residue of moral idealism and social optimism inherent in their outlook. The "most characteristic tendency" in American thought, declared the Episcopal clergyman and writer Samuel Osgood in 1857, "is what might be called an ideal realism, or a disposition to bring ideal convictions to bear upon practical realities." Osgood knew that most American realists, despite their claims to scientific detachment and hard-edged candor, manifested a powerful moral impulse in their efforts to represent the objects and concerns of their contemporary world. For some, that moral stimulus grew out of the evangelical reform tradition; for others it was a more secular force promoting social better-ment; for a few, such as Henry James, it was essentially an aesthetic morality emphasizing the artist's obligation to depict accurate "pictures of my time" so as to enlarge the public's sense of the world.[8]

Whatever its essential mode or motive, American realism revolved around a moral axis. A young Willa Cather made the connection explicit when she reviewed several plays at the end of the century. "They are realistic with the realism of good," she explained, "and remind one that realism after all may not be an absolute synonym for evil." In fact, most of the leading American thinkers and artists willingly mixed moral optimism with their aesthetic realism. Walt Whitman carried on a vibrant inner dialogue between the real and the ideal, as did Louisa May Alcott, Louis Sullivan, Winslow Homer, William James, and many others.[9]

Their mingled loyalties and fertile tensions suggest that labels such as realism and idealism are relative terms that gain clarity—and utility—when discussed in reference to individual figures and through comparison with one another. "Absolute realism and absolute idealism do not exist," proclaimed a writer in the *North American Review*. Nevertheless, he insisted, individuals could "be graded according to the percentage of realism or idealism contained" in their works. In this context adjectives assume a crucial delimiting significance. Readers will encounter in the following pages genteel realists, romantic realists, sentimental realists, impressionistic realists, muckraking realists, and naturalistic realists, among others. The interpretive issue is one of degree, of proportion, of the *relative* importance of one or the other outlook to a given individual or the larger cultural scene at a particular moment.[10]

The most conspicuous social ideal promoted by American realists was the creation of a more democratic culture. By giving coherent representation to the diverse racial and ethnic groups and social classes making up their rapidly changing nation, realists sought to promote societal toleration and consensus. Literature and the fine arts, they argued, need not simply be elevating or distracting luxuries for the few; a truly democratic aesthetic could exercise minds, open eyes, and enlarge sympathies. Such an inclusive vision acknowledged the fact of a pluralistic America and fought to secure a common cultural ground uniting an increasingly diverse and fractious public.

William Dean Howells, for instance, urged writers and artists to include in their works unrepresented groups, locales, and points of view. Industrialists and laborers, patricians and paupers, bankers and farmers, reactionaries and anarchists, blacks and whites, natives and immigrants, pietists and agnostics, men and women, old and young, beautiful, ordinary, vulgar, criminal, and ugly, were all to receive literary and artistic attention. At the same time that realists expanded the *scope* of cultural representation, they also sought to broaden their *audience* beyond the genteel elite that had long circumscribed American taste. Whitman asserted in *Democratic Vistas* (1871) that literature and the arts should not be created by and for "a single class alone, or for the parlors or lecture-rooms, but with an eye for the practical life" of the "middle and working strata."[11]

There is a charming innocence to this ambitious ideal of a democratic culture. The "aesthetic of the common," as Howells called it, harbored inherent problems. In giving artistic representation to long-ignored social classes and groups, realists often adopted a patronizing tone that turned subjects of description into objects of condescension. The "aesthetic of the common" also raised a daunting conceptual dilemma: how could the writer, artist, or architect presume to represent the "real world" in all its polyglot variety? A multicultural society does not stand still to have its portrait painted on canvas, its character rendered in prose, or its spirit embodied in stone and steel structures. Holding a mirror to a culture in motion produces a blurred image, and many of the realists' mimetic efforts fell short of their ideals. In sum, this "aesthetic of the common" represented a very complicated phenomenon. Understanding its historical development entails probing epistemological assumptions, exposing internal contradictions, and analyzing social causes.

Among the factors giving rise to a realistic frame of mind, none was more potent than the question of gender ideology. During the course of the nineteenth century, long-held stereotypes about men and women clashed with changing social roles and new opportunities for artistic expression. Women not only agitated for voting rights and legal equality; they also exercised an increasingly prominent role in the workplace, in religious and charitable activities, school faculties and library staffs, and the literary and art marketplaces. Several prominent female activists maintained that women had a special mission and mandate to exercise leadership in the realms of art and morality. "The higher interests of society must be cared for by women," declared Elizabeth Peabody, the Boston teacher, writer, and reformer. "That is, literature, art, and all the virtues and graces that make society progressive spiritually, morally, and intellectually."[12]

Yet many people charged that the growing female presence in artistic life brought with it a dangerous "effeminacy." American high culture, observed a writer in the *Atlantic Monthly*, "lacks the impulse and virility necessary for its own propagation. It is too dainty for a land like ours." During the nineteenth century, literature and the fine arts seemed to become increasingly artificial and "unreal," overly refined, and out of touch with practical matters. For a man to practice "art" or to be "artistic" risked being labeled "feminine" or "effeminate." Those worried by such developments often blamed the growing influence of women. "There is no doubt," the flamboyant English poet Oscar Wilde told Americans, "that within a century from now the whole culture of the New World will be in petticoats."[13]

The "sissification" of American culture alarmed many women as well as men. Concerned female writers, artists, and critics argued that "masculine" qualities were neither the exclusive property nor the inevitable birthright of men. In fact, many wondered why women could not be more like men in outlook and expression. Willa Cather, for instance, repudiated conventional femininity and all its trappings, especially its propensity for sentimentalism. She lamented that women remained "so horribly subjective" and displayed "such a scorn for the healthy commonplace." She even questioned "why God ever trusts talent in the hands of women, they usually make such an infernal mess of it." Cather remained skeptical that women writers would ever achieve greatness until they gained access to "real" experiences outside the home and developed a hardier disposition. "When a woman writes a story of adventure, a stout sea tale, a manly battle yarn, anything without wine, women, and love, then I will begin to hope for something great from them, not before."[14]

Cather and others believed that the cluster of values associated with traditional femininity—spiritual piety, moral purity, blissful domesticity, and overheated emotionalism—not only damaged the psyche of women but also impeded the development of a potent cultural life. Margaret Fuller, the brilliant Boston feminist, advised women at mid-century to nurture the "masculine" side of their nature and produce a more vigorous literature dealing with contemporary concerns. She herself refused to "write like a woman, of love and hope and disappointment," but instead resolved to write "like a man, of the world of intellect and action." Artist Mary Cassatt felt the same need to defy conventional expectations and become a "masculine and assertive" painter. She wanted her brush to convey force rather than sweetness, truth rather than sentimentality. Any "talk of womanliness," Cassatt contended, "left her cold."[15]

The supposed "effeminization" of social and artistic life became a prominent theme in realistic art and fiction. In Henry James's *The Bostonians* (1886), a tormented Basil Ransom declares that the "whole generation is womanized; the masculine tone is passing out of the world; it's a feminine, nervous, hysterical, chattering, canting age, an age of hollow phrases and false delicacy and exaggerated solicitudes and coddled sensibilities." To counteract such "pantywaist" tendencies, Ransom prescribes an infusion of "masculine character, the ability to dare and endure, to know and yet not fear reality, to look the world in the face and take it for what it is."[16]

In this way, both men and women seized upon cultural realism as an antidote to syrupy sentimentalism and as a stimulant to help revitalize "masculine" virtues. Recall Huckleberry Finn's strenuous efforts to escape the pious domesticity of the Widow Douglas and Miss Watson. Or consider Walt Whitman's claim that American culture suffered from a surplus of "delicacy, refinement, elegance, prettiness, propriety, criticism, analysis." In his view, the "shoats and geldings" dominating the country's cultural life could hardly be expected to produce poetry adequate to modern needs. Theodore Dreiser also rebelled against the "tea and toast crowd" governing American taste and morals, but he was more explicit in his chauvinism. Women, he maintained, should rest content with fulfilling their "natural capacities" rather than pursue "a lot of schemes of reorganizing the world and putting love on a higher basis." During the second half of the nineteenth century, Dreiser, Whitman, Twain, and many other male writers and critics transformed their sense of besieged masculinity into a campaign for a more energetic and "realistic" American culture. Their chest-thumping crusade at times displayed both comic and ridiculous elements, but its hormonal energies helped propel and shape the realistic enterprise.[17]

TODAY, OF COURSE, many sophisticated theorists who deny the existence of objective reality dismiss realism as a dull and naive notion unworthy of sustained attention. Yet realists during the Gilded Age were considered cultural radicals. Their candor in treating previously ignored or suppressed social facts shocked prevailing sensibilities and generated lively debates about the purpose and nature of artistic expression. "Recently there has sprung up what purports to be a new theory of the proper method of writing novels and of painting pictures," Professor

William Roscoe Thayer told his Harvard students in 1889. The "apostles of realism," he charged, were determined to "abolish our idols and our ideals." Thayer contemptuously dismissed their "futile piling up of facts." True *reality*, he countered, "is broader, and deeper, and higher than we could ever infer from the Realists' philosophy."[18]

Thayer's defensive comments indicate that the controversy over realism involved issues of profound—and perennial—significance: Should writers and artists transcend contemporary social realities or confront them? Should they imitate visible surfaces, probe for the vital spirit beneath appearances, seek to reform social evils, or express the artist's emotional turbulence and imaginative power? Not surprisingly, people fastened upon quite different answers to these questions. To revitalize the diversity of their responses, I have tried to recreate their historical context, retrieve their intellectual energy, and take their ideas seriously.

Most accounts of nineteenth-century cultural life claim that a realistic consciousness first surfaced at the end of the Civil War, but I trace its roots to the 1850s. During that decade, orthodox religious believers and natural philosophers felt the first unsettling tremors of evolutionary science. At the same time, debates between "idealists" and "realists" filled the literary journals. It was also during the 1850s that Walt Whitman published *Leaves of Grass*, Winslow Homer began his career as a magazine illustrator, Horatio Greenough launched his campaign for a more functional American architecture, and mass-circulation newspapers and illustrated weeklies introduced readers to the fact-devouring emphases of metropolitan journalism. Over the next half-century the rage for "facts" and the appeal of mimetic representation grew in scope and intensity. Avowed realists appeared in virtually every art form and academic discipline, ranging from drama and music to philosophy and the social sciences.[19]

With the dawn of the twentieth century, however, a new "modernist" sensibility challenged the fact-worshipping spirit of the age and questioned whether the arts could or should "mirror" contemporary society. In the face of new revelations in physics and psychology and in the aftermath of the horrifying carnage of the Great War, many people lost confidence in their ability to *know*, much less represent in words, paint, or steel, a common reality. Said a journalist in 1920: "We appear to have passed beyond the region of physics into that of metaphysics, where paradox and magic take the place of solid fact." To be sure, a realistic impulse has remained a prominent force in cultural life, but since 1920 or so it has grown more complex and self-aware. It has also lost much of its cohesion and faith as a collective enterprise.[20]

The following thirteen chapters are divided into five parts. In the first section, Chapter 1 surveys the landscape of American thought and culture during the antebellum period. It is intended primarily for those readers unfamiliar with the various modes of "idealistic" sensibility that flourished between 1830 and 1860.

Chapter 2 describes different manifestations of the new realistic consciousness that emerged during the 1850s. The second part consists of three chapters that range across the second half of the nineteenth century in an effort to evaluate the major *causes* of the rise of realism. They deal, respectively, with the impact of the Civil War, the epistemological claims and astonishing accomplishments of scientific materialism, and the emergence of urban-industrial civilization and new media such as the metropolitan newspaper, mass advertising, and photography.

In Part III, Chapters 6, 7, and 8 focus on the maturation of a realistic consciousness in three separate fields after the Civil War: literature, art, and architecture. The four chapters making up Part IV deal with developments at the turn of the century. Chapter 9 discusses the ways in which post-Civil War social reformers adopted a more realistic and less "sentimental" outlook. Chapter 10 surveys the emergence of a "naturalistic" form of literary realism. Chapter 11 offers in-depth analyses of Stephen Crane, Frank Norris, and Theodore Dreiser. The last chapter in the fourth part, Chapter 12, assesses the contributions of the "Ash Can" school of painters. The book's final part—composed of Chapter 13 and the epilogue—examines the emergence of "modernism" and the ways in which it challenged many of the premises associated with common-sense realism.

As a broad historical synthesis, this book draws upon both archival sources and secondary works and trespasses across a variety of intellectual disciplines and art forms. Developments in sculpture, drama, and music deserved attention as well, but doing justice to those fields would have made the book prohibitively long. In staking out the wide boundaries of the realistic impulse and indicating its major causes and forms of expression, I have been brutally selective in casting the main characters and have at times treated them with ruffian brevity. "When we cannot look at everything," as Henry James once recognized, "we must look at what is most characteristic." In this sense, most of the individuals featured in the following chapters are quite familiar; several, however, are known only to specialists. While not claiming to be encyclopedic in my coverage, I have tried to be comprehensive. Of course, such an overview of a complex and often incoherent cultural phenomenon has its pitfalls. Historical context provides only a partial explanation of shifting currents of intellectual inquiry and artistic expression. Yet I hope that this book will give readers a more textured understanding of the realistic movement and entice them to explore or re-examine in a new light some of the individual works discussed in the following pages. If so, this decade-long project will have been well worth the effort.[21]

Part I

Setting the Stage

Understanding the rise of realism requires first understanding what it rose up against. A writer in the New York Times *announced that realism "ranks itself as one of the opposites of idealism." Quite so. But defining idealism was—and is—no easy matter. "There are," Ralph Waldo Emerson once acknowledged, "many degrees of idealism." During the first half of the nineteenth century, four types of idealistic sensibility—genteel, domestic, transcendental, and brooding—dominated American "high" culture.[1]*

Although quite diverse in motive and method, idealists shared a basic conviction that fundamental truths rested in the unseen realm of ideas and spirit or in the distant past rather than in the accessible world of tangible facts and contemporary experiences. This meant that literature and the arts should deal with the beautiful or the exotic rather than the common aspects of life. A literary critic observed at mid-century that the arts in America were so thoroughly idealistic in orientation that they had come to constitute a "world apart, above the world of mere material combinations." More than anything else, this excessive preoccupation with otherworldly concerns and historical subjects provoked the rise of a realistic sensibility. Novelist Hjalmar Boyeson, one of the most ardent advocates of literary realism, claimed that American culture during the first half of the ninteenth century displayed such "a sovereign disregard of reality" that a reaction was inevitable.[2]

1

Antebellum Idealism

In 1913 the Spanish-born Harvard philosopher George Santayana used the phrase "genteel tradition" to characterize the formidable elite that had governed nineteenth-century American taste and sensibility. The nation's most prominent writers, artists, architects, educators, ministers, and moral philosophers, he noted, shared refined tastes, delicate sensibilities, and magisterial demeanors. Well born and well informed, beyond money worries and brimming with erudition and authority, they viewed themselves as guardians of conventional religious beliefs, aesthetic standards, and moral values. The cantankerous historian Henry Adams, both a product and a critic of the genteel tradition, remembered that the New England "social hierarchy" of the first half of the nineteenth century was one "in which respectability, education, property, and religion united to defeat and crush the unwise and vicious."[1]

To be sure, some patricians exercised a benevolent paternalism toward those below them on the social scale. Others, however, played the role of moral and political police, and at times their snobbery assumed sublime proportions. Sidney George Fisher, a Philadelphia lawyer, essayist, and art critic, lamented in 1840 that the "demon of democracy is abroad & triumphant and will drive us to the devil before very long." He felt "socially superior to a man who is not a gentleman by *birth*." Increasingly aware of their isolation and vulnerability amid the gusty whims of a leveling age, the nation's intellectual and moral aristocrats were motivated as much by their fears of the new as by their hopes for renewal. Many of them saw in the arts an instrument for reinforcing the values of their own social class in the midst of spreading vulgarity and ignorance.[2]

The members of the genteel elite were usually Congregationalist, Unitarian, or high-church Episcopalian in religion; Federalists or Whigs in politics; graduates of Harvard, Yale, or Princeton; and regular contributors to the staid *North American Review*. They displayed a serene certitude regarding cultural values and rigorously enforced what the Harvard professor-poet James Russell Lowell called the "orthodox creed of the ideal." That is, they endorsed the eighteenth-century neo-classical assumption that the arts should esteem the beautiful, revere the past, reinforce social stability, and uphold a chaste morality.[3]

Such genteel idealism placed little faith in facts alone. The Harvard professor and poet Oliver Wendell Holmes, for example, believed that a person of real cultural refinement transcended a mundane emphasis on facts. "All fact collectors, who have no aim beyond their facts," he wrote, "are one-story men. Two-story men compare, reason, generalize, using the labors of the fact-collectors as well as their own. Three-story men idealize, imagine, predict; their best illumination comes from above through the sky light." Lowell, too, preferred to explore aesthetic delights far removed "from the plane of the actual and trivial." Throughout his early career, he professed a deep-seated reluctance to confront common life itself. Great art, he preached, "always idealizes" by enticing people away from "the association of everyday life." He also encouraged writers to avoid discussing anything that might shock or disgust polite sensibilities. His editorial credo established the standard for the genteel tradition: "let no man write a line that he would not have his daughter read."[4]

Other idealists derived their inspiration more from Jerusalem than classical Athens. Ministers and moral reformers such as Horace Bushnell and Catharine

Washington Allston, *Elijah in the Desert*, 1818. *(Courtesy, Museum of Fine Arts, Boston.)*

Beecher combined high culture with intense piety, using a divine measuring rod in evaluating artistic merit. In melding good manners with traditional Christian values, they added an emotional warmth to the cool civility of the neoclassical elite. These pious idealists practiced what Holmes called "sentimental religion," a deeply felt spiritual impulse that manifested itself in personal devotion and strenuous efforts at social improvement. Like their Puritan ancestors, the antebellum religious leaders professed that true "reality resided in a region beneath appearance and beyond manipulation."[5]

Whatever their animating perspective—neo-classical or Christian—the new nation's cultural aristocrats took little interest in the commonplaces of everyday life. The very brightness of their idealism blinded them to the essential beauty latent in their actual surroundings. "Give me the writers," Lowell pleaded, "who take me . . . away from my neighbors! I do not ask that characters should be real; I need but go into the street to find such in abundance." Likewise, artist Washington Allston painted intensely emotional scenes derived from biblical stories or ancient history in an effort to evoke the mystery of realms beyond human understanding. He mixed theology with his paints, arguing that the true poetry of art occurs "where the *real* ends and the *ideal* begins."[6]

The Cult of Domesticity

During the nineteenth century, the genteel elite—as well as an emergent middle class—developed an ardent faith in the civilizing power of moral women. Females were widely assumed to be endowed with greater moral sensibility and religious inclinations than men. Such pedestaled notions of women helped nourish a powerful "cult of domesticity" which assigned to women the role of self-denying guardians of the hearth and soul. As the more complex economy of the nineteenth century matured, economic production was increasingly separated from the home, and the absence of men who left to work long hours in the city transformed the middle-class home into a "separate sphere" governed by mothers. In 1842 a writer in *Parents' Magazine* commented that the typical middle-class American father "was eager in his pursuit of business, toils early and late, and finds no time to fulfill his duties to his children." William Dean Howells testified to such a situation when he recalled that in his Ohio household his "mother represented the family sovereignty; the father was seldom seen."[7]

In the artistic realm, most women during the second quarter of the nineteenth century preferred edifying paintings and poems or uplifting "romances" that trumpeted "feminine" virtues. As early as the 1840s anxious men began to complain about the loss of "manliness in thought." By mid-century such complaints had formed a chorus. "Like subtle lighting," a literary critic reported, "the womanly nature is penetrating the life of the age." Written mostly by women for women readers, hundreds of popular romances appealed directly to the reader's sentiment and conscience. They typically included melodramatic incidents, despicable villains or immoral suitors, wrenching deathbed scenes, and climactic marriages or spiritual conversions.[8]

To be sure, some of the popular female writers openly rebelled against what Louisa May Alcott called the "twaddle" about a separate "woman's sphere." Alcott, for instance, chafed at the recurring image of women as "clinging vines" seeking "man's chivalric protection." So did Fanny Fern (Sara Willis Parton), the Boston-born author of *Ruth Hall* (1855) who became a columnist for the *New York Ledger* in 1856. She contended that a woman should have "the right to love and hate persons and things as well and as strongly as a man." Eventually, Harriet Beecher Stowe also came to resent and resist patriarchal rule. "The more I think of it," she told a friend, "the more absurd this whole government of men over women looks."[9]

At mid-century, however, most female writers preached that the life of feeling took priority over the outer world and that women should accept and even glory in their submissive role as self-denying angels ministering to a male-dominated world. "True feminine genius," declared Grace Greenwood (Sara Jane Clark), "is ever timid, doubtful and clingingly dependent; a perpetual childhood." Literary critics would later denigrate the popular female authors for coating their fiction and verse with a meringue of piety and inventing improbable characters and melodramatic situations, preachy narrators, self-righteous heroines, harrowing personal trials and tragedies, and predictable happy endings. Such charges have considerable validity. Yet in focusing on the stylistic weaknesses of sentimental fiction, male critics cloaked a more fundamental objection: women writers often dealt with alternative realms of meaning. Women were not so much evading reality as they were focusing on different realities from those favored by men. By writing lovingly about the homely activities and emotions common to their "separate sphere," women writers challenged masculine assumptions about what subjects were worthy of imaginative representation.

The South Carolina romancer William Gilmore Simms, for example, discouraged male writers from dealing with the "ordinary events of the household" or "the snug family circle." Such subjects made for inferior fiction and were "best handled by the other sex." He and other male authors preferred robust stories of masculine adventure and camaraderie. Yet in exploring with caring eyes the everyday occurrences of domestic life, many of the popular women writers anticipated the major concerns of the realistic movement. They often included in their fiction accurate, detailed descriptions of common social experiences: home life, church activities, factory and mill conditions, quilting bees, excursions to town or city. And they embraced in their fiction social outcasts who had been rarely represented: prisoners, slaves, children, and the poor.

Although loath to admit it, many of the most prominent male "realists" would later build their own fiction upon this feminine foundation.[10]

The writings of female sentimentalists featured tears, prayers, hymns, and intimate relationships between mothers and daughters because such emotions and experiences were palpably *real* to them; they themselves dealt with familial grief, practiced spiritual devotion, and relished fantasy worlds. Perhaps antebellum readers preferred stories about self-denying heroines living in an imaginary realm of chivalric love and adventure or acting in domestic dramas of religious conversion and moral triumph because the secular forces outside the ideal home and church were too troubling to confront or too complex to resolve. Pious sentimentalism and rural nostalgia—the two most prominent characteristics of such literature—offer instruments of escape, and the increasing tempo of life, the vicissitudes of the emerging middle class, as well as the growing political, economic, and social tensions provoked by slavery, westward expansion, and commercial development made memories of simpler and more spiritual times increasingly attractive.[11]

Transcendental Idealism

At their best the genteel tradition and the cult of domesticity endowed early nineteenth-century cultural life with stability, piety, and cultivated intelligence; at their worst they served to stifle creativity and ignore festering social tensions. This rigid resistance to change helped generate a revolt among a younger generation of artists and thinkers. There is a point at which dignity and dogmatism become mere dullness and provoke rebellion, and during the 1830s that point was reached, as a wave of democratic individualism washed across the barricades of patrician conservatism. For many rebellious young men and women, it was time to begin thinking afresh.

The second quarter of the nineteenth century was the much celebrated (and exaggerated) "age of the common man," when, as Herman Melville wrote in *Moby-Dick,* the "great Democratic God" picked up "Andrew Jackson from the pebbles." As politicians extended voting rights to propertyless males, cultural life began to feel the intoxicating effects of the egalitarian spirit and the unbridled celebration of democratic values. Cultural rebels began to argue that America must produce art and literature that would give unapologetic representation to common things and common folk.

The most profound manifestation of this new democratic culture was the peculiar version of romanticism practiced by the Transcendentalists, that eclectic coterie of inspired poets and philosophers centered in Concord and Boston. During the 1830s and 1840s, this fluid group of geniuses and cranks launched a vigorous rebellion against the barnacled religious, economic, and social status quo. Most of them began their spiritual pilgrimage in the bosom of the Unitarian Church, but they eventually abandoned the "corpse-cold" denomination in search of a more vital faith. In their view, the genteel elite had become decorous and smug, cramped and cramping. To startle the patricians out of their parlor complacency, the Transcendentalists resolved to *transcend* the limits of reason and cultivate inner states of consciousness. Were there not, after all, more experiences than reason and logic could box up and explain—moods, impressions, feelings; mysterious, unknown, and half-seen things? Were these not facts as well?[12]

The Transcendentalists embraced a philosophical idealism derived from Plato and Kant. Stated most simply, their "sublime creed" asserted that the mind could penetrate the deceptive surface of things in order to discover the universal spirit underlying and embodying all matter. They thus strove to "quit the world of experiences" in order to engage in an intuitive search for essences. The "Real" for them existed beneath and beyond the realm of sensory experience. It was something to be imagined rather than seen or reasoned. "The idealist," Emerson explained, "takes his departure from his consciousness, and reckons the world [merely] an appearance."[13]

The Transcendentalists used the lamp of personal inspiration to illuminate changing states of consciousness rather than simply hold a mirror to the "facts" of the physical world and human society. "Mind is the only reality" for the transcendental idealist, Emerson asserted. Longing for the unbounded and indefinable, full of perfectionist illusions about the possibilities of human nature and the American social experiment, he and other kindred spirits aspired to rise from what Orestes Brownson called "the dull Actual" to the "rich and magnificent world of the Ideal lying beyond" and "above things."[14]

Yet what makes the Transcendentalists so compelling—and baffling—are their contradictory impulses. While searching for a mystical union of their souls with the Universal Being, they also celebrated the essential dignity and beauty residing *in* common things and common experiences. They believed it possible to unite the common and the divine, to produce a fertile synthesis of the real and ideal. In his famous Phi Beta Kappa address at Harvard in 1837, entitled "The American Scholar," Emerson linked the growing interest among younger writers and artists in depicting their contemporary surroundings to the era's democratizing political tendencies, and he enthusiastically sanctioned their efforts: "I ask not for the great, remote, the romantic . . . I embrace the common, I explore and sit at the feet of the familiar, the low." This romantic celebration of the actualities of common life, however tentative, vicarious, or celestial, greatly expanded the range of cultural expression. By embracing the aesthetic significance of their everyday surroundings, the Transcendentalists nudged the boundary of idealism toward the border of realism.[15]

Yet while promoting the use of common subjects in literature and art, Emerson added that such heretofore neglected material should be "poetized" to illuminate the "sublime presence of the highest spiritual cause." This key qualifying statement clarifies his contradictory effusions. He never endorsed the *literal* representation of the homeliest trifles. Instead of taking commonplace facts at face value, he wanted writers and artists to *idealize* the "near, the low, the common," to transmute realities into intuitive spiritual truths, to find the miraculous and poetic in the common life of street, mill, railroad, and farm. The ideal poet, he wrote, converted "the daily and hourly event of New York, of Boston, into universal symbols."[16]

Although willing to wade vicariously in the shallows of everyday life, Emerson refused to submerge himself for fear that the undertow would drown his genius. He once confessed that he felt a "sore uneasiness in the company of most men and women" and detected in his own character an "absence of common sympathies" and a "want of animal spirits." Emerson's occasional collisions with city crowds and commercial activities bruised and intimidated him. In a fit of candor, he admitted that he "had no skill to live with . . . such men as the world is made of." Despite his fraternal rhetoric, he remained at heart a detached individualist; he never developed a genuine respect for the working class or the destitute poor. In *Nature* (1836), for instance, Emerson declared that "you cannot freely admire a noble landscape, if laborers are digging in the field hard by." He later called German and Irish immigrants "rueful abortions that squeak and gibber in the streets." They were a herd of unthinking "bugs and spawn" with undeveloped mental capacities.[17]

Emerson wanted above all to be a prophet and seer rather than a "mere" reporter of common realities or a fellow traveler with common folks. He repeatedly asserted that the "ideal is truer than the actual." In this regard Emerson resembles the albatross in Baudelaire's poem: his wings were so large they prevented him from walking. His essential idealism repeatedly frustrated his realistic inclinations. To his credit, Emerson recognized the "clangor and jangle of contrary tendencies" within his warring psyche. "God has given me the seeing eye," he admitted, "but not the working hand." Burdened by frail health all his life, he craved the balsam tonic of a more robust outlook but never found it. "We love to paint those qualities which we do not possess," he admitted, and his own deficiencies help explain why he spent so much time glorifying people of a practical bent.[18]

"I like people who can do things," Emerson said, and Henry Thoreau, fourteen years his junior, could do many things well—carpentry, masonry, painting, surveying, gardening. The enigmatic son of a pencil-maker father and abolitionist mother exuded a spirit of uncompromising integrity, manly vigor, self-reliant simplicity, and tart individuality that Emerson found captivating. After graduating from Harvard and serving a brief stint as a teacher, Thoreau settled down to eke out a living as a surveyor and craftsman in his father's pencil-making operation. But he spent most of his time drinking in the beauties of nature. Short and sinewy, the blue-eyed Thoreau mastered the woodland arts. He loved to muck about in swamps and fields, communing with mud turtles and loons as well as his inner self. Emerson admired his "muscle," his pugnacious determination, his native intelligence, and his keen powers of perception. Thoreau, he confessed, was "far more real . . . than I."[19]

Thoreau was certainly more at home in the natural world than Emerson. Armed with jackknife, spyglass, diary, and pencil, he found the woods and fields alive with fascinating mysteries, spiritual meaning, and elemental truths. Thoreau undertook his twenty-six-month bivouac at Walden Pond in the mid-1840s in order to kidnap himself beyond the pinch of society and "entertain the true problems of life." In *Walden* (1854) he emphasized that "we are enabled to apprehend what is sublime and noble only by the perpetual instilling and drenching of the reality that surrounds us." But Thoreau defined "reality" in a transcendental sense. He was willing to exchange facts for "vague and misty forms." He loved nothing more than to close his eyes and float into the recesses of his inner consciousness, becoming in the process "a subjective—heavily laden thought, in the midst of an unknown and infinite sea." Knowledge, he taught, "does not come to us by details, but in flashes of light from heaven." This mystical vision proved far more meaningful to him than active immersion in the village life of Concord: "I find the actual to be far less real to me than the imagined."[20]

Brooding Idealism

The genteel, domestic, and transcendental forms of idealism dominated the intellectual landscape during the 1830s and 1840s, but three of America's most powerful and enduring writers fell outside such categories. Edgar Allan Poe, Herman Melville, and Nathaniel Hawthorne rebelled against the static conservatism of the genteel tradition, displayed a jealous disgust at the popular female sentimentalists, and castigated the Transcendentalists for their irrepressible optimism and airy metaphysics. Poe, for instance, characterized Emerson's philosophy as mere "twaddle" and expressed a desire to hang him.[21]

Hawthorne, Melville, and Poe found their guiding motive in the deep recesses of the tormented soul rather than in the wonders of nature or what Hawthorne called the "white sunshine of actual life." Although each wrote fictional works remarkable for their accurate details and realizing effects, their outlooks were essentially idealistic—or romantic—in their common desire to probe the heart of the self and explore the hidden spheres of feeling and spirit. They all demanded a wide latitude for the imagination. Melville said in *The Confidence-*

Man (1857) that fiction, like religion, should "present another world" to people while Hawthorne once assured his wife that "the grosser life is a dream, and the spiritual life is a reality."[22]

Poe, a self-described "citizen of somewhere else," argued that the object of poetry was not truth but beauty, and he curtly dismissed those who wrote fiction about everyday life. "That a story is 'founded on fact,'" he asserted in 1836, "is very seldom a recommendation." Poe found a welcome outlet for his own pursuit of the beautiful and bizarre in his "unworldly" fiction and verse. Adept at exploring the terrifying and hallucinatory peripheries of consciousness, he eagerly exploited the seductive lure of horror, addressing most of his poems and tales "to the dreamers and those who put their faith in dreams as the only realities." The gripping accounts of Poe's spectral raven, Washington Irving's headless horseman, the ghosts and spooks populating painter John Quidor's canvases, as well as the melodramatic plots of dime novels and popular stage plays, appealed directly to a perennial human desire for vicarious escape from the commonplaces of life.[23]

Melville also developed a keen interest in subjective explorations of the dark "world of mind." He could not rest content with a simple equation of realism with visual facts. The true artist must always look for the "little lower layer" of experience beneath surface appearances. After completing *Typee* (1846) and *Omoo* (1847), his embroidered travel stories set in the Polynesian islands, Melville confided to his English publisher that he was developing an "incurable distaste" for "my narrative of facts." He felt "irked, cramped & fettered by plodding along with dull common places," unable to explore fully his interest in the marvelous and the mysterious. To that end he began writing profoundly original works drawing upon his own maritime experiences and his complex metaphysical vision into the "soul of man."[24]

Moby-Dick (1851), for instance, begins as a realistic adventure tale. Its richly textured descriptions of people and places and its detailed scientific discussion of the whale as a species convince the reader that this is the "real thing." As the chapters unfold, however, the novel becomes more of a cosmic allegory than a mimetic representation of the whaling industry. The monomaniacal Captain Ahab cannot rest content with the world of appearances. He decides that "all visible objects, man, are but as pasteboard masks." The ultimate objective must be to "strike through the mask!" so as to grasp the "ungraspable phantom of life."[25]

Much of the genius of *Moby-Dick* resides in its hybrid quality: the book incorporates aspects of virtually every literary genre. Melville's close friend, Nathaniel Hawthorne, also combined romance with reality to produce several literary masterpieces. To preserve his own imaginative freedom, he, too, dissolved the actual world into shadows. In the preface to *The Scarlet Letter* (1850), Hawthorne related how his contact "with the materiality of this daily life" and "its petty and wearisome incidents" while working as Surveyor of the Custom House in Salem, Massachusetts, had threatened to deaden his creative abilities. Because America lacked an ancient and densely layered social order, he decided to explore that "neutral territory somewhere between the real world and fairyland, where the Actual and the Imaginary may meet, and each imbue itself with the nature of the other. Ghosts might enter here, without affrighting us."[26]

Although later literary realists stood in awe of Hawthorne's craftsmanship, they regretted his aloofness from his own social world and his preference for what he called "cloud land." To be a realist in the late nineteenth century meant to highlight the web of social relations and mores rather than focus on the self-absorbed ruminations of the isolated self. William Dean Howells, after reading several of Hawthorne's works in the 1860s, concluded that they "were so far from time and place that . . . I could not imagine anything approximate from them; and Hawthorne himself seemed a remote and impalpable agency." Henry James agreed with Howells's assessment. While admitting that Hawthorne "was the greatest imaginative writer we had," he stressed that "his principle was wrong. . . . Imagination is out of place; only the strictest realism can be right."[27]

The Relevance of the Ideal

In a perceptive essay on Hawthorne, the British novelist Anthony Trollope emphasized the idealistic quality of American thought and artistic expression during the first half of the nineteenth century. American works, he observed "generally are no doubt more given to the speculative — less given to the realistic" than their English counterparts. Trollope said that he and other self-proclaimed English literary "realists" sought to draw pictures "as like to life as possible, so that readers should feel they were dealing with people whom they might probably have known." Hawthorne, however, dealt with people "barely within the bonds of possibility."[28]

Trollope thus fastened upon the ruling theme of antebellum American culture. Throughout the second quarter of the nineteenth century, the most prominent writers and artists were more concerned with navigating imaginative eddies or preaching pious sentiments than with exploring the rapids of common life. Few focused their attention on the emerging new realities of mills and railroads, cities and banks, wage labor and class tensions. For them, classical standards and rural traditions, transcendent ideals and romantic mysteries, evangelical piety and Victorian moralism were compelling truths, and they refused to limit their attention to the present material world and its social imperatives.

Such a principled refusal to face crass material facts or mundane imperatives can be both splendid and necessary. Although sharply criticized by later realists, the idealistic and romantic writers, painters, and designers of the early nineteenth century reflected the era's peculiar cultural dynamic and satisfied its felt needs. They affirmed that concerns of the spirit *did* matter, that there was more to reality than what met the eye. The explicit moralism so evident during the Jacksonian era and after also proved to be a potent weapon in the crusade against slavery, in the empowerment of women, and in the cause of many other worthwhile social reforms. For all of the Olympian aridity attributed to the genteel tradition, several of its virtues remain appealing. Today we have nothing comparable to the orderliness, the composure, the sheer delight in the life of the mind and the respect for sophisticated civility and craftsmanship promoted by men of such magisterial eloquence and mannerliness as Lowell and Allston. If they seem remote from *us*, they spoke to the central values of *their* time. An infant American Republic still firmly grounded in the certainties of Christian religion found in such diverse forms of idealism an appropriate and redeeming cultural voice.[29]

2

New Paths

During the 1850s the tide of cultural idealism crested and slowly started to recede as profound new forces began to transform the landscape of social concerns and artistic taste. The decade before the Civil War witnessed rapid urban-industrial development, a sharp financial panic, wrenching sectional tensions over slavery and western expansion, and dramatic scientific advances. All of this helped to focus public attention on what a writer in *Putnam's Monthly* called the growing "predilection for the real and the practical." The leading idea of the day "is the *true.*"[1]

Such "practical" interests helped generate a self-conscious new realistic outlook in social thought and the arts that developed into a full-fledged movement by the end of the decade. Little by little, here and there, efforts at more candid and precise representation of contemporary life began to surface in cultural expression. Public discourse began to lose its sonorous quality and ethereal emphasis as people developed a consuming passion for "actual" knowledge about their increasingly diverse society.

Abolitionists, for example, found a potent weapon in first-hand accounts of slavery by blacks who escaped from bondage or managed to buy their freedom. White activists touted the slave narratives as unaltered testimonies, but they in fact frequently rewrote passages and fabricated events so as to excite the reader's interest and sympathy. Frederick Douglass, for example, reported that his abolitionist publishers told him simply to "give us the facts," and they would "take care of the philosophy." Such collaboration between well-intentioned white editors and fugitive slaves usually resulted in hybrid works. They mixed verisimilitude (first-person narrative, professions of objectivity, documentary evidence—letters, bills of sale, newspaper clippings) with characteristics of sentimental fiction (garish asides, shrill polemics, and melodramatic incidents). The result was a potent new genre in which blacks strove to write themselves into American consciousness.[2]

Whatever their actual degree of veracity, the slave narratives both fed and increased the public appetite for accurate information concerning life in the "real world." Equally instrumental in this regard were the many new magazines catering to the broadening pool of urban middle-class readers. *Putnam's Monthly, Knickerbocker, Harper's Weekly, Graham's Monthly*, and *Frank Leslie's Illustrated News*, among others, focused on the daily concerns and experiences of city life. In 1853 Charles F. Briggs, an ex-sailor turned writer of popular novels about Manhattan life, became the editor of *Putnam's Monthly*, a new journal of literature and the arts based in New York City, and he aggressively used the publication to promote a metropolitan cultural sensibility. "Local reality," he asserted, "is a point of utmost importance," adding that readers were weary of fiction that "paints only the gentle, the grieving, and the beautiful." Writers and artists in the coming "age of realism," Briggs concluded, must renounce the "extravagances of sentiment and action" and "unheard-of adventures and impossible exploits." They instead should provide "veritable and veracious" representations of everyday life.[3]

Where romantic idealists found inspiration in virgin woods and atop craggy mountains, Briggs and other editors, writers, and artists celebrated the grimy cityscape. "God made man to dwell in cities as well as the plains," declared a writer in New York's *Knickerbocker* magazine. Creative Americans must therefore learn to represent the city's sprawling vitality in their art.[4]

During the 1850s many writers of fiction began to exploit the metropolis. Herman Melville, for instance, turned his attention to the urban scene in his novels *Pierre* and *Redburn*, as well as several short stories published in *Putnam's Monthly*. So, too, did George Foster, a now forgotten New York journalist who published several revealing sketches of slum life while roaming the streets and by-ways of the city as a reporter for the *New York Tribune*. He gathered up "the fragments, the refuse, of every-day life" so as to reveal the degradations of the city to those insulated from such coarse actualities. Foster claimed that his "authentic" portrayal of the urban underclass, suffused as it was with equal doses of prudery and prurience, could be put to great social use. "It is such facts that society most needs—facts which show the actual conscience, color and dimensions of the cancer that lies eating at its very vitals."[5]

A Realistic Architecture

At the same time that journalists and editors were celebrating the vibrant new urban culture emerging in the United States at mid-century, several commentators were criticizing the slavish dependence of American architects on classical or medieval styles. The *Christian Examiner*, for instance, maintained that instead of "attempting to breathe life into the dry bones of a by-gone age, our architects should strive to make their buildings subservient to the uses and wants of to-day."[6]

Those promoting an indigenous American architecture found an ardent advocate in the Boston-born sculptor Horatio Greenough. During the 1840s, he urged artists to stop imitating classical and medieval styles and instead deal candidly with the practical needs of everyday life. Greenough found far more beauty in an American clipper ship, railroad bridge, or cotton mill than in neo-classical buildings because such utilitarian conveyances and structures were organically related to their planned uses. He applauded those architects and engineers who designed structures to serve their functions rather than to imitate historical styles. In discovering the "majesty of the essential," they eliminated superfluous adornment and liberated American architecture and design from a slavish adherence to outdated models.[7]

After returning to the United States in 1851, Greenough incorporated his new "functionalist" theory into a book entitled *The Travels, Observations, and Experience of a Yankee Stonecutter.* Emerson found it a "very dangerous book, full of all manner of reality & mischievous application." Yet "it contains more useful truth than anything in America I can readily remember." Emerson encouraged his friend's campaign for a utilitarian and organic theory of design, even offering to "fife in his regiment."[8]

Emerson's enthusiasm for Greenough's aesthetic program revealed his grow-ing resignation to the imperatives of a new cultural perspective. The "ghastly reality of things" continued to challenge his transcendent idealism. He now acknowledged that his outlook had been too cerebral, too detached from com-mon concerns and activities; it must now root itself more in practical realities. Persons "of romantic character," Emerson conceded, could no longer avoid facing the facts of daily life. "We cannot trifle with this reality. No picture of life can have any veracity that does not admit the odious facts." Writers and artists must admit the power of modern scientific method and the popular attrac-tion of practical values: "Let us replace sentimentalism by realism."[9]

Although a middle-aged Emerson grudgingly conceded the rise of a new realistic sensibility, others welcomed it as the first step in the development of an indigenous *American* culture. "In the century wherein we exist," proclaimed the editor of the *Cosmopolitan Art Journal*, "we have but a solitary standard of merit—utilitarianism." The "worshippers of the ideal" must give way to the "triumph of the real."[10]

The most fervent celebrant of the "triumph of the real" during the 1850s was a writer hardly known today: Charles Godfrey Leland, a clever journalist, folklorist, and essayist who became the spirited editor of Philadelphia's struggling

Graham's Monthly late in 1856. Leland brought to *Graham's* a fresh conviction that America had reached a transitional stage in its cultural history. The "old paths of thought" were being "broken up," Leland announced, and he urged young writers to escape from the confines of polite convention and "write about what you *see* and not what you *read*." *Graham's*, he pledged, would print no more effeminate "handkerchief poetry" or mawkish fiction. "Puling pathos, sentimental transcendentalisms, and vague reverie" would not be tolerated. Instead he solicited only virile writers who displayed a "keen appreciation and observation of life."[11]

Walt Whitman

Leland discovered in Walt Whitman the epitome of the vigorous new cultural outlook he advocated. By rooting romantic idealism in an affection for everyday realities, Whitman became the most potent catalyst for change in nineteenth-century American culture. Of course, Whitman displayed a rich passion for extremes, but, as T. S. Eliot observed, he saw "no chasm between the real and the ideal." More than any other member of his talented literary generation, Whitman successfully fused romantic and realistic values. In the process he became one of the most provocative figures in American cultural history. "If there is a distinctive 'American realism,'" novelist John Updike has written, "its metaphysics are Whitman's."[12]

In March 1842, Whitman, then a twenty-three-year-old editor of the *New York Aurora*, attended a lecture at the New York Society Library. It proved to be a remarkable occasion, for Emerson was the speaker, and his address, entitled "Nature and the Powers of the Poet," spoke directly to Whitman's own emerging literary personality. At one point in his remarks, Emerson declared: "We have had yet no genius in America, with tyrannous eye," who sees "in the barbarism and materialism of the times" the source of inspired poetry. Yet he predicted that the world's next great poet would be an American who "visits without fear the factory, the railroad, and the wharf."[13]

Whitman characterized Emerson's lecture as "one of the richest and most beautiful compositions . . . we have ever heard anywhere, at any time." Thirteen years later *Leaves of Grass* made its controversial appearance, and Whitman rushed an autographed copy to Emerson. He found the unusual poems remarkably powerful. The new volume, he wrote Whitman, constituted "the most extraordinary piece of wit & wisdom that America has yet contributed." Convinced that Whitman might be the sturdy new American poet he had been expecting, Emerson applauded "the beginning of a great career, which yet must have a long foreground somewhere."[14]

Whitman's new career as poet did have a "long foreground." He was born in 1819 (the same year as Herman Melville) on a central Long Island farm. At age four, while "still a little one in frocks," he moved west with his family to Brooklyn, then a furiously growing harbor and factory town. Walt Whitman lived an open-air, barefoot childhood. His was a boy's world—fresh, free, grainy, toughening. He learned carpentry from his father, and his older playmates showed him how to swim and sail, dig for clams, gather gull eggs, and spear eels. Apprenticed first to a Brooklyn newspaper editor at age twelve, he learned the "art and mystery" of printing over the next four years. In 1834, when the Whitman family—now boasting eight children—moved back to rural Long Island, the strapping fifteen-year-old stayed in Brooklyn and made frequent trips to Manhattan.

Almost daily Whitman took the ferry from Brooklyn to New York and back. The mast-hemmed harbor and the magnificent panorama of the rising city captivated his imagination. Upon arriving at "million-footed Manhattan," Whitman would become a carefree saunterer, taking in the city's activities with the curiosity of a tourist and the memory of a poet. The living tableau of urban life provided him with a constant stimulus, flooding his mind with images and encounters that would later be recreated in his poems. At the same time, his newspaper work reinforced his desire to give voice to contemporary concerns. "The true poem," he proclaimed, "is the daily paper."[15]

In the spring of 1841, the twenty-two-year-old Whitman moved to Manhattan and became a compositor-reporter for the New York World. He joined a local debating society and developed what would become a life-long love for the theaters and music halls along Park Row, Chatham Square, and the Bowery. In his spare time he read Emerson and wrote a liquor-inspired temperance novel, several flat short stories, and some undistinguished poems. He also castigated the nation's conventional literary standards and sanctimonious morality. Public taste, he claimed, had grown dependent on the "tinsel sentimentality" and "dish-water senility" of imported British fiction and poetry. Fresh, "virile" American writing could not gain a hearing.[16]

These were the years in which Whitman accumulated most of the experiences and images destined to appear in Leaves of Grass. The democratic ethos of the city provoked his self-appointment as the poet of the contemporary American scene and spirit. As a participant, observer, reporter, and poet all at once, Whitman displayed a compulsive appreciation for the pulsating energies of the physical world and the human hive. He did more than observe the spectacle before him. He bathed in the invigorating stream of "simple humanity" that represented America's lusty energy and democratic potential.[17]

This willing immersion in the whirl and boil of everyday life differentiated Whitman from Emerson and other Transcendentalists who preached the attractions of a democratic culture from a comfortable distance. He never believed that communion with "first principles" required "transcending" the physical or material world. To him, ultimate meaning resided not in the shadowy realm of the "Over-Soul," but in the everyday experiences of the here and now. Nor did he consider the inner life superior to the outer shell of material necessity. Whitman saw body and soul, ideal and real, spiritual and material, as equally important aspects of the good life. Unlike many of the Transcendentalists, he chose to entangle himself with the coarse hemp of human society. Where Emerson stressed that he "loved man, but not men," Whitman placed human-kind at the center of his own universe. A vague distinction perhaps, but an illuminating one. It explains why Whitman, unlike Emerson and Thoreau, could plant himself so comfortably amid his urban surroundings. The poet, he demanded, "must flood himself with the immediate age as with vast oceanic tides."[18]

Whitman also differed from his Transcendentalist counterparts in his attitude toward nature. Although he loved the countryside and shore, he discarded the pastoral myth as mere pap. The rural life that he knew entailed incessant hard work, a poor diet, and a paucity of educational and cultural resources. "No matter what moralists and metaphysicians may teach," he declared, people living in the countryside do *not expand and improvise so well morally, intellectually, or physically*" as city dwellers. Whitman stressed that he "did not fear the age of steam." The city provoked in him "a continued exaltation and absolute fulfillment."[19]

In developing his own literary philosophy, Whitman displayed a claustro-phobic resistance to the constraints of both genteel conservatism and domestic sentimentalism. Nothing about life shocked him, and prevailing Victorian mod-esties held no claim upon his social views or artistic expression. In 1850 he announced: "We have no patience with the merely imaginative when we can get the real." Whitman, with all his innocent arrogance, dismissed most Amer-ican poets as mere "confectioners and upholsterers of verse." He thought Lowell too delicate, Whittier too syrupy, Longfellow too refined, and Poe too morbid and shadowy. The true poet, he felt, must be "manly," must break the bonds of the dainty and the demure in order to transform the "immediate age" into verse and bring its teeming actualities to life with force and passion. He himself sought "To exalt the present and the real,/To teach the average man the glory of his daily walk and trade." He would give voice to the inarticulate and unheard, would speak for the lowest of the low, the "deform'd, trivial, flat, foolish, despised." Such an egalitarian idealism grounded in a prosaic realism reflected Whitman's unique blend of celebratory romanticism and earthly prac-ticality. In this union of genius and humanity resided the true strength of his vision. "Away with old romance!" he shouted. "I am the poet of reality."[20]

An extravagant proposition, but Whitman carried it off, with monumental results for American literary history: in 1855 he privately published *Leaves of Grass*. Just as *Uncle Tom's Cabin* loosened the shackles of slavery, *Leaves of*

Grass helped liberate American culture from its timid restraints and thereby altered the channel along which literature and the arts flowed. At last, it seemed, America had produced a robust writer unafraid of exposing coarse facts yet serenely optimistic about the future of the American dream. Fanny Fern declared in the *New York Ledger:* "Walt Whitman, the effeminate world needed thee." It demanded a man "who dared speak out his strong, honest thoughts in the face of pussillanimous, toadying, republican aristocracy."[21]

Of course, genteel critics blanched and scowled at the candid effusions in *Leaves of Grass*. One reviewer called it an "intensely vulgar, nay, absolutely *beastly* book." Another sneered at Whitman's "realistic" creed, his affection for banal particulars, and his desire to serve as a "camera to the world, merely reflecting what he sees." An outraged James Greenleaf Whittier supposedly threw the book into the fire, while James Russell Lowell promised to keep it out of the hands of Harvard gentlemen.[22]

What caused such intense reactions? Whitman's brassy egotism, profanity, and impiety, his erotic vehemence and explicit sexual references (copulation, masturbation, homoeroticism), shocked many conservators of decorum and morality. Almost as scandalous was his defiance of established conventions of rhyme, meter, and subject matter. Bad grammar, faulty punctuation, gutter slang, and violent transitions did little to endear his verse to the literary establishment. In beating "the gong of revolt," Whitman asserted that the poet must express "the momentous spirit and facts" of daily life. "What I tell I tell for precisely what it is." To that end he invited readers to "stand by my side and look in the mirror with me."[23]

In compiling his montage of facts, chants, confessions, and reflections, Whitman extolled things "as they are": farms, factories, and cities, utilitarian concerns and scientific phenomena, "eating, drinking, and breeding," material objects, popular entertainments, common laborers, farmers, miners, slaves, and prostitutes. Full of vernacular raciness and bristling with brawny declarations and tactile images, Whitman's rough-hewn, sensual prose-poems lyricized the common aspects of life. He resolved to be

> . . . the poet of commonsense and of the demonstrable and
>   of immortality;
> And am not a poet of goodness only. . . . I do not
>   decline to be the poet of wickedness also.

Whitman gave voice to the forbidden subjects of vice as well as human lust: "Copulation," he maintained, "is no more rank to me than death is."[24]

True, much of Whitman's poetry is bombastic and repetitive. Neither a systematic nor a consistent thinker, he catalogued things rather than analyzed ideas, and his universal hurrahs and overstocked inventories can grow tiresome.

For all of his faults and idiosyncrasies, however, he induced a new gusto and injected vital new facts into American poetry. "Give me initiative, spermatic, prophesying, man-making words," he wrote in his journal. Opposed to both a realism that sought "merely to copy and reflect existing surfaces" and an idealism that ignored prosaic or unpleasant truths, he displayed a rare ability to give "vivification to facts," to implant some bone and sinew into literature, to infuse the ordinary with "life" and "spirit" by unveiling the transcendent meaning inherent in the most dense particulars.[25]

In sum, Whitman served as the ferry linking antebellum romanticism to late nineteenth-century realism. There was something elemental in his wild, unkempt character, something generous and compelling in his universal sympathies, something invigorating in his voracious appetite for life in all its forms. Braiding idealism with materialism, embracing and caressing all around him, the undiscriminating Whitman showed the power and possibility of an inclusive poetry moored to everyday realities and unspoken truths. Many of the most talented among a younger generation of philosophers, writers, artists, and architects would draw their nutrients from Whitman's fertile version of romantic realism and his vision of a truly democratic culture. Philosopher William James and architect Louis Sullivan, for instance, saw in Whitman a contemporary prophet. William Dean Howells referred to the "good, gray poet" as the great "liberator." Writer Hamlin Garland labeled him the "genius of the present," adding that Whitman taught people to appreciate "the common things of the present, undaunted by the vulgar words and stern realities of our day."[26]

Transatlantic Literary Influences

Whitman's controversial effort to fuse poetry and prose reflected a broader trend at mid-century: the reading public was deserting conventional poetry in favor of modern fiction. In an illuminating short story entitled "The Nervous Man," John Greenleaf Whittier suggested that poetry had become too detached from practical concerns. Poetry "was something woven of my young fancies, and reality has destroyed it." In a later essay Whittier repeated his concern that American culture had become emasculated. It showed "no evidence of manly and vigorous exertion. . . . We are becoming effeminate in everything—in our habits as well as our literature."[27]

New currents in European literature reinforced the appeal of such a "masculine" realism. In France during the 1830s, Honoré de Balzac had rebelled against romantics who spent their time "smoking enchanted cigarettes." A prodigious, compulsive writer, Balzac professed an unabashedly cold and skeptical materialism. "Tell me what you have," he contended, "and I will tell you what you think." Such a materialistic philosophy led Balzac to crowd his pages with meticulous descriptions of houses, chairs, curtains, pictures, glassware and other things, leading Henry James to observe that "we often prefer his places to his people." With a scientist's probing eye and detached temperament, Balzac dissected rather than embellished his subjects. As he told his novelist friend George Sand, "You look for a man as he ought to be; I take him as he is."[28]

While Balzac's brutal candor and materialistic point of view jolted most American readers, several aspiring American writers marveled at his prodigious achievements. Henry James called him "the father of us all," and Theodore Dreiser attributed his own conception of life as a "spectacle" as well as his "first taste of what it means to be a creative writer" to his excited reading of Balzac.[29]

By the time Balzac died in 1850, his trailblazing efforts had helped create in France a full-fledged cultural movement devoted to the candid representation of contemporary life. "At last," Fernand Desnoyers exclaimed in 1855, "Realism is coming." Such controversial developments attracted excited attention in the United States. The *North American Review* claimed that the new school of French writers "are called Realists, and Realism is the object they are supposed to be trying to attain."[30]

The French realists had counterparts throughout Europe and in Great Britain. In Russia, Fyodor Dostoevsky, Ivan Turgenev, and Leo Tolstoy published stories remarkable for their severe objectivity and exact description. Meanwhile, English novels of contemporary social life by Charles Dickens, William Makepeace Thackeray, Charlotte Brontë, George Eliot, and Anthony Trollope displaced the historical romances of Walter Scott and his many imitators. Although gentler, more discreet, and frequently more sentimental than their French or Russian counterparts, these Victorian novelists shared a desire to couple realistic detail and everyday circumstances with a firm assertion of individual morality.[31]

Feminine Realism

Hawthorne perceived the growing taste for literary realism in the United States at mid-century when he observed that readers now "terribly insisted upon" fiction that dealt with their immediate world. He himself yearned to compose thickly worsted novels of contemporary life like those by Thackeray and Balzac. "It is odd enough," he confessed, "that my own individual taste is for quite another class of novels than those which I myself am able to write."[32]

While Hawthorne lamented his inability to write a fiction rooted in commonplaces, women authors took the lead in developing a distinctive brand of literary realism. In 1853 Caroline Kirkland, a prominent professional writer, wrote a long review-essay for the *North American Review* in which she assessed the state of feminine fiction. Novels of contemporary life, she reported, were beginning to displace "romances" dealing with "an ideal world." Where fiction in the past sought "merely to delight or to exalt the imagination" and was unrelated to "humdrum" activities, women writers were focusing on everyday topics and concerns. Nowadays, she observed, "there is no truth but literal truth." Kirkland announced that the "most popular" women writers had exchanged the cultivation of "tenderness, piety, imagination, and fancy" for "keen observation, powerful satire, knowledge of the world, strong common sense, and—though last not least—democratic principles."[33]

Although still marked by a strain of pious moralism and nostalgic sentiment, the local-color fiction written by women writers at mid-century questioned many prevailing social assumptions and literary stereotypes. In the process it took a bold first step toward greater realism. For the most part, the first generation of female realists described the familiar experiences of local life in a common manner, shifting attention from the grand, remote, and fantastic themes of much romantic fiction. For example, Alice Cary insisted that "the fictitious narrative has rarely the attractive interest of a simple statement of facts." She offered readers stories drawn from life in her "little village" rather than "a flight in the realms of fancy." For Cary, a primary motive for such "real-life" stories was "to dispel the erroneous belief" about "the necessary baseness of the 'common people' which the great masters of literature have in all ages labored to create." This required writers to adopt a close-up vision, focusing on the details and fragments of everyday life. [34]

Like Cary, Rose Terry (Cooke) felt equally at home amid the familiar routine of small-town New England domestic life. The daughter of a prominent Hartford family and a devoted member of the Congregational church, she graduated from the Hartford Female Seminary at age sixteen. She then began teaching first in rural Connecticut and later at a church school in New Jersey. In the late 1840s, when her sister died, Terry returned home to serve as surrogate mother for her nieces. While doing so she found time to write pious poetry and moralizing short stories. [35]

Over the next decade or so, however, Terry's style matured and her subject matter changed. Her stories in the *Atlantic Monthly* (eight were published in the first eleven numbers), *Harper's Monthly*, and *Putnam's Monthly* stripped the veneer off delicate parlor fiction and revealed the careworn face of village life in decline. Like the small towns of the New England hinterland, her fiction is populated mostly with women, since thousands of men had opted for work in mill towns, seaport cities, California gold fields, or western prairies. Unlike the domestic sentimentalists, however, Terry often assaulted conventional gender stereotypes. Instead of being self-denying angels or frail ornaments encased in corsets, most of her women characters are plain-looking spinsters or raw-boned wives mired in drudgery, their faces lined with care and exhaustion, manifesting a tenacious resolve against the elements and relentless domestic demands. [36]

Harriet Prescott Spofford, a prominent writer during the late nineteenth century, remembered that she and her friends found Terry's stories to be engaging "transcripts of genuine life, the interest interwoven with pure wit and humor, sweetness and tenderness." Terry's best stories avoided the improbable situations then common in "feminine fiction." Lowell praised her "vigorous prose" and called her one "of the most successful sketchers of New England character, abounding in humor and pathos."[37]

Rebecca Harding

Rebecca Harding (Davis) joined Rose Terry in charting a new course for American literature from romanticism to realism. She, too, introduced readers to tough, coarse, and sometimes grim new realities. Born in Pennsylvania in 1831, Harding lived in Wheeling, Virginia (now West Virginia), from age five until her marriage to Clarke Davis in 1863, then spent the rest of her life in Philadelphia. Wheeling was a coarse mill town consisting of two poplar-lined streets nestled between the slow-moving Ohio River and the steep Appalachian ridges.[38]

Harding was a rapt observer of town life in Wheeling. From her bedroom window she watched faceless crowds of Irish and Cornish mill workers trudging to and from work. Such scenes, she realized, harbored the compelling raw materials for a story, and she decided to paint the "reality of soul-starvation, of living death, that meets you every day under the besotted faces on the street."[39]

All very well, but a risky business, this decision to craft a realistic story set in a grimy industrial town. Would a genteel reading public balk at such alien and discomforting material? After months of writing and rewriting, Harding submitted "Life in the Iron-Mills" to the *Atlantic Monthly* and, to her surprise, the editor accepted it in January 1861. Still uncertain of her efforts, she asked that her name not be printed with the story.[40]

"Life in the Iron-Mills" appeared in April 1861 and created an instant sensation. Few American writers had depicted the conditions of working-class life or industrial labor with such power or candor, and several readers guessed that the anonymous author of such a virile story must be a man. Like Rose Terry, Harding realized the pioneering nature of her efforts. The grim story opens with an unsettling invitation: "I want you to hide your disgust, take no heed to your clean clothes, and come right down with me—here into the thickest fog and mud and effluvia. I want you to hear this story. There is a secret down here, in this nightmare fog, that has lain dumb for centuries. I want to make it a real thing for you."[41]

To make the "real thing" visible and palpable to her readers, Harding used her keen sensory perception and attention to detail. She highlights the difficulty of seeing life clearly and implicitly chastises the reader for ignoring the reality of social injustice:

> A cloudy day: do you know what that is in a town of iron-works? The sky sank down before dawn, muddy, flat, immovable. The air is thick, clammy with the breath of crowded human beings. It stifles me. I open the window, and, looking out, can scarcely see through the rain the grocer's shop opposite, where a crowd of drunken Irishmen are puffing Lynchburg tobacco in their pipes. I can detect the scent through all the foul smells ranging loose in the air.

A "slow stream" of mill workers flows underneath the narrator's alert eyes, displaying "dull besotted faces bent to the ground, sharpened here or there by pain or cunning; skin and muscle and flesh begrimed with smoke and ashes; stooping all night over boiling caldrons of metal, laired by day in dens of drunkenness and infamy."[42]

Harding then introduces Hugh Wolfe, a young furnace-tender already grown haggard and consumptive amid the grinding toil of the rolling mill. The Welsh immigrant puddler differs from his peers in that he has had some schooling and possesses an artistic talent for carving figurines from korl, the pink refuse from smelted pig-iron. A quiet, morbid man, he attracts the attention of several gentlemen visitors who take an interest in his carvings and encourage the mill owner to help cultivate Wolfe's talent. The owner, however, shows no interest. He has washed his "hands of all social problems—slavery, caste, white or black." In a desperate attempt to escape the soul-numbing routine of the mills, Deborah, Wolfe's hunchbacked cousin who ardently loves him, steals some money, and Hugh is falsely accused of directing the crime. Both are arrested. Hugh, restrained by *iron* manacles, is sentenced to nineteen years of hard labor, and the day before leaving for the penitentiary, he commits suicide in his cell. Deb serves out her prison term and then is befriended by Quakers who invite her to live in their neighboring community.[43]

The coarse subject matter, vernacular dialect, and harrowing tone of "Life in the Iron-Mills" stunned readers. In offering "no hope that" the injustices heaped upon industrial laborers "will ever end," Harding denied readers both the balm of a happy ending and the usual platitudes about the regenerative effects of Christian piety. Instead she provided the "outline of a dull life, that long since with thousands of dull lives like its own, was vainly lived and lost." For an America still wedded to rosy notions of the new industrialism beginning to dot the once pastoral landscape, "Life in the Iron-Mills" represented an enlightening glimpse of the horrors associated with a factory system. As one reviewer recognized, her fiction provided "a welcome exception to the common run of sensational or sentimental novels."[44]

Courbet and Millet

During the 1850s American painting also began to witness the first stirrings of revolt against neo-classical idealism. Here, too, European developments exercised a direct influence on American culture. As early as the late 1840s, young French painters led by Gustave Courbet rebelled against the "the ideal of the conventional" being taught in the salons and the Academy, and they called for a revival of the realist tradition embodied in seventeenth-century Dutch and Flemish art, a tradition that included common folk as subjects and depicted real people in real places doing real things.[45]

After the jury of the Universal Exposition refused to exhibit Courbet's paintings in 1855, he bitterly announced that the "label of realist has been imposed on me," and he proudly confessed his determination to depict "the manners, ideas, and appearance of my time as I see it." In his view, the "ultimate aim of realism is the negation of the ideal." True art could only exist "in the representation of *real* and *existing* objects." Other prominent French painters such as Jean François Millet shared Courbet's realistic philosophy and social commitments. They, too, sought to democratize art with the sincere representation of plain folk—coal heavers, gleaners, nuns, goosegirls—drawn from their familiar surroundings. When Walt Whitman saw Millet's *The Sower*, he announced that Millet "is my painter: he belongs to me: I have written Walt Whitman all over him."[46]

Several young American painters who studied in France fell under the spell of the French realists. Vermonter William Morris Hunt, for example, became a devoted follower of Millet. After being dismissed from Harvard during his junior year for rebellious behavior, Hunt had traveled to Europe, touring museums and studying under the best European masters in Rome, Düsseldorf, and Paris. In 1853 he settled in the village of Barbizon, where Millet and like-minded young French painters were living. There Hunt developed the same hatred of affected style and elegance, the same love of common appearance and the ways of humble folk, that inspired Millet and the other so-called Barbizon painters. "When I came to know Millet," he recalled, "I took broader views of humanity, of the world, of life. His subjects were real people who had work to do."[47]

After returning to the United States in 1855, Hunt set up a studio at Newport before finally settling in Boston in 1862. Among his Newport students was a young philosopher-to-be named William James, whose brother Henry also frequented the studio. The James boys delighted in Hunt's charismatic eccentricity. Tall and sinewy, with a long nose, gray beard, and penetrating eyes, he pelted his students with aphorisms about art. "Paint firm and be jolly," he told them. When he spied a student improving a model's appearance, Hunt ordered him to "Stick in the faults! A man's nose may be too long; but it belongs to him, and God made it." He taught students to respect "everything that *is, because it is!*" Hunt became both a beloved and dominant figure in the art community, and he was especially instrumental in educating Americans about the turbulent new trends in French painting and alerting them to the attractions of "realistic" art.[48]

The Pre-Raphaelites and John Ruskin

At mid-century a small group of young painters and poets in England known as the Pre-Raphaelite Brotherhood was also defying prevailing artistic conventions and promoting its own version of realism. In 1848 John Everett Millais, William Holman Hunt, and Italian-born Dante Gabriel Rossetti revolted against the dogmatic standards of the Royal Academy of Art. A sterile conservative tradition, they charged, had been built around a continuing attempt to imitate the theatrical grandeur of Raphael, the early sixteenth-century Italian master, who had depicted holy figures as dignified, lovely, and above the tawdry realities of everyday existence. The British insurgents, however, preferred a simpler and more sincere religious art characteristic of the medieval Florentine and Flemish painters who preceded Raphael.[49]

These "Pre-Raphaelites," all earnest and deeply reverent individuals, promoted "truth to nature": the detailed representation of the literal appearance of things. They painted with microscopic precision in full sunlight so as to represent objects in their most clearly illuminated state. Their ambition was to substitute scientific accuracy for emotional expression and imaginative license. The *Westminster Review* commended the Pre-Raphaelites for "refusing to let a mist of the so-called ideal and beautiful blind it to the rough realities of the world we live in."[50]

The outspoken young British art critic John Ruskin also applauded the Pre-Raphaelites for being true to nature as well as to their own times. In 1843 this dogmatic evangelist of taste began publishing a series of influential books which preached a common theme: nature was superior to art; realism was preferable to imagination. Painters, he felt, had become too egotistical; they were more concerned with displaying their mental virtuosity than with accurately representing their subjects. Ruskin chastised artists who preferred to paint "things as they are *not.*" The truly great painters, he insisted, "mirror" nature just as they see it, "rejecting nothing, selecting nothing, and scorning nothing."[51]

God was in the details, Ruskin believed, and he praised the Pre-Raphaelites for filling their canvases with concrete particulars rendered with excruciating precision. Artistic greatness, he stressed, depended upon an "intense sense of fact," and the more facts the better. Ruskin added, however, that verisimilitude

William T. Richards, *In the Woods*, 1860. *(Bowdoin College Museum of Art.)*

was only a means to an end, and the end was ultimately moral rather than mimetic. Visible facts—the petals of a flower, the veins of a leaf, the grain of a rock, or the purling of a brook—were not to be equated with artistic truth. Ruskin wanted realistic methods applied to uplifting subjects only, and he demanded that Nature's moral symbols be represented as well as its visible details. Like most of the Pre-Raphaelites, he remained a philosophical idealist and religious medievalist, convinced that art, while founded on the reality of natural fact, must nevertheless be filtered through the spiritual imagination of the artist to arouse "noble emotions." The greatest painters presented "sifted truth."[52]

A New Path

The notion of "sifted truth" found a receptive audience among young American painters, writers, and intellectuals. In 1855 a writer in *Putnam's Monthly* declared that the "unreal art" of Washington Allston and other romantics was giving way to a younger group of painters "of things real, and of which they know." In 1863 several of these young "realists" organized the Society for the

Advancement of Truth in Art and began publishing their own periodical called *The New Path*. Among the charter members, all in their twenties, were Thomas Charles Farrer, an Englishman who had studied under Ruskin before emigrating to America in the late 1850s; Clarence Cook, a New York art critic; Clarence King, a writer, geologist, and art critic from Newport; Russell Sturgis, Jr., and Peter Wight, both New York architects; and Charles Herbert Moore, a New York landscape painter. Wight explained that they "never called themselves Pre-Raphaelites. They preferred to be called only 'Realists.'"[53]

In the *New Path's* first issue in May 1863, the editors claimed that they were fighting "for truth and reality" in all the arts. American painters, writers, architects, and sculptors must find their "own free way" to represent their contemporary surroundings: "Time and future generations will ask of our art and our literature, 'Is this the way the people of the nineteenth century worked, dressed, and acted?' " It was "the artist's first duty," Charles Moore concluded, "to be true to the real." The *New Path* announced that there were now two contrasting schools of art evident in the United States: "One is sentimental, dreamy and struggling after that it calls the ideal. The other is hard-working, wide awake, and struggling after the real and true." The editors named young Winslow Homer as the best representative of the new realism and predicted that "the Realists will carry the day."[54]

Yet idealism in its various guises remained a potent force in American cultural life. George Inness, the celebrated painter of poetic landscapes, contemptuously dismissed the work of American "realists" as "puddling twaddle." Public taste, he lamented, "has been led to desire what *is called* the real in landscape, that is to say, the local and particular, and not the universal or the ideal." Inness's anxious defense of artistic idealism reveals that the battle lines of a profound cultural debate were being established at mid-century. By the 1860s a self-consciously realistic perspective had organized for a massive assault on prevailing aesthetic assumptions. A writer in 1858 observed that it was too early to predict the fate of what he called the "new school of fact," for realism remained in its childhood, "but a childhood of so huge a portent that its maturity may well call out an expectation of awe."[55]

Part II

The Generative Forces

From its first tentative stirrings in the 1850s, realism matured into a full-fledged cultural force during the second half of the century. Not surprisingly, its rapid development provoked excited speculation about its causes. In 1886 the Boston literary critic Thomas Sergeant Perry asked his friend William Dean Howells what had generated the realistic movement they both had been zealously promoting for the past twenty years. Howells feebly replied that realism just "came," and it seemed "to have come everywhere at once." One can sympathize with Howells's inability to discern the causes of a whirlpool in which he was still swirling, but the question remains: Why did the pendulum of taste swing toward realism during the second half of the nineteenth century?[1]

On the most general level the answer is obvious: the rise of realism resulted from a transformed social, intellectual, and moral landscape. The "traditions and usages of past ages are broken, or at least discredited," declared the outspoken Yale social scientist William Graham Sumner, and modern America required new modes of thought and expression. The following three chapters explore how historical events, intellectual developments, and social convulsions combined during the second half of the nineteenth century to focus attention on practical concerns and everyday realities.[2]

3

Touched with Fire

mong the primary agents of social and intellectual change during the nineteenth century, none proved more potent or traumatic than the Civil War. It was, of course, the pivotal event of American political and social history. Yet it also served as the hinge in the nation's cultural development, a turning point after which intellectual life and artistic expression were perceptibly different. A writer in *Century* magazine claimed that the war "brought a change in the disposition of the intellectual forces of America—one that was to be looked for, and in many respects to be desired."[1]

Many scholars have argued that the Civil War's murderous requirements were alone sufficient to generate the realistic movement in thought and the arts. "The Civil War," wrote Lars Ahnebrink, "gave a fatal blow to romanticism and Emersonian idealism." This explanation, however, is too tidy; it oversimplifies the complex—and often contradictory—effects of the nation's bloodiest struggle. Romanticism suffered numerous blows as a result of the Civil War, but none proved fatal. Realism, meanwhile, experienced its own triumphs and setbacks amid the fratricidal struggle.[2]

From the moment Lee surrendered to Grant, the actual Civil War has been concealed under layers of comforting legends and nostalgic sentiments, and this haze of myth complicates the task of sorting out the war's various effects. As Justice Oliver Wendell Holmes, Jr., predicted, there would always be "two Civil Wars," the "war in fact" and the "war in retrospect." Songs, statues, novels, prints, and paintings, as well as movies and television shows, have portrayed the Civil War as a panorama of colorful uniforms, heroic deeds, noble purposes, and sword-and-roses courtships.[3]

The war contained all of these elements. Georgia's poet-soldier Sidney Lanier, for example, spent most of his time in uniform as a scout watching Union troop movements near Petersburg, Virginia, playing the flute for his

George C. Lambdin, *Kissing the Sword*, 1865. *(Berry-Hill Art Galleries, Inc., New York.)*

commander, courting local belles, or visiting the public library. "Our life," he recalled, "was as full of romance as heart could desire. We had a flute and a guitar, good horses, a beautiful country, splendid residences inhabited by friends who loved us, and plenty of hair-breadth 'scapes from the roving bands of Federals."[4]

Yet underneath the Civil War's romantic veneer lurked grim realities: mass killing, maiming, and civilian travail that sobered many participants and onlookers. In this sense the war served as a double-edged sword. For some the wrenching event reinforced romantic and sentimental tendencies. For others it provoked—or heightened—a more realistic outlook toward life and culture.[5]

Patriotic Ecstasies

War is the most romanticized and the least romantic of human endeavors. From a distance, a battlefield is all spectacle and glory; yet the actual combat experience of the Civil War proved otherwise. The editor of *Harper's Monthly* remarked that "while the hum of war is heard far away, we see only the romantic aspect, and feel only the stirring excitement." The Civil War turned into a devastating affair—both physically and psychologically. For many combatants, the horrors of battle and the tedious degradations of camp life and sea duty corroded ideals and punctured illusions. The loss of an arm or leg stripped combat of its excitement, and the unrelenting carnage caused even the most calloused observers to blanch in disbelief. As Abraham Lincoln confessed in his second inaugural address, no one expected the war to become so "fundamental and astonishing." The struggle's prolonged agonies led many Americans, at least for a while, to call into question the comforting sentimentalism prevalent in affairs of the mind. "The cannon-ball," said a reporter, "shoots away veils; it opens eyes, and ears, and mouths."[6]

In the spring and summer of 1861, however, few Americans realized what a chastening furnace the war would become. The firing on Fort Sumter spawned innocent visions of glory and gallantry as men eagerly left farms, factories, and shops to join the ranks. "To me, in my boyish fancy," remembered Wilkinson James, the younger brother of William and Henry, "going to war seemed glorious indeed." He and others felt a sudden urgency of purpose absent from their peacetime regimen. "The war is making us all tenderly sentimental," observed Mary Chesnut, wife of a Confederate congressman. She viewed the coming war as a vicarious fantasy, "all parade, fife, and fine feathers."[7]

Prominent New Englanders predicted that the war would do more than end slavery and nullify secession; they saw in the struggle a catalyst for societal regeneration and masculine revival. Oliver Wendell Holmes, Sr., called it "our Holy War" and promised that the "war fever" would afford "our poor Brahmins" a welcome opportunity to harden their flabby muscles and effete sensibilities. So he urged the hesitant to join the crusade: "Listen young heroes! Your country is calling! Time strikes the hour for the brave and true!"[8]

Walt Whitman also gloried in the coming clash of arms. "I welcome this menace," he wrote. "I welcome thee with joy." Even young Louisa May Alcott felt a rush of martial passion. She longed "for a battle like a warhorse when he smells powder." Alcott later added that "I've often longed to see a war, I've often longed to be a man, but as I can't fight, I will content myself with working for those who can."[9]

Alcott's neighbor and hero, Ralph Waldo Emerson, also expressed a militant fury at the outset of the fighting. War, he wrote, offered a potent "tonic" that would "restore intellectual & moral power to these languid & dissipated populations." The war to preserve the Union and end slavery would have many other "good purposes." It "is a realist, shatters everything flimsy & shifty, sets aside all false issues, & breaks through all that is not real itself." He therefore encouraged its grim reaping. "Let it search, let it grind." Emerson then revealed how little he knew about the actualities of modern warfare when he asserted that the battlefield was no more dangerous than the home: "The child," he wrote, "is in as much danger from the staircase, or the fire-grate, or a bath-tub, or a cat, as the soldier from a cannon or an ambush." Nathaniel Hawthorne, who knew better, told a friend that Emerson "was breathing slaughter."[10]

After visiting Boston and nearby Concord in June 1862, Rebecca Harding testified to the transcendental detachment with which the philosophical idealists discussed the distant but glorious war. She stayed at Wayside, the home of Nathaniel and Sophia Hawthorne, and her hosts invited Emerson and Bronson Alcott to meet the lady responsible for writing "Life in the Iron-Mills." Harding remembered that she "had just come up from the border where I had seen the actual war; the filthy spewings of it, the changes in it . . . for brutish men to grow more brutish, and for honorable gentlemen to degenerate into thieves and sots. War may be an armed angel with a mission, but she has the personal habits of the slums." The Hawthornes' celebrated friends, however, seemed unaware of such realities. Alcott and Emerson, Harding concluded, "knew no more of war as it was, than I had done [as a child] in my cherry-tree when I dreamed of bannered legions of crusaders."[11]

Harding decided that these philosophers were childishly ignorant of life. While "they thought they were guiding the real world, they stood quite outside of it, and never would see it as it was." For all of their savage fervor, Alcott and Emerson lacked the "back-bone of fact." Hawthorne, by contrast, harbored a humane skepticism about the spreading conflict. After allowing his guests to preach about war's virtues for more than an hour, he had heard enough. Like an elderly priest bored by too long and tangled a confession, he suddenly rose and interceded: "'We cannot see that thing at so long a range. Let us go to dinner.'"[12]

Sobering Experiences

Harding and Hawthorne knew that wars maim ideals as well as combatants. The Civil War proved to be a relentless and random reaper. It became the first total war, stubbornly protracted and involving whole societies across eighteen states and territories. Some three million men served in the military and 630,000 died of wounds or disease. In the Union army one of every nine soldiers died in the war; in the South the toll was one in every four. Fifty thousand of those who survived lost one or more limbs. The impact of such ghastly slaughter penetrated well behind the lines: homes were turned into headquarters, churches into hospitals, and civilians into engaged partisans. Hawthorne recognized as much: "There is no remoteness in life and thought, no hermetically sealed seclusion . . . into which the disturbing influences of the war do not penetrate."[13]

Gender roles especially felt the conflict's transforming impact. While men fought the war, women supported and endured it. Initially the call to arms reinforced heroic images of female self-sacrifice and domestic skills. Women north and south sewed uniforms, composed uplifting poetry and songs, managed plantations or businesses, organized relief societies to raise money and supplies, and nursed the wounded and sick.[14]

But the war's gruesome toll eventually eroded martial enthusiasm. For many young women forced to shoulder new household burdens, the conflict's hardships accelerated their maturity and denied them an adolescence. In January 1865 a southern woman confided that the war "commenced when I was thirteen, and I am now seventeen and no prospect yet of its ending. No pleasure, no enjoyment—nothing but rigid economy and hard work—nothing but the stern realities of life." The death of loved ones was especially sobering. In 1868 a Tennessee widow named Mary Brown wrote a letter to her cousin in which she reflected on the war's impact: "In many respects I feel like a different mortal, have learned that life is earnest, life is real." She and other bereaving women came to look on the war with what Emily Dickinson called a "chastened stare."[15]

Of course, the Civil War also transformed the outlook of thousands of men. Many of the soldiers and sailors saw the event primarily as a ritual of masculine achievement rather than as a moral or political crusade. Henry James, Sr., interpreted the onset of fighting as a natural transition in the life of the nation: "from youth to manhood, from appearance to reality, from passing shadow to deathless substance." After visiting his son Wilkinson, who was recuperating from a wound received during the Union attack on Fort Wagner, James wondered at "so much manhood so suddenly achieved." When another son, Bob, talked of quitting the army because of poor health, his father dismissed the idea as a sign of "passing effeminacy" and an "unmanly project." He told his wavering son "to be a man, and force yourself like a man to do your whole duty."[16]

Most of America's promising young writers and artists did not participate in the Civil War, and this fact often served as a psychic wound that dogged their consciences and influenced their art long thereafter. William Dean Howells, who lived in Italy as consul to Venice during the war, admitted that he had always felt bound to make "some excuse for turning his back on his country in her hour of need." By avoiding the most profound event of the century, he and others were forever denied the badge of martial experience confirming their masculine identity. Henry James, Jr., confessed that his relation to the war consisted of "seeing, sharing, envying, applauding, pitying, all from too far-off." In many ways, the dreadful trials and masculine camaraderie of the Civil War, whether experienced or avoided, helped steer some people toward a more "masculine" posture and thereby buttressed the "realistic" assault on "feminine" sentimentalism and decorous idealism. As a writer noted in 1863, the "war has done much to make the people more manly, thoughtful, serious, earnest."[17]

Holmes — Patrician Patriot

Oliver Wendell Holmes, Jr., experienced a similar rite of passage amid the war's dreadful rigors. An ardent young abolitionist imbued with a keen sense of patrician duty, he was commissioned a first lieutenant shortly after graduating from Harvard in the summer of 1861. Holmes found camp life in Maryland an enlightening experience, as he mingled with diverse young men from across the nation. "All these things you see," he wrote his parents, "give reality to the life, but I don't expect any fighting for the present." The fighting erupted sooner than Holmes expected. A few weeks later, at the battle of Ball's Bluff, a spent bullet knocked him down. Helped to his feet, he again charged forward, only to have another ball rip through his chest. "I felt as if a horse had kicked me and went over," he remembered. Taken to a field hospital, he watched a surgeon calmly amputate the finger of an unanaesthetized soldier. Holmes had now seen the cruel face of war and felt its painful spite. Still, he took some comfort in his own "manly" efforts under fire. He reassured his mother that the wound was not mortal and that he had "felt and acted very cool" in performing his duty.[18]

By the summer of 1862, Holmes, now a captain, was back with his unit as it participated in the series of battles known as the Seven Days. Tidewater Virginia's heat, humidity, and thick undergrowth made the business of killing especially enervating. The intense fighting leached away Holmes's quixotic bravado and gave rise to a more calloused perspective. He no longer could afford the luxury of "'unproductive emotion" in the midst of such prolonged hardships and mortal threats. "It is singular," he wrote after the battle of Fair Oaks, "with what indifference one gets to look on" the "fly blown and decaying" bodies, "of men shot in the head, back, or bowels." Far from being a romantic messenger, he added, a bullet "has a most villainous greasy slide through the air." Three months later, on September 17, 1862, a bullet slid through the back of Holmes's neck in Antietam's bloody cornfield. It missed his windpipe and jugular, then split the seams of his collar before exiting. When his father found him on a hospital train in Pennsylvania, their terse greetings spoke eloquently to the transforming impact of war: "How are you, Boy?" Holmes asked. "Boy, nothing," his son replied.[19]

While recuperating at home, Holmes shocked young ladies eager for first-person accounts of his heroism when he irritably announced that war "is an organized bore." Young Holmes had changed, his family and friends whispered. After two months convalescing, Holmes returned to the front. Then, in the spring of 1863, near Chancellorsville, a bullet struck him in the heel, shattering the bone and tearing the ligaments. It took eight months for him to return to the ranks, and, soon thereafter, he witnessed the furious butchery at Spotsylvania, Cold Harbor, and Petersburg. At Spotsylvania he saw "the dead of both sides . . . piled in the trenches five or six feet deep—wounded often writhing under superincumbent dead." This was too much; he had seen enough bloodshed and incompetence; the narcotic of patriotism and the adrenalin of combat had worn off; he could no longer accept the "butcher's bill." Holmes now began to display the pathos that prolonged war gives to the sensitive combatant. He wrote his parents that he was going "to leave at the end of the campaign . . . if I am not killed before." When they urged him to stay at his post, he retorted with disenchanted brevity: "I started in this thing a boy. I am now a man and I have come to the conclusion . . . that the duty of fighting has ceased for me." No longer "the same man" who had originally enlisted, he was "not so elastic as I was and *I will not acknowledge the same claims upon me under those circumstances* that existed formerly."[20]

Soon thereafter, twenty-three-year-old Oliver Wendell Holmes, the seasoned, sickened campaigner, returned to Cambridge, entered the "black and frozen night" of Harvard Law School, and immersed himself in "a thick fog of [legal] details." While a law student, he visited Emerson, the man who before the war had been his "revered master." But this encounter disillusioned Holmes. The aging Emerson "aired transcendental ideas about life and death, conscience and duty" that the veteran soldier now found utterly annoying. Having crossed "the threshold of reality," he could no longer relate to Emerson's cosmic utterances or relentless innocence.[21]

With almost grim dedication—some said obsession—Holmes embarked upon what would become a distinguished career in jurisprudence. His war experience permanently affected his outlook. He had been "soaked in a sea of death" and had "touched the blue steel edge of actuality." In 1884 Holmes, by then a judge, delivered an eloquent Memorial Day address in which he spoke for his generation of veterans: "Through our great good fortune, in our youth our hearts were touched with fire." The flames were chastening as much as exhilarating; they burned through many illusions. Holmes had become ruthlessly unromantic. Eleven years later he noted that the "cannon's roar" had "stilled the song" of patriotic glory. War, he had discovered, "is horrible and dull," anything but romantic, and in this sense it led him to look afresh at life as something throbbingly real. [22]

The war reinforced Holmes's already developing skeptical materialism. Neither flowery rhetoric nor religious belief engaged his loyalties any longer; he now sought knowledge of society by examining the concrete actions of people rather than their words or conventions. The law, he repeatedly stressed, was "not the place for the artist or the poet." It derived its real meaning—and vitality—from neither abstract logic nor absolute and eternal truths but in the "actual forces" of contemporary experience. An early convert to a gentlemanly Darwinism, Holmes insisted that the law must evolve with society; it must "fit the facts," and the facts were constantly changing. [23]

Life, Holmes concluded, represented "a roar of bargain and of battle." It paid no attention to polite sentiment, metaphysical speculation, or the presumed certainties of a formal legal tradition. As he once asked the dean of the Harvard Law School, "Doesn't the squashy sentimentalism of a big minority of our people about human life make you puke?" Holmes considered utopians not only naive but dangerous. Only an outlook rooted in "palpable realities" could survive in the fluid and treacherous modern world. Realism for Holmes had come to mean the acceptance of life's "predatory" quality and its many contingencies and ambiguities. "I like to see someone insist," he wrote, "that the march of life means a rub somewhere." The experience of war had thus transformed Holmes's outlook, leading him to see how the friction of reality, with its grainy "actualities and immediacies," eroded the abstractions of idealism. [24]

Drum Taps

At the outbreak of hostilities Walt Whitman had urged "relentless war" to end secession, but he, too, changed his exuberant tone after witnessing the war's grim reaping. Working tirelessly as a volunteer nurse in military hospitals in Brooklyn and later in Washington opened his middle-aged eyes to what he called a "new world." He told his mother that he had "seen *war-life*, the real article."

Armed with oranges, peppermints, and limitless sympathy, Whitman talked with the convalescents and the dying, wrote or read letters for them, shared their miseries, and fought their loneliness. The sight of amputated limbs piled high, the soul-wrenching cries of the maimed and the dying, and the nauseating stench of overcrowded and gangrene-infested hospitals took the foam off Whitman's martial fervor and yanked him out of a stagnant egoism. "Mother," he wrote home, "I feel the reality more than some because I [am] in the midst of its saddest results so much." The war became "the very centre, circumference, umbilicus, of my whole career."[25]

For Whitman the Civil War proved to be not "a quadrille in a ball-room," but a "strange, sad war." Its "stern realities" planted seeds of doubt that haunted him like specters:

> Year that trembled and reel'd beneath me!
> Your summer wind was warm enough, yet the air I breathed froze me,
> A thick gloom fell through the sunshine and darken'd me,
> Must I change my triumphant songs? said I to myself,
> Must I indeed learn to chant the cold dirges of the baffled?
> And sullen hymns of defeat?

Whitman's wartime letters and notebook entries never mention the romantic side of war, only the human waste. In 1863 he told his mother that "one's heart grows sick of war, after all, when you see what it really is—every once in a while I feel so horrified and disgusted—it seems to me like a great slaughterhouse & the men mutually butchering one another."[26]

The war's "mortal reality" gave greater depth, poignancy, and clarity to Whitman's poetry. He told a friend that *Drum-Taps*, his collection of fifty-three war poems, was "realistic" because it expressed "the pending action of this *Time & Land we swim in*," and avoided all "superfluity." He also claimed to have exercised greater "control" over his ego. Many of Whitman's war poems are tersely descriptive rather than lyrical in tone. More solemn and disciplined than the exultant songs in *Leaves of Grass*, they report what he saw, letting the images and facts speak for themselves. "The Wound-Dresser" reveals his leaner style:

> From the stump of the arm, the amputated hand,
> I undo the clotted lint, remove the slough, wash off the matter and
>   blood,
> Back on his pillow the soldier bends with curv'd neck and side-
>   falling head,
> His eyes are closed, his face is pale, he dares not look on the bloody
>   stump,
> And has not yet looked on it.

The final poems in the collection are bittersweet, celebrating the Union victory but regretting the ghastly cost. In "Turn, O Libertad," Whitman depicts the war as a liberating event for writers and artists as well as for slaves. He urges Americans to turn away from "the chants of the feudal world," to give "up the backward world," and turn to the modern world.[27]

At the same time, however, Whitman continued to stress the importance of enduring ideals. He celebrates: "The announcements of recognized things, science,/The approved growth of cities and the spread of inventions." These "stand for realities—all is as it should be." But equally real are "our visions, the visions of poets, the most solid announcements of any." After the Civil War, a chastened but not cynical Whitman remained a uniquely powerful hybrid of individual mysticism and democratic realism.[28]

Camera Images

The hardening effects of participation in the Civil War reflected in the experiences of Holmes and Whitman were widely felt. A Union surgeon serving with the Army of the Cumberland confessed in a letter to his wife that he was "sick and tired of bloodshed. Weary and worn out with it. . . . I have amputated limbs until it almost makes my heart ache to see a poor fellow coming in the ambulance to the Hospital." After describing in gruesome detail his daily surgical routine, he stressed that the "horror of this war can never be half told. Citizens at home can never know one fourth part of the misery brought about by this terrible rebellion."[29]

How could anyone convey the agonies of such a war to those behind the lines? Of course, it was impossible. Yet the new medium of photography provided civilians with more accurate images of the Civil War than of any other previous conflict. Frenchman Louis Daguerre's discovery in 1839 that sunlight could be used to make precise images on chemically treated metal plates had captured the popular imagination in the United States. Overnight a new medium was born, and people were astonished, overwhelmed, enthralled by the camera's terrible objectivity and scientific exactitude. "A camera! A camera! cries the century," Emerson noted in his journal. By the onset of the Civil War, the quality of photography had improved enormously, and Matthew Brady's platoon of skilled assistants, as well as the hundreds of other photographers covering the war, provided Americans—north and south—with thousands of images revealing the visual facts of war.[30]

People fell silent upon seeing the black-bordered photographs of "grim-visaged war" displayed in Brady's New York gallery during the fall of 1862. The editor of the *New York Times* said that Brady had revealed "the terrible reality and earnestness of the war. If he has not brought bodies and laid them in our

Matthew B. Brady or staff, *Dead Confederate Soldier with Gun, Petersburg, Virginia,*
1863. *(Library of Congress.)*

door-yards and along [our] streets, he has done something very like it." Although
photographers occasionally repositioned their subjects or retouched negatives to
heighten the emotional impact, the public at the time readily accepted the
images as distinctively "true." In fact, one visitor to a gallery naively proclaimed
that "it is impossible for photography to lie."[31]

Although the entrepreneurial Brady mistakenly received credit for many of
the 7000 Civil War photographs, two of his talented associates, Alexander Gardner and Timothy O'Sullivan, were responsible for the best pictures. The Scot-
tish-born Gardner served as Brady's assistant for seven years before starting his
own gallery with his son in 1863; O'Sullivan then joined Gardner. Like Brady,
Gardner highlighted photography's documentary function, but he went even
further in declaring the new medium's superiority to written description. In his

Field Where General Reynolds Fell, Gettysburg, July 1863. (Chicago Historical Society.)

Photographic Sketchbook (1866), he contended that "words may not have the merit of accuracy; but photographic presentments of them will be accepted by posterity with an undoubting faith."[32]

One of Gardner's better-known photographs is "A Harvest of Death at Gettysburg," a grisly scene of sun-bloated, slain "heroes," their limbs stiffened and their mouths agape. In an accompanying commentary, Gardner concluded that such "a picture conveys a useful moral: It shows the blank horror and reality of war, in opposition to its pageantry. Here are the dreadful details!"[33]

The elder Oliver Wendell Holmes, who had long been fascinated by the camera's scientific and aesthetic implications, became the most perceptive analyst of war photography. In the summer of 1863, he visited Brady's gallery and lingered over the gripping pictures of the Antietam battlefield, awed by their clarity and power. Camera views, he concluded, provided such a "frightful amount" of detail that "the mind feels its way into the very depths of the picture." Those who wanted "to know what war is," Holmes wrote in the *Atlantic*, must "look at these illustrations."[34]

The Antietam photographs particularly impressed Holmes because they recorded scenes that he had witnessed. After the battle on September 17, 1862, he had received a telegram reporting that his son had been wounded. Frantic for details, Holmes boarded a train and headed for Maryland. The scenes of death and destruction along the way overwhelmed him. "It was a pitiable sight, truly pitiable."[35]

Upon reaching Sharpsburg, Holmes learned that his son was not mortally wounded and had been sent north, so he decided to linger and visit the battle-field. Although most of the dead had been buried, sobering evidence of fighting on a horrible scale remained painfully visible. One mass grave containing over a thousand bodies had been covered so hastily that hands and feet protruded above ground. The air stank with rotting flesh, and the fetid ground ached with a magnificent sadness. The debris of mass combat littered the horizon: fragments of bodies, bloated horse carcasses, clothing, haversacks, blankets, drums, canteens, cartridge boxes, scraps of food, and patches of curdled blood. "It was like the table of some hideous orgy left uncleared," he gasped. Just as death stiffened bodies, the sight of death stiffened sensibilities. Holmes, who had earlier viewed the war as a metaphysical abstraction, now realized that it was far more real and ghastly than he had imagined from the comfort of his Cambridge home.[36]

Several months later Holmes purchased copies of Brady's photographs of the same bloody landscape at Antietam. "It was," he wrote, "so nearly like visiting the battle-field to look over these views, that all the emotions excited by the actual sight of the stained and sordid scene, strewed with rags and wrecks, came back to us, and we buried them in the recesses of our cabinet as we would have buried the mutilated remains of the dead they too vividly represented." The camera's "terrible mementoes" portrayed war as "a repulsive, brutal, sickening, hideous thing." Holmes reported to a Boston audience that he now knew "what war means." It is "a grim business."[37]

The War in Print

Since newspaper and magazine reproduction of photographs had yet to be mastered, most civilians followed the course of the war through the eyes of journalists. "When the times have such a gunpowder flavor," Henry Wadsworth Longfellow declared, "all literature loses its taste. Newspapers are the only reading. They are at once the record and the romance of the day." To satisfy the insatiable demand for war news, editors sent hundreds of reporters to the front. Some wrote dispatches soaked in sentimentality, others wrote brazen propaganda, but a few provided strikingly authentic accounts of the fighting. One correspondent maintained that battle summaries should be "written without fear, favor, or affection." Accounts of combat "should be as minute and detailed as possible."[38]

Several editors and reporters sought to educate the public about the realities of the war. In the summer of 1862, for instance, *Harper's Weekly*, then boasting

120,000 subscribers, concluded: "We are learning war. It is a dreadful lesson, in a fearful school." Charles Leland saw in such gruesome truths a weapon to help wean the American public from its deeply entrenched sentimentalism, not only about war but about life itself. The conflict's "all-smashing hammer of *facts*," he wrote in the *Knickerbocker*, would help rid the nation of a "vast amount of sickening folly and disgusting mock romance."[39]

Hand-drawn illustrations reinforced the written accounts of the war printed in newspapers and journals. Dozens of sketch artists, or "specials," joined the armies in the field and sent their pencil drawings to New York, where engravers traced them onto wood blocks for printing in the weekly illustrated magazines. The best combat artists—Winslow Homer, Alfred and William Waud, Charles Reed, Arthur Lumley, Theodore Davis, and Edwin Forbes—became increasingly accurate and nonpartisan in depicting the protracted war. Like Brady's photographers, they stripped soldiering of its romantic veneer and revealed its underlying tedium, humor, sentiment, and pain.

Of the many Civil War artist-correspondents, Winslow Homer was the most talented. In fact, he was the most striking American artist of his time, single-mindedly devoted to the realistic depiction of his diverse surroundings. When asked if he ever took the liberty of modifying what his eyes saw, he snorted: "Never! Never! . . . I paint it exactly as it appears." Honesty, precision, freshness—Homer constantly evokes such terms. Essentially self-taught and adamantly unacademic, he had a discerning eye, steady hand, and flinty personality. Homer's sketches and paintings have a singular tone, an austere, uncompromising devotion to fact, and a subdued emotion that provoke sympathy and respect. Forced to make his living through his art, he walked the tightrope of popular sentiment without falling into the net of sentimentality. What enabled him to keep his balance was his recognition that real life brimmed with poetry; it needed no false emotion to endow it with interest.[40]

When the Civil War erupted, the Boston-born Homer was in his midtwenties, living in New York, and contributing regularly to *Harper's Weekly*. In the fall of 1861 he accompanied the Army of the Potomac into battle, and he later made several more trips to the front. Upon returning to his New York studio, he would supplement his sketches and notes by viewing the war photographs on display at New York galleries.

Homer's drawings appeared often in *Harper's Weekly*. His studies of individual soldiers grew more candid and detached over time, conveying in the process the sense of confusion in battle and the agonies of combat—as well as the tragic results. Homer also discovered that armies spend much more time training and waiting than fighting. His drawings provide a wonderfully accurate record of camp life—its cosmic boredom, tensions, drudgery, loneliness, filth; its tired, shabby men, but also its periodic joys.

In *Home Sweet Home*, a scene of two Union soldiers listening pensively as a band plays "Home Sweet Home," Homer evoked a common sentiment without being maudlin. *Harper's Weekly* observed that there "is no strained effect in it, no sentimentality, but a hearty, homely actuality, broadly, freely, and simply worked out." A writer in the *New Path* made the same point in announcing that "Mr. Homer is the first of our artists . . . to tell us any truth about the war." *Home Sweet Home*, he emphasized, was "too manly-natural to be called sentimental." Reviewers of Homer's war paintings repeatedly invoked adjectives such as "manly," "vital," "rugged," and "vigorous" to characterize his emerging style.[41]

As the fighting came to a close, Homer produced his best war painting, *Prisoners from the Front*. Modeled after actual soldiers, it shows a spit-and-polish Union officer accepting the surrender of three grubby, grizzled Confederates of markedly different ages and outlooks. The victorious captain's mien, however, reveals more regret than haughtiness. People who saw *Prisoners from the Front* marveled at its cropped composition, muted colors, complex psychology, and evident authenticity. "Those are real men," exclaimed a writer in the *Nation*. Another art critic observed that the painting "has attracted more attention" and "won more praise than any genre picture by a native hand that has appeared of late years."[42]

Yet for all of Homer's "healthy realism," there were many other artists, North and South, determined to romanticize the war. A writer in *Godey's Lady's*

Winslow Homer, *Prisoners from the Front*, 1866. *(The Metropolitan Museum of Art. Gift of Mrs. Frank B. Porter, 1922.)*

Book stresses the didactic and edifying function that art must play in any war effort. Battle paintings and drawings should "refine the mind, enrich the imagination, and soften the heart of man." The mass-produced Currier and Ives prints of the war frequently followed such a formula. Produced not at the scene of battle but in Manhattan, these panoramic battlescapes portray straight lines of smartly dressed soldiers advancing shoulder to shoulder, moving effortlessly in measured tread through pastoral countryside, seemingly eager to engage in fierce hand-to-hand combat. Such sublime conflict bleached of bloodstains and fear contradicted the experiences of soldiers themselves. When asked if he had ever charged shoulder to shoulder, one veteran snorted: "No, God don't make men who could stand that."[43]

War in Fiction

In both North and South, the novels written during and immediately after the war were partisan and formulaic. Most included heart-rending scenes of lovers' parting, heroes on deathbeds, and gallant acts of self-sacrifice. They pictured the happy march but not the killing fields, miserable camps, or veterans with wooden legs and empty sleeves. Only rarely did the ghastly toll of death and destruction intrude into these sanitized war romances and melodramas. This is not surprising. As Mary Chesnut confessed in 1864, she liked to read "pleasant, kindly stories now. We are so harrowed by real life." Such attitudes lingered long after the last battles. In the postwar South, defeat fed a powerful appetite for self-delusion. The "downfall of empire," explained one writer, "is always the epoch of romance."[44]

Among the most popular of the southern writers who self-consciously idealized the "lost cause" was John Esten Cooke. When Virginia seceded in April 1861, Cooke, a successful writer of charming historical romances and the son of a prominent Richmond family, immediately joined the Confederate army. He participated in several raids, became a favorite of both J. E. B. Stuart and Stonewall Jackson, and quickly rose to the rank of captain. He also grew tired of the hellish strains of combat. "I wasn't born to be a soldier," he later told a friend. Fighting was the "work of brutes and brutish men." In such a war, "where men are organized in masses and converted into insensate machines, there is really nothing heroic or romantic or in any way calculated to appeal to the imagination."[45]

Yet Cooke the professional writer unabashedly manipulated the actuality of his war experience to excite reader interest and promote sales. While privately reporting that the war was anything but romantic, he declared in *Hilt to Hilt* (1869) that "it is almost impossible, indeed, to exaggerate the wild romance" of the Civil War. *Surry of Eagle's-Nest* (1866), *Mohun* (1868), and *Hilt to Hilt* blend contrived incidents and larger-than-life personalities with stilted dialogue and melodramatic love stories directed at the stereotypically emotional and gullible female reader. Cooke's battles are always "splendid spectacles" described from the lofty perspective of the generals and their staffs; the actualities of war rarely march across the pages. No shirkers or cowards serve in Cooke's literary army, and soldiers display no fatigue or boredom.[46]

Cooke's Civil War romances of masculine grace and chivalry were not believable—except to those seeking self-deception. As one irritated reviewer noted, Cooke "was afflicted with a pair of rose-colored goggles of enormous magnifying powers." Yet many readers of the day suffered from, or gloried in, the same affliction, and Cooke's writings became quite popular, especially in the South. Such highly embellished treatments of the nation's bloodiest conflict led Walt Whitman to worry that the "real war would never get into the books."[47]

There was, however, one intrepid novelist of the time who depicted the "real war." Born into a wealthy Connecticut family in 1826, John W. De Forest spent some six and a half years in the Union army. During the war he served according to his own count "forty-six days under fire." The experience transformed him. "How changed we all are!" he wrote his family. By the end of the conflict, De Forest had "learned to draw a broad distinction between the words discomfort & suffering."[48]

While in the army, De Forest quickly recognized the literary potential of his experiences and observations. He wrote lengthy letters to his wife and kept a detailed journal, both of which he later used to write *Miss Ravenel's Conversion from Secession to Loyalty* (1867), the most authentic contemporary novel about the Civil War. De Forest prized candor because he was determined to counter the "trashy, misleading" accounts of the conflict written by a "host of ignorant romancers." He would "show just what war is."[49]

Miss Ravenel's Conversion highlighted the war's actual nature, but De Forest wanted to go beyond writing simply a war story. He yearned to produce a masterpiece, a realistic novel dealing with "the ordinary emotions and manners of American existence," a book whose scope and fidelity would compare favorably with the works of Balzac. At the same time, however, he believed that the feminine reading public demanded a love story. Years later he confessed that his novels always included "a boy and girl in love" because "it was the only kind of plot a writer could get the public interested in." He therefore centered *Miss Ravenel's Conversion* on a romantic triangle that enabled him to portray the texture of civilian life in both North and South as well as to incorporate aspects of the military scene. His grandiose scheme required the literary abilities of a Tolstoy or a Stendhal, abilities that De Forest lacked. *Miss Ravenel's Conversion* is not "the great American novel" he aspired to write, but it is an interesting, competent example of nascent literary realism at work.[50]

The novel tells the story of tall, slender, blonde-haired and blue-eyed Lillie

Ravenel, a young woman of greater charm than beauty who moves with her widower father, a good-hearted, principled medical professor, from New Orleans to New England just before the firing on Fort Sumter. Dr. Ravenel had been forced to leave Louisiana because of his antislavery and Unionist convictions; Lillie, however, remains loyal to her native South and its "noble" cause. In "New Boston," the fictional city modeled after De Forest's own New Haven, she meets two suitors: Edward Colbourne, a sincere, athletic, somewhat dull and priggish young lawyer, and Colonel John Carter, a rowdy, hard-drinking, Virginia-born graduate of West Point. Colbourne raises a company in Carter's unit, which then joins in the expedition against New Orleans. Once the city is captured, Lillie and her father return to their former home in Louisiana. Confederate sympathizers thereupon assault Dr. Ravenel, and the incident causes Lillie to abandon the rebel cause. Meanwhile, she marries the coarse, yet magnetic and sensual Colonel Carter and bears him a son, only to learn that her heroic husband is a rake engaged in a tryst with Mrs. Larue, a sultry Creole temptress who believed that widows should enjoy the same forbidden pleasures as widowers. An indignant Lillie thereupon returns to New England with her infant. After Carter is killed in battle, Captain Colbourne, toughened and matured by his war experiences, returns home and marries Lillie.

Despite its conventional happy ending, *Miss Ravenel's Conversion* represented the most successful attempt yet to produce a realistic American novel with a large canvas. De Forest later recalled that writing the book exposed him to "the value of personal knowledge of one's subject and the art of drawing upon life for one's characters. . . . From *Miss Ravenel* on I have written from life and have been a realist." The book's geographical sweep is enormous, ranging from New England to New Orleans with glances at life in Washington as well as on southern plantations. Politicians, reformers, soldiers, and African Americans come to life in sharply realized scenes. Although clearly sympathetic to the North, De Forest is remarkably critical of his native region. He questions the motives of some abolitionists, reminds readers that there were cowards and skulkers in every unit, points out the many examples of graft among government officials, and flays New Englanders for their prudery.[51]

De Forest also assaulted the romantic literary tradition. When Colbourne and Lillie are reunited in New Orleans after a long separation, there are no fulsome speeches or breathless sighs. Instead, they meet in a stiff, formal manner. De Forest explains: "This is not the way that heroes and heroines meet on the boards [on stage] or in some romances; but in actual human society they frequently balk our expectations in just this manner. Melodramatically considered real life is frequently a failure." In dealing with slavery, De Forest also punctured sentimental illusions. While repeatedly stressing the evils of the "peculiar institution," he recognizes that there are bad as well as good blacks, just as there were bad and good whites. "Uncle Tom is a pure fiction," Dr. Ravenel concludes. "There never was such a slave, and there never will be."[52]

Miss Ravenel's Conversion boasts drama aplenty but no melodrama. Its prose is relatively plain and terse, devoid of the grand metaphors and lyrical adjectives favored by Cooke and other romancers. The realism of *Miss Ravenel's Conversion*, however, is most evident in its treatment of the war itself. One of the novel's most salient themes is the hardening effect of war upon American idealism. "The old innocence of the peaceable New England farmer and mechanic," De Forest notes, "had disappeared from these war-seared visages, and had been succeeded by an expression of stony combativeness, not a little brutal, much like the look of a lazy bull-dog." The veteran soldiers have lost the spit-and-polish spirit of the early weeks of the war. They are now hard-swearing and hard-drinking men, not the lavendered cavaliers portrayed in Cooke's "Dixie romances." Their uniforms are filthy, frames gaunt, and feet swollen; their outlook is stern and grave. Even iron-nerved campaigners feel the chill of cowardice amid the deafening roar of bursting shells. Fear felt under fire, De Forest reveals, poses unexpected questions.[53]

De Forest knew from experience that battles were mostly scenes of chaotic horror, not splendid tapestries. "Nothing is more confusing, fragmentary, incomprehensible," he later told Howells in language anticipating Stephen Crane's Henry Fleming, "than a battle as one sees it. And you see so little, too, unless you are a staff officer and ride about, or perhaps a general." What he saw of major battles bore little resemblance to his romantic preconceptions. In his own first combat, he had expected "a bayonet struggle, not knowing then that hand-to-hand combats exist mainly in newspapers." De Forest's graphic battlefield accounts read like coroners' reports. Note the effect of sober observation and the conspicuous lack of sentimentality in the following description:

> A dozen steps away, rapidly blackening in the scorching sun and sweltering air, were two more artillerists, stark dead, one with his brains bulging from a bullet hole in his forehead, while a dark claret-colored streak crossed his face, the other's light-blue trousers soaked with a dirty carnation stain of life-blood drawn from a femoral artery.

De Forest evoked other sensory images of the battlefield. War, he recognized, has peculiar sounds such as "the sharp, multitudinous *whit-whit* of close-firing; the stifled crash of balls hitting bones and the soft *chuck* of flesh wounds mingled with the outcries of the sufferers."[54]

Many reviewers applauded the book. A critic in *Harper's Weekly* announced that the "air of reality through the whole story is so pronounced that it is hard not to feel, as you close the book, that it is a tale of actual incidents and persons."

Harper's Monthly praised the book as "the best American novel published for many a year." Although critical of De Forest's frequent authorial intrusions and his pandering to the romantic expectations of female readers, a neophyte writer named Henry James commended the "very lively" novelist for providing "some excellent description of campaigning," description so precise as to have been "daguerreotyped from nature."[55]

American literature, it seemed, finally had its own realistic novel. So thought a young William Dean Howells, who reviewed *Miss Ravenel's Conversion* in the *Atlantic Monthly* and cited the novel as "the first to treat the war really and artistically." He found the book a welcome relief from the typical fictional accounts of the war written by women. The "heroes of young lady-writers in the magazines," he observed, "have been everywhere fighting the late campaigns over again, as young ladies would have fought them." But De Forest's "soldiers are the soldiers we actually know, — the green wood of the volunteers, the warped stuff of men torn from civilization and cast suddenly into the barbarism of camps, the hard, dry, tough, true fibre of the veterans that came out of the struggle." Such characters struck Howells as being "like people in life."[56]

The critical acclaim, however, did not translate into popular success. Readers balked at the gruesome candor of *Miss Ravenel's Conversion*. War romances such as Cooke's were much preferred, and De Forest never achieved the literary fame that many critics predicted. Howells later attributed this to the skittishness of the predominantly female audience. De Forest, he claimed, was "a man's novelist," too boldly truthful for feminine tastes. The immoral behavior and amoral outlook of Mrs. Larue and Carter, as well as the graphic war scenes, were disconcerting to Victorian sensibilities. As a result, De Forest failed to attract the wide audience he sought. "Finer, not stronger workmen succeeded him," Howells remarked in a veiled confession of his own limitations, "and a delicate realism, more responsive to the claims and appeals of the female over-soul, replaced his inexorable veracity."[57]

The Residue of War

As the Civil War retreated into history, the battlefields became breeding grounds for myth. Over the years a crust of romance formed around the Civil War, making it stirringly unreal, a war of euphemistic clichés, misleading iconography, and forgotten horrors. The soul-numbing war depicted in De Forest's writings, Homer's illustrations, and Gardner's photographs soon disappeared, and in its place emerged a misty image of chivalric combat and magnolia romance, especially in the postwar South. Virginian Ellen Glasgow said that such "evasive idealism" afflicted most of her region's intellectuals and artists. In the South, she reported, there were only "a few scattered realists, as lonely as sincerity in any field, who dwell outside the Land of Fable inhabited by fairies and goblins." The genteel tradition and the romantic spirit thus continued on into the Gilded Age, more defensive and doubting than before the Civil War, to be sure, but remaining powerful presences in literature and the arts.[58]

Yet the Civil War also provided the impetus for at least some writers, artists, critics, and members of the reading and viewing public to look at life through clearer lenses. One commentator observed in May 1865 that the nation now displayed a "sterner temper." Two months later the *Nation* declared that the gauzy romanticism of the antebellum era was incapable of representing the experience of the Civil War. "No ideal nobleness, no invented sacrifices, no romantic adventures, equal the realities we have known." The war also enhanced a sense of cultural nationalism, as many younger writers, painters, and architects decided that they must develop a closer relationship to actual life. A critic in the *Nation* cited with approval the "increased interest in things American" among the literary and art communities. Another art critic remarked that "our people refuse to look upon art as a thing apart from their daily life. . . ." Patrons wanted to see their own commonplace, everyday surroundings, "the things they love," represented in the arts. An editor attending the annual Academy of Design Exhibition in 1865 sensed "the greater reality of feeling developed by the war. We have grown more sober, perhaps, and less patient of sheer artificiality."[59]

4

A Mania for Facts

L ong before the outbreak of the Civil War, a brawny new phase of
scientific development laid siege to the citadels of philosophical idealism and theological orthodoxy. The modern scientific spirit led
many people to equate belief in facts with the substance of belief itself.
Such a "positivist" perspective dismissed philosophy as mere speculation, religion as superstition, and held that carefully verified sense perceptions constituted
the only admissible basis of human knowledge. By producing what one writer
called a "mania for facts," a new generation of professional scientists served as
the first shock troops for the realistic crusade in thought and the arts. "This is
a world of reality," lamented a literary critic at the end of the century, "and
romance breaks against many hard facts." The sentimental tradition, she
claimed, had been forced to give way to a "scientific age" and a "positivist
gospel" which valued only "facts that can be seen or heard or weighed or
measured."[1]

The prestige of positivistic science increased enormously during the nineteenth century as researchers explored electromagnetic induction, the conservation of matter, the laws of thermodynamics, and the relationship between
heat and energy. Breakthroughs in chemistry led to new understandings of the
formation of compounds and the nature of reactions, fossil discoveries opened
up new horizons in geology and paleontology, and greatly improved microscopes
enabled zoologists to decipher cell structures. The growing sophistication of
such research fostered a new conception of scientific knowledge as a field for
educated specialists. By the 1840s the term "scientists" had replaced "natural
philosophers" as the label for those trained to unlock the mysteries of nature.[2]

Striking new technological developments—the telegraph, railroad, and electric dynamos and lights—and spectacular achievements in civil engineering such
as the Brooklyn Bridge provided conspicuous physical evidence of the trans-

forming effects of science. Henry Adams, destined to become the most perceptive and perplexed analyst of science's cultural significance during the nineteenth
century, marveled at the awe-inspiring force of the modern scientific spirit. "I
tell you these are great times," he wrote his brother Charles in 1862. "Man has
mounted science and is now run away."[3]

That same year, Charles Leland, now editor of the *Knickerbocker*, predicted
that American culture will "fall in with the realizing tendency of the age, with
its practical onward march, and be raised to the ranks of Science or of accurate
knowledge." The "stupendous power of Science" would rid the arts of "every
old-time idea, every trace of old romance and art, poetry and romantic or
sentimental feeling" and wash away "the ideal, the morbid, the sickly and
visionary." He urged writers and artists to relish with the scientist the "delights
of the actual" and "go to work in this . . . commonplace, vulgar, tremendous
world" and extract its "warm truth from real life."[4]

Scientific Materialism

Fired by the investigative impulse, tempered with distrust and edged with curiosity, the much-discussed new scientific spirit pursued the "what is" of life with
relentless intensity. "All facts are exposed," Emerson observed in 1844. "The
age is not to be trifled with: it wishes to know who is who, and what is what.
Let there be no ghost stories [any] more." Modern empirical scientists taught
people not to wonder at natural phenomena but to experiment and reason, to
discern and dissect, to classify and compare. "Facts," remarked a literary critic,
"please the many simply by reason of their force and sheer reality." Henry James
echoed the sentiment when he reported in 1871 that "the analytic instinct rises
supreme." The scientist's "explanatory gaze" and unrelenting emphasis on "firm
particulars," he predicted, must necessarily affect the arts as well.[5]

In cultivating their "explanatory gaze," scientists opened up a gulf of doubt
about many inherited truths and spiritual convictions. Investigators now claimed
to see through the "false" assumptions of both philosophical idealism and conventional religious beliefs. As early as 1847, a writer in the *North American
Review* claimed that this "culminating era of material philosophy and science"
had supplanted idealism's "world of chimeras and fantastic forms by a world of
stiff, stubborn, angular facts, which she can neither bend nor mold."[6]

Chauncey Wright, the mathematical genius and charismatic Harvard tutor
who served as the mentor of both William and Henry James, emerged as one
of the most outspoken proponents of this new positivist creed. American
thought, he repeatedly declared, must descend from the clouds and focus on
the concrete actualities of everyday life. "Behind the bare phenomenal facts,"
Wright declared, "there is *nothing*." Modern scientific study, he maintained

with naïve exuberance, was absolutely objective and impersonal, for it dealt solely with "questions of facts," without reference to how "we ought to feel about the facts." Wright and others preached such "scientific realism" with the enthusiasm of evangelists. In 1857 Emerson reported that the "scientific whim is lurking in all corners." He then made a prescient prediction: "We are coming on the secret of a magic which sweeps out of men's minds all vestige of theism and beliefs which they and their fathers held and were framed upon."[7]

The Impact of Darwin

Almost three years after Emerson's prophecy, in November 1859, Charles Darwin published On the Origin of Species. It stood the world on its head. The New York Times reported that the book contained "arguments and inferences so revolutionary" that they promised "a radical reconstruction of the fundamental doctrines of natural history." Even though Darwin had essentially crystallized ideas that had been fermenting for three-quarters of a century, his emphasis on the material causes of natural phenomena rather than providential design had a shattering impact on Western thought. "I have just finished reading [Origin of Species]," Chauncey Wright wrote a friend in early 1860, "and . . . I have become a convert."[8]

On the Origin of Species represented twenty years of painstaking research and reflection. The book overflows with references to the authenticating stature of facts. Darwin directs the reader to "astonishing" facts, "remarkable" facts, "grand" facts, and "well-established" facts, all of which combined to reveal plants and animals engaged in a brutish "struggle for life" against rivals, predators, and a stingy environment. Darwin's provocative thesis that the "modification" of species occurred through a ceaseless process of "natural selection" challenged the biblical story that all animal species had originated in an act of divine creation which had forever fixed their forms. In Darwin's world, new species were not "special creations" of God; they emerged randomly from the struggle for existence. Natural selection, he implied, was arbitrary, capricious, and devoid of ultimate meaning—a long, gradual process of intense competition and hereditary development without divine plan or purpose.[9]

Readers were stunned. With one fell swoop and a wealth of corroborating evidence, Darwin had challenged established beliefs about nature and cherished assumptions about providential design and life processes. "If this be truth," growled one college president, "let me live in ignorance." As time passed, however, more and more people accepted many aspects of evolutionary naturalism. Fifteen years after Origin of Species appeared, Titus Coan, a literary and art critic, announced that Darwinism "had taken possession of the human mind, and will not be dislodged. Our views of life, of art, of the world, all our conceptions, are modified by this master-thought."[10]

As unsettling new scientific facts hardened into accepted truths, religious faith dissolved into skepticism. "Genuine belief seems to have left us," Walt Whitman commented in *Democratic Vistas* (1871). Of course, faith in God did not disappear in the face of evolutionary naturalism, but the post-Darwinian God became for many people a more distant and diffuse Being than the immanent, providential Deity of antebellum days. "Religion is not extinct," William Dean Howells observed in 1878, but "it is constantly losing ground." For many who remained professing believers in the Darwinian era, God no longer served as the center and circumference of life; firm convictions softened into flaccid platitudes. In his novel *A Modern Instance* (1882), Howells wrote that religion "had largely ceased to be a fact of spiritual experience, and the visible church flourished on condition of providing for the social needs of the community." An uneasy agnosticism became intellectually fashionable during the second half of the nineteenth century. "The world has grown tired of preachers and sermons," attorney Clarence Darrow reported in 1894; "to-day it asks for facts."[11]

The Vogue of Spencer

Herbert Spencer, a brilliant English civil engineer turned cosmic philosopher, did more than any other thinker to sanctify the authority of modern scientific method and evolutionary naturalism in the United States. By repeatedly asserting that "objective facts are ever impressing themselves upon us," Spencer emerged as the era's consummate positivist. In 1852 he coined the telling phrase "survival of the fittest," which Darwin later borrowed, to explain how some people succeeded and others failed when competing for life's bounties. Competition, he decided, ensured the continuing progress of the human race. It weeded out the weak and incompetent and preserved the superior members of society.[12]

Spencer's empirical spirit, rugged individualism, and optimistic faith in material and social progress generated excited attention in the United States. A world of indestructible matter operating in certain space and time bolstered faith in human knowledge and enabled Victorians to accommodate even Darwinism to their optimistic outlook. The wealthy industrialist Andrew Carnegie recalled in his autobiography that he had been mired in an intellectual and moral quandary until he read Darwin and Spencer. "I remember that light came in as a flood and all was clear. Not only had I got rid of theology and the supernatural, but I had found the truth of evolution."[13]

The foremost American popularizer of evolutionary naturalism was Edward

L. Youmans, a self-taught chemist who became a popular lecturer and writer on scientific subjects. During the 1860s he developed a crusading commitment to preach the "gospel of evolution" and to promote scientific professionalism. For the next twenty years, he touted articles and books written by the English evolutionists, raised funds to support their research, and wrote favorable reviews of their publications. In 1872 Youmans assumed the editorship of the *Popular Science Monthly* and molded the new magazine into a widely read forum for the promotion of evolutionary science. In the first issue he proclaimed that "the progress of science" is "undoubtedly the great fact of modern thought." Through the application of scientific method, he promised, thought and the arts would "be brought into the exactest[sic] harmony with things." Such statements reveal how positivists assumed that the self-confidently representational outlook of empirical science could be easily transferred to the arts.[14]

Youmans repeatedly called for the integration of modern science and practical skills into college and university curricula. Instead of memorizing archaic knowledge and learning "dead" languages, students should be taught "facts, laws, relations and realities of the world of experience." Youmans's plea for "realistic" instruction struck a resonant chord. In the 1870s Charles W. Eliot, the president of Harvard, revamped the university's classical curriculum so as to require students to take more science courses. In the laboratory, he explained, the student "scrutinizes, touches, weighs, measures, analyzes, dissects, and watches things. By these exercises his powers of observation and judgment are trained." Oliver Wendell Holmes, Sr., who taught in the Harvard medical school, announced that Eliot's curricular reforms had overturned the university "like a flapjack."[15]

Other colleges implemented similar curricular changes. Yale's William Graham Sumner, the outspoken Spencerian sociologist, suggested that courses in idealistic philosophy be eliminated altogether. "Philosophy," he charged, "is in every way as bad as astrology. It is a complete fake . . . unworthy of serious consideration." Moreover, the predominance of sentimental "gush" in popular literature "means that people are trained away by it from reality. They lose the power to recognize truth." Real life was "grim and hard," Sumner insisted, and educational leaders must cast off the pieties of an earlier era and expose students to unidealized facts.[16]

The Social Sciences

After the Civil War the authoritative premises of scientific positivism invaded every bastion of the nation's intellectual life. "This scientific current," a writer in the *North American Review* concluded, "is moving more or less all schools of thought." New specialized disciplines founded on modern scientific method appeared in rapid sequence. During the second half of the nineteenth century, psychology, sociology, economics, politics, and anthropology gained recognition as social *sciences* and became separate departments in colleges and universities. Sociologist Lester Frank Ward exclaimed in 1883 that the new academic disciplines must adopt the same priorities as the natural sciences: "Tangible facts" and "material objects" were the "materials for the intellect to deal with" and were the true "objects of knowledge."[17]

Edward Ross, a pioneering sociologist at Stanford University, agreed that the study of American life must be grounded in the objective interpretation of hard data. Americans, unlike Europeans, "have steadily sought the light of fact and reason, and shunned the fog" of romantic "sentimentalism and idealism." His own efforts to analyze American social forces reflected a desire to "see things as they are and to report what I see." Ross had an absolute "horror of the subjective." He coveted "clear outlines," "firm colors," and "hard numbers" that "make life seem solid and real."[18]

What really mattered for Ross and other positivists was what could be counted, measured, weighed. Truth resided in quantifiable facts. The decennial publication of the census beginning in 1840 attested to this rising interest in statistics as the most valuable source of knowledge. Carroll D. Wright, the director of the Massachusetts Bureau of Statistics of Labor, stressed that the "point aimed at always in the collection of labor statistics is truth," and the results must be "fearlessly stated, without regard to the theories of the men who collect the information." Those possessed of such statistical knowledge, Wright claimed, "live in a light never before granted us, and the facts give us this light and lead us to the true conclusion."[19]

Wright and many other social scientists sought to use statistics as a new instrument of social reform. In part this reflected the failure of conventional moral suasion and evangelical preaching to uplift human behavior. Beginning in the 1840s, sanitary reformers and urban missionaries realized that the unruly denizens of city slums were not listening to pious appeals for clean living and upright morality. Some argued that this reflected middle-class ignorance of new social realities. The ministers of New York, asserted reformer Charles Loring Brace in 1853, "know scarcely anything about the masses." To inform people about what came to be called the "social question," many activists embraced a statistical point of view that resulted in an outpouring of data about contemporary social conditions. A writer in the *Popular Science Monthly*, for instance, said that "the new statistical method" provided "realistic thinkers" and reformers with a powerful new tool to be used against the prevailing assumptions of "sentiment and authority." He especially welcomed the "introduction of scientific evidence into problems over which theologians and moralists have long claimed exclusive jurisdiction." If the facts were revealed and faced, his logic went, reformers could adopt more effective means to attack social ills.[20]

With uncritical zeal, Wright and other statisticians and social scientists developed an abiding faith in numbers as the key to understanding, improving, and controlling a turbulent urban-industrial society careening out of control and fraught with volatile ethnic and class tensions. "Fact-worship," noted one observer, "is the popular worship of our times." This reverence for facts also embodied the gender stereotypes of the age. Numerical facts, supposedly "masculine" in their objectivity and potency, seemed to offer a formidable new weapon in the assault on "feminine" idealism. So thought economist Francis Amasa Walker, a Union general who after the Civil War became chief of the Federal Bureau of Statistics and Superintendent of the Census. The postwar era, Walker insisted, demanded "stern and practical inquiry" into "the facts of society." He associated antebellum idealism with "transcendentalism and sentimentalism," subjective "female" states of mind that were "bankrupt" in modern America. "The romance of political youth is gone," he wrote in 1868; "the stern duties of manhood remain. We must regard liberty no longer as a female, but as a fact."[21]

Walker and other social scientists assumed that a "manly" disposition entailed the dispassionate analysis of social facts. By contrast, a "feminine" perspective avoided or idealized harsh truths. Albert G. Keller, a Yale sociologist, expressed such stereotypes when he congratulated a colleague for planning "a 'He-Book,' on Sociology. Lord knows there are enough . . . 'She-Books.' If some scientific virility can be injected into this sadly emasculated subject, the sociologists will not need to climb a tree every time a real scientist heaves into sight." Keller saw in his colleague William Graham Sumner a "hard hitting" social scientist who would "appeal to anyone who, in these times, prays: 'O God, give us a man.'" Sumner displayed his "robust" scientism when he declared that sociology must be "built on this thirst for reality." Ideals, he contended, "are necessary phantasms. They have no basis in fact."[22]

Such bald statements reveal again how men were presumed to be commonsensical, vigorous participants in public life while women were seen as naturally sentimental and domestic beings, adept at nurturing children and cultivating morals but out of place amid the realm of public or practical concerns. In a children's story entitled "Fancy's Friend" (1868), Louisa May Alcott acknowledged the prevailing images of male and female attributes. She describes the allegorical character "Aunt Fiction" as a "graceful picturesque woman, who told stories charmingly, wrote poetry and novels, [and] was much beloved by young folks." Her counterpart, "Uncle Fact," is a "grave, decided man" who "knew an immense deal; and was always taking notes of things he saw and heard, to be put in a great encyclopedia he was making. He didn't like romance, loved the truth, and wanted to get at the bottom of everything."[23]

SCIENCE'S COCKSURE ADVANCE across every field of knowledge helped inspire and legitimize the realistic movement in thought, literature, and the arts. "The scientific tone of mind, the modern critical spirit," concluded literary critic and biographer Gamaliel Bradford, "is distinctly realist." The insistent hammering of science's empirical approach challenged writers, painters, architects, and other thoughtful Americans to use first-hand experience and direct observation to depict the people, places, and events of contemporary life. "In an age which is pre-eminently an age of observation and science," an architect explained, "it is natural that the arts of observation should be dominant, and that realism in art should be stimulated by the habit of observation." American writers and artists, another commentator advised in 1884, "must accept every scientific method . . . applicable to our ends." This "Age of Realism," he concluded, had resulted from "a single cause—the spread of modern scientific ideas."[24]

A Younger—and More Skeptical—Generation

In 1870, Oliver Wendell Holmes, Sr., devoted his Phi Beta Kappa address at Harvard to the impact of scientific materialism on thought and culture. "The atittude of modern Science," he declared, "is erect, her aspect serene, her determination inexorable, her onward movement unflinching." But the measured tread of positivism, he feared, threatened to displace idealists like himself. "We must bestir ourselves," he advised his peers in 1870, "for the new generation is upon us. Each generation strangles and devours its predecessor."[25]

In fact it was among a younger generation of writers, thinkers, and artists who came of age after mid-century that the skeptical outlook and empirical method of modern scientific realism proved especially appealing. Oliver Wendell Holmes, Jr., who as a young boy had been "set on fire" by Emersonian idealism, entered Harvard in 1857 and by his junior year had joined the rebels who were stealthily reading Darwin and Spencer, much to the chagrin of the faculty and administration. Holmes doubted "if any writer of English except Darwin has done so much [as Spencer] to affect our whole way of thinking about the universe." He recalled that his own advocacy of a hard-headed empiricism, which rejected all social panaceas and "put everything to the test of proof," led to numerous clashes with his father. Even though the elder Holmes had himself been "brought up scientifically," his son emphasized, "there was with him, as with the rest of his generation, a certain softness of attitude" toward areas challenged by modern science. There were questions that his father "didn't like to have asked . . . so that when I wanted to be disagreeable I told him that he straddled."[26]

As a lawyer and jurist the younger Holmes refused to straddle; he attacked those who clung to the delusive certainties of formal logic and past conventions. The Civil War veteran advocated the rigorous application of scientific method to the law, arguing that "the only proper method by which truth may be discovered is by observation and experiment." The "man of the future," he believed, "is the man of statistics and the master of economics," and he dreamed of an America "in which science is everywhere supreme."[27]

Similar intellectual tensions between generations emerged in the James household. William and Henry James, their friend George Santayana remembered, "were as tightly swaddled in the genteel tradition as any infant geniuses could be." Inherited wealth enabled Henry James, Sr., the father of William and Henry, to lead a life of philosophical leisure and practice a rarefied spirituality. Henry Jr. recalled that his effervescent father, while "delighting ever in the truth," was "generously contemptuous of the facts" when discussing the family's economic circumstances. In the James household, the practical affairs of life "played in our education as small a part as it perhaps ever played in any." The elder James wanted his extraordinary boys to "*be* something" rather than simply be trained to *do* something.[28]

As William and Henry developed into young men of genius and were increasingly exposed to modern scientific thought, they grew impatient with their father's unquestioning belief in God and his blithe dismissal of "the facts" of contemporary life. William cultivated a nimble, spidery mind, forever musing, discovering, and chipping away at inherited conventions. What set him apart from his peers and makes him such a compelling figure was his freedom from the sententious Victorian cant so prevalent among the emotionally parched exemplars of the genteel tradition. He especially disdained the "enormous emptiness" of the abstractions preoccupying his father and his circle of "woolyminded" friends, continually "killing themselves with thinking about things that have no connection with their merely external circumstances." He resolved to "come in contact with the stern realities of life."[29]

Pragmatic Realism

To wean himself from his father's introspective idealism, William James clasped the empirical authority of science in a firm embrace. He read Herbert Spencer's *First Principles* in 1860 at age eighteen, and he rejoiced in the English theorist's emphasis on "irreducible and stubborn facts." A few months later, as the Civil War erupted, William succumbed to his father's persistent nagging and abandoned the "feminine" world of art in favor of the "masculine" study of science at Harvard's Lawrence Scientific School. Like many of his peers, he developed a gendered understanding of the pursuit of knowledge. "The scientific-academic mind and the feminine-mystical mind," James concluded, "shy from each other's facts, just as they fly from each other's temper and spirit. Facts are there only for those who have a mental affinity with them." James told his family that his "vocation was . . . quite positively and before everything for Science, physical Science, strenuous Science."[30]

By 1867 William James had become such an ardent champion of modern scientific method that he savaged an ethereal article his father had written. Its spongy argument, he wrote home, was so soaked in abstraction that it made no sense to anyone except the author: "I cannot logically understand *your* theory." The elder James archly replied that "it is very evident to me that your trouble in understanding it arises *mainly* from the purely scientific case of your thought just at present, and the temporary blight exerted thence upon your metaphysic wit." Like his friend Oliver Wendell Holmes, Sr., Henry James, Sr., decided that professional scientists were temperamentally incapable of perceiving intuitive spiritual truths. William, his brother Henry, and Wendell Holmes, Jr., however, came of age during the vogue of Darwin and Spencer, and they could not so easily dismiss scientific materialism's insights and methods. William attributed his father's opposition in part to his living "in such mental isolation that . . . you must see in your own children strangers to what you consider to be the best part of yourself."[31]

For the moment William James severed all ties to his father's fuzzy metaphysics. In 1869, having received his medical degree, he confessed to a friend that "I'm swamped in an empirical philosophy." Exhilarated by the rush of change and new intellectual challenges, he found the attractions of the "rugged and manly school of science" irresistible, and he rejoiced over the "great unsettlement" in the "philosophic atmosphere of the time."[32]

In the early 1870s, however, James began to have second thoughts about positivism's brittle certainties and rigid determinism. After spending two years studying physiology in Germany, he suffered a series of health problems—eye trouble, backaches, general nervousness—that culminated in a prolonged bout with depression punctuated by nightmarish hallucinations and thoughts of suicide. Several developments conspired against his sanity and serenity. The death of a beloved young cousin, lingering guilt over his failure to participate in the Civil War, and his inability to fasten upon a clear vocational direction apparently triggered his despondence. Should he be a physician? psychologist? philosopher? Awash in uncertainty, he confessed to his father that "my strongest moral and intellectual craving is for some stable reality to lean upon."[33]

What finally lifted James out of the "pit of insecurity" was a sheer act of will. An unexpected Harvard faculty appointment in 1872 bolstered his spirits and convinced him that he could no longer afford the luxury of perpetual anxiety. So he plunged into his work as professor and scholar, resolving in the process to end his "morbid" ways and discover the "real me." His "first act of free will," James announced, "shall be to believe in free will." Most people, he now concluded, derived meaning not from abstract speculations but from doing, acting, and struggling with real concerns that involved "the sweat and blood and tragedy of this life."[34]

James thus started his life afresh, fronting essentials and taking nothing for granted, determined to rid his thinking of all preconceptions and abstractions. In the process he became both a revered professor of psychology and philosophy at Harvard and a scholar of international reputation. A man of lively spirits and vivacious curiosity, he displayed a singular charm and an elastic "sense of reality" that served as an enlivening example of willed authenticity. If not the greatest American philosopher of his time, he was certainly the most influential.

James recognized that most people were not philosophers; they faced stern choices each day and could not wait for final solutions or indulge in endless speculations. To help make philosophy relevant to such common folk, he developed his pragmatic theory of truth. Philosophers, James noted, had traditionally defined truth in two ways: either the coherence of statements with an all-encompassing system, or the correspondence of statements with the objects they sought to represent. Yet neither approach, he decided, helped people get at the truth of actual life. The true and the real could never be discovered through speculation alone, for the "truth" of an idea was not a static quality inherent in things. "Truth *happens* to an idea," he wrote. "It *becomes* true, is *made* true by events." The mind thus "*engenders* truth *upon* reality. Our minds are not here simply to copy a reality that is already complete. They are here to complete it, to add to its importance by their own remodeling of it."[35]

Armed with his pragmatic concept of truth, James urged "transcendental" thinkers to abandon their "sentimental and priggish" theorizing and instead address that "element in reality which every strong man of commonsense willingly feels" when in vigorous contact with the "teeming and dramatic richness of the concrete world." Idealistic philosophy, he charged, "is out of touch with real life, for which it substitutes abstractions. The real world is various, tangled, painful." Yet "white-winged" thinkers continued to ignore the "complexity of fact" by treating life as solely "noble, simple, and perfect." The disorderly, intractable realities of daily living defied such easy optimism. "No abstract concept," he stressed in 1879, "can be a valid substitute for a concrete reality."[36]

By making philosophy more applicable to the realm of daily activity and common concerns, James may have done more than any other American *thinker* to influence cultural trends and democratize epistemology during the last quarter of the century. His pragmatic method, although fraught with ambiguity, bridged the gap between positivist materialism and philosophical idealism by giving the individual a vote in deciding what was "true" or "real" when all other instruments of proof had failed.[37]

James's pragmatism found a devoted disciple in John Dewey. The young philosopher from Vermont dedicated his first book to James and considered James's *Principles of Psychology* (1890) his greatest scholarly inspiration. Dewey, too, decided that the mind and reality were not separate entities but parts of a common process. "Things," he stressed, ". . . are what they are experienced as." In linking the mind with the external world of experience, Dewey agreed with James that "the presuppositions and tendencies of pragmatism are distinctly realistic."[38]

Philosophy, Dewey contended in terms echoing James, "suffers by being made vague and unreal," and he encouraged his students to see thought as an instrument of concrete action rather than speculative musing. Dewey developed a zealot's faith in what he termed "the scientific habit of mind." The syllabus for his introductory philosophy course at the University of Michigan included the following description: "Science and philosophy can only report the actual condition of life, or experience. Their business is to reveal experience in its truth, its reality. They state what *is*." By the end of the century, however, Dewey had decided with James that there was more to the scientific spirit than a narrow preoccupation with evident facts. The more important element of the scientific creed was an open-ended commitment to investigation—wherever it might lead. Dewey's own investigations led him to conclude with James that the spectator theory of knowledge—whereby a person simply mirrors a true Reality—was misleading and incomplete. Thought did not merely reflect a reality "that *is*"; it could *change* reality by interacting with "actual experience" and thereby earn "the realistic stamp of contact with life." Philosophy "recovers [significance for] itself," he later wrote, "when it ceases to be a device for dealing with the problems of philosophers and becomes a method . . . for dealing with the problems of men."[39]

Dewey and James saw a direct connection between their pragmatic method of philosophy and literary realism. The "importance and the endurance" of art, Dewey maintained, depended upon "its hold upon reality." James, too, admired "realistic writers" who turned away from a "spurious literary romanticism" in order to confront the real issues of contemporary life. He yearned for the time when "philosophers may get into as close contact as realistic novelists with the facts of life." Writers returned the compliment. Howells, for instance, praised James's *Principles of Psychology* (1890) for its "poetic sense of facts" and its "absolute devotion to the truth."[40]

Such comments document the way in which a scientific rationalism tempered by respect for the shaping power of individual consciousness came to serve as the intellectual foundation for late nineteenth-century cultural realism. Far from simply being naive positivists about the solidity and accessibility of facts, the most gifted realists in fiction and art decided with James and Dewey that mirroring a stable reality was an illusion. The true realist, as James concluded, was "not simply a mirror floating with no foot-hold anywhere, and passively reflecting an order that he comes upon and finds simply existing." True, there was a reality "out there" independent of the mind, but in practice people "perceived" and represented that "reality" in different ways.[41]

5

Goods and Surfaces

The commanding reality in the United States during the nineteenth century was an ever-accelerating industrial revolution that transformed the very nature of social life and generated a conspicuous new urban consciousness and culture. As sociologist William Graham Sumner remarked, industrialization "controls us all because we are all in it. It creates the conditions of our own existence, sets the limits of our social activity, regulates the bonds of our social relations . . . and reforms the oldest and toughest customs." With each passing year, the myriad effects of urban-industrial development led more and more people to jettison traditional folkways associated with rural life in favor of new urban environs, a more secular outlook, and enticing new economic and social opportunities. People living amid the marvels and miseries of metropolitan America saw their habits of mind profoundly altered. In the process, many came to expect literary and artistic representation of their new urban culture. Henry James noted that "modern manners, modern nerves, modern wealth, and modern improvements, have engendered a new sense . . . recognizable in all the most characteristic productions of contemporary art."[1]

In the preface to the 1855 edition of *Leaves of Grass*, Walt Whitman asserted that the poet and artist must assimilate "the factories and mercantile life and labor-saving machinery." Artist John Ferguson Weir experienced a similar enchantment in the presence of modern industrial technology. In late 1864 Weir visited the West Point Foundry at Cold Spring, New York, where cannons were being produced for the Union war effort. The "virile activities of that busy place" and the "swarthy smiths with brawny arms, arms of muscle large and strong," so dazzled him that he resolved to depict their matter-of-fact heroism on canvas. Two years later he completed *The Gun Foundry* and exhibited it at the National Academy of Design. It portrays several visitors observing grimy workers casting a cannon, their strenuous labors illuminated by the lurid glow of a huge cauldron of molten iron. Critics labeled the painting one of the most original of the year. "The work is really getting done," a writer in the *Nation* observed; "the men are really busy, and sensibly busy . . . the picture is real, and that is less often true of American pictures than could be wished."[2]

More than any other mechanical innovation during the nineteenth century, the railroad served as the most ubiquitous symbol of the new industrial age. Nathaniel Hawthorne saw in the railroad the spirit of commercial concerns intruding upon the pre-industrial imagination. The approaching locomotive, he wrote, "comes down upon you like fate, swift and inevitable." With its unsettling whistle it brought "the noisy world into the midst of our slumberous peace." Still other painters, poets, and novelists sought to reconcile the presence of "the machine in the garden" by integrating the railroad into their artistic works, hoping to acknowledge industrial advances without sacrificing pastoral virtues. Yet they fought a losing campaign. Harriet Beecher Stowe reported in 1869 that the "simple, pastoral germ-state of society is a thing forever gone. . . . The hurry of railroads, and the rush and roar of business" had displaced the agrarian myth forever.[3]

Of course technology alone did not produce the widespread "incorporation" of American life during the nineteenth century. Primary responsibility lay with the industrial and financial capitalists whose investments put such machines to work, with the managers and efficiency experts who turned offices and plants into machines, and with the foremen who monitored laborers tending looms and furnaces. The emergence of a new corporate work culture based upon mass production and mass marketing aroused concern as well as awe. In 1864 the *New York Times* expressed heightened anxiety about entrenched corporate power:

> The growing wealth and influence of our great corporations . . . is one of the most alarming social phenomena of our time. . . . Our public companies already wield gigantic power, and they use it like unscrupulous giants. . . . They are generally . . . close monopolies, and all attempts to escape from their tyranny by starting rival or supplementary enterprises, are every day becoming more hopeless. They control the Legislatures, and crush all schemes that seem in the least likely to interfere with their schemes.

Others worried about the corrupting effect of new commercial priorities and bureaucratic structures. "Civilization is becoming more and more material," one literary critic complained, "and idealism is everywhere being crowded to the wall." Industrial development, the settlement of the West, unrelenting foreign immigration, and the heightened scramble for goods and status all gave a rough new texture to an American society "swimming in plenty," as Whitman wrote.[4]

Perhaps more than any other factor, this pervasive materialism helped spawn the realistic movement in thought and the arts. "The Greeks were preeminently an aesthetic and ideal people," a minister wrote. "We are proverbially practical and real." Whether such a worldly orientation actually increased during the nineteenth century is impossible to prove, but Americans certainly thought so. Edmund C. Stedman, the refined New York poet, stockbroker, and literary historian, highlighted "the realistic tendencies of the present time" and attributed them in large part to "the so-called materialism of our century."[5]

The City of Realism

This materialistic spirit drew most of its energy from the mushrooming commercial cities springing up across the nation. During the nineteenth century, the United States experienced the fastest rate of urban growth in the world. Between 1820 and 1860, as manufacturing production more than tripled, the urban population grew five times as fast as the rural. A writer in *Putnam's Monthly* exclaimed that *"the great phenomenon of the Age is the growth of great cities."* Sleepy villages were transformed into bustling towns, and small cities blossomed into sprawling metropolises. In 1860 there were but sixteen cities with populations over 50,000; by 1910 there were well over eighty. Between 1870 and 1920 almost 11 million Americans left farms and rural villages for the cities, and even more newcomers arrived from abroad. By 1920 less than 30 percent of Americans still worked in agricultural occupations, and the total population had soared to 106 million.[6]

The modern city's frenetic pace, its sprawl, disorder, beauty, and energy combined to give it a commanding presence and almost sensual allure. The crowded metropolis was both a dramatic spectacle and a menacing dynamo. As the vortex of production and consumption, it controlled the pace of national life and tied most people to its secular priorities and collective culture.

In his poems, Whitman celebrated the diverse collectivity making up New York City. He gloried in the "many-item'd Union" and his "million-headed city." Nothing inspired him more than "the majesty and reality of the American scene *en masse*." By offering an epic scene of mass democracy at work and on the move, the new commercial cities nurtured the realistic impulse to represent the face and mood of contemporary life. Writer Fanny Bates asserted that "the people of the cities" now "live more in realities than imagination." Realism in the arts, she added, especially appealed to "the people of the cities, whose lives are crowded with a variety of interests, and whose desire for change and new experience is constantly gratified."[7]

The overwhelming fact and force of city culture meant that moralists, writers, and artists could no longer wish away urban-industrial life in a fit of agrarian

Matthew B. Brady, *Broadway, North from Spring Street*, 1867. *(Museum of the City of New York.*

nostalgia. With all its faults and horrors, and despite the laments of "mild-mannered rural philosophers," a journalist contended, the "busy, swarming metropolis exists, and will continue to exist, inviting criticism, condemnation, reformation, but defying abolition, and making recognition of some sort imperative."[8]

As people grew accustomed to the permanence and excitements of city life, art patrons, book buyers, builders, and critics demanded that literature and the arts focus on metropolitan themes and subjects. Between 1840 and 1870, more than three times as many books were written about city life as about rural and frontier culture. "This is the age of cities," proclaimed writer Hamlin Garland; "we are now predominantly urban and the problem of our artistic life is practically one of city life."[9]

Theodore Dreiser acknowledged the same urban imperative. Born in Indiana, he grew up in a rural setting that celebrated the pastoral myth. "The very soil smacked of American idealism and faith," he recalled. After touring the nation's major cities as a young journalist, however, Dreiser discarded his "romantic" rural perspective: "I had seen Pittsburgh!" Dense with mythic meaning, the city of steel's "roar of life" and brilliantly illuminated thoroughfares exuded a seductive allure and an imperial power. "To go to the city," Dreiser realized, "is the changeless desire of the mind. To join in the great, hurrying throng . . . becomes a desire which the man can scarcely resist."[10]

Money enticed young rural and foreign migrants to urban-industrial centers and in the process helped undermine conventional moral values and social stability. Realists did not endorse this development, but they did insist that the nation's compulsive preoccupation with money be represented in literature and the arts. "The true meaning of money," Dreiser observed in *Sister Carrie*, "yet remains to be popularly explained and comprehended." He understood that money was more than a tangible medium of exchange: it symbolized infinite possibilities, fantasies, mobility, and power; it represented a complex new system of industrial and financial capitalism; and it served as the ticket for social advancement and the standard of self-esteem. "Human beings," Dreiser decided, "are innately greedy—avaricious. They dream of fine clothes and fine homes and of rolling about luxuriously in carriages; and to make their dream come true they are ready and willing to stoop to anything."[11]

Perhaps the most salient topic in realistic fiction was the getting, keeping, spending, and owing of money. In Howells's *The Rise of Silas Lapham* (1885), for example, Bromfield Corey asserts that money "is the romance, the poetry, of our age. It's the thing that chiefly strikes the imagination." Such overweening material ambition was not limited to men. Lily Bart, the tragic heroine of Edith Wharton's *House of Mirth* (1905), declares that she "must have a great deal of money" to be happy among the frantic, frivolous posturings of fashionable society. At the other end of the social spectrum, Dreiser's Carrie Meeber, struggling to make her way in Chicago, professes the same ambition: "Ah! money, money, money! What a thing it was to have!"[12]

Many observers claimed that the lust for money and its social rewards coincided with the emergence of the middle class. This mercurial term first appeared in the 1830s and gained currency in the 1850s, as newly affluent Americans viewed themselves as a distinct social group between the ragged and the rich, upwardly mobile people possessed of their own anxieties, ambitions, and activities. The middle class, explained the Chicago writer George Ade, meant those people "who work either with hand or brain, who are neither poverty-stricken nor offensively rich, and who are not held down by the arbitrary laws governing that mysterious part of the community known as society."[13]

Members of this self-assertive middle class, flush with new wealth and social ambition, yearned to see their own values and experiences portrayed in the arts. This is not surprising. All bourgeois societies—fifteenth-century Flanders, seventeenth-century Holland, and nineteenth-century Europe and America—have created an expanding market for mimetic art that consecrates the faces, scenes, and events of everyday life. People of uncertain social standing understandably want the arts to mirror and celebrate their common experiences. Walt Whitman was among many willing to oblige the new bourgeoisie. He remarked that the unsophisticated middle class was the "most valuable class in any community," for it represented "the real, though latent and silent bulk of America, city or country." As such it presented to the writer or artist "a magnificent mass of material, never before equalled on earth."[14]

Spreading affluence also increased social tensions and ripened perceptions of class stratification. During the second half of the nineteenth century, the number of very rich Americans skyrocketed. In 1861 there were only a few dozen millionaires in the country; by 1900 there were more than 4000. Wealth brought the bourgeoisie leisure, power, and prominence, but it did not always earn them social respect among the older patrician elite. Philadelphian Henry Tuckerman reported that an uneducated and unrefined bourgeoisie was forcing its way into social, civic, and cultural circles long dominated by the gentry, displacing distinguished families of demonstrated "integrity, hospitable living, and public spirit." In a clumsy effort to win prestige and gain attention, the *nouveaux riches* often flaunted their new wealth. They wore elegant clothes, rode in liveried carriages, and built gaudy brownstone mansions. Tuckerman claimed that the "new rich" brought with them the "devotion to appearances" characteristic of the age.[15]

Then, as now, city life placed a premium upon visible signs of social status. In the city, noted Robert Park, a pioneering urban sociologist, "the art of life is largely reduced to skating on thin surfaces and a scrupulous study of style and manners." Realistic novelists often highlighted the friction between families of established wealth and those of new money. In Howells's *The Rise of Silas Lapham*, the tension is particularly acute for Lapham, the parvenu rustic who has made a fortune in the paint business. Upon first meeting the Boston aristocrat Bromfield Corey, whose son Lapham's daughter wants to marry, Silas feels "the struggle of stalwart innocence not to feel flattered at the notice of sterile elegance." While visiting the Corey family amid their decorous gentility,

Lapham brags incessantly about his money and resolves to buy the social respect
and aplomb the Coreys enjoy. Until his moral rise through his financial fall,
when he loses his cherished million, Silas continues to think, in his wife's words,
that "his money can do everything." That assumption also bothers Bromfield
Corey, who complains to his son that "the suddenly rich are on a level with
any of us nowadays. Money buys position at once."[16]

A *Spectatorial Vision*

The urban world of social thriving and social acting provided the theater of
cultural realism. It offered poetry, romance, tragedy, and a palpitating force. In
his diary, painter John Sloan described the sensual attractions of the city scene
during a walk through Manhattan's East Side: "Life is thick! colorful. I saw
more than my brain could comprehend, a maze of living incidents—children
by thousands in the streets and parks." Here was plenty of *life* to watch and
record on canvas and in prose.[17]

Like Sloan, other artists and writers loved to stroll city streets, window-
shopping and people-watching, consuming with remembering eyes what Henry
James called the rousing "spectacle of the world's presence." The urban scene,
he stressed, unleashed a "flood of the real" upon the observing consciousness.
City life was "a thing to be looked at, seen, apprehended, enjoyed with the
eyes." Yet he took care not to allow the urban mass to absorb his individuality.
He observed the events of the streets from a comfortable distance. In his autobiography, James reported that the lobby of a grand hotel offered the best
"chance to dawdle and gape" at the spectacle of public life and manners.[18]

This spectatorial emphasis appears often in James's fiction. In *The American*,
Christopher Newman, a retired American businessman in search of an attractive
European wife to "possess," sits "for three days in the lobby of his hotel, looking
out through a huge wall of plate-glass at the unceasing stream of pretty girls in
Parisian-looking dresses, undulating past." James initially assumed that such a
spectatorial sensibility was an innocent commonplace. In later years, however,
he recognized that looking is rarely an indifferent activity; it often involves a
system of class control as privileged observers classify and categorize others on
the basis of their dress and physical appearance. In *The American Scene* (1904),
he acknowledged that "his luxurious perch of spectatorship" was an "awful
modern privilege" that could become a "monstrous thing" because it denied "to
so many groups of one's fellow creatures any claim to a 'personality.'"[19]

During the nineteenth century, opportunities for people "to see and be seen"
increased enormously. The introduction of public transportation, parks, bal-
conied apartments, congested tenements, public recreations, spectator sports,
office towers, and plate-glass windows gave people an intimate glimpse of others

and their possessions. The mathemetician-philosopher Charles Peirce recorded in his diary his own pleasure at people-watching while riding in a street car: "Detained there, with no business to occupy him, one sets to scrutinizing the people opposite, and to working up biographies to fit them." The raging popularity of new etiquette books testified to the heightened sensitivity to social appearances among the *nouveaux riches*. As one adviser counseled, "You should . . . observe at once all the people in the room, their motions, their looks, and their words; and yet without staring at them, and seeming to be an observer."[20]

But the social elite had no monopoly on people watching. "My favorite pastime" as an impecunious young man, Dreiser remembered, "was to walk the streets and view the lives and activities of others." His "great interest was in life as a spectacle," perhaps because he considered his own life so incomplete and shameful. Whatever the reason, he spent enchanted hours gaping at the festival of urban experience. So, too, did his fictional characters. In *Sister Carrie*, for instance, the traveling salesman Charles Drouet takes Carrie Meeber to a restaurant and chooses "a table close by the window, where the busy rout of the street could be seen. He loved the changing panorama of the street—to see and be seen as he dined." For her part, Carrie uses her "gift of observation" to view people through "the open windows" of shops, offices, and factories, imagining what "they dealt in, how they labored, to what end it all came." Likewise, Eugene Witla, the painter in Dreiser's *The "Genius"* (1915), "was always looking out of the window" at the "seething masses of people!"[21]

The spectatorial sensibility associated with realism derived much of its legitimacy from the scientific spirit. As an architectural critic observed, "realism in art" was "stimulated by the habit of observation." Just as scientists took visual facts and transformed them into knowledge, realistic writers and artists assumed that to see was to know. The gift for detailed and discriminating observation, for finding the exact words or pigments to represent some experience of the eye, became the most valued attribute of the realistic artist. The Chicago writer George Ade focused a short story entitled "From the Office Window" on a middle-class office worker whose most prized perquisite was a window with a view and "a seat where he could look out and see the world move up and down." The daily vignettes unfolding before his eyes provided "a perpetual source of joy."[22]

Such social scrutiny—perhaps surveillance is the better word—occasionally took on almost predatory qualities. John Sloan derived subjects for his drawings and paintings from his habit of spying on people from his New York studio or from city sidewalks or benches in Madison Square. He confided in his diary that he was "in the habit of *watching every bit* of human life I can see about my windows, but I do it so that I am not observed at it. I 'peep' through real interest, not being observed myself." He also "peeped" as a means of compensating for his own lack of worldly experience. During a downtown stroll he once "shadowed a poor wretch of a woman on 14th St. —watched her stop to look at billboards, go into Five Cent Stores, take candy, nearly run over at 5th Avenue, dazed and always trying to arrange hair and hatpins." Occasionally, Sloan the observer became the observed, when he saw "eyes between the slats of shutters" on West 24th Street. [23]

A Consumer Culture

One of the most striking elements of this new urban culture and its visual imperatives was the sight of fashionable women on shopping expeditions. "The lady-element of Broadway," noted a journalist, "is one of its most dazzling features." Walt Whitman took special pleasure in loitering along Broadway, "looking at the beautiful ladies, the bustle, the show, the glitter, and even the gaudiness." Such shopping spectacles testified to the dizzying expansion of a consumer culture during the nineteenth century. Spreading affluence and the mass production of goods meant that millions of Americans could acquire items that earlier generations had considered luxuries. Increasingly standardized production generated new patterns of mass marketing while growing social stratification gave heightened importance to conspicuous consumption. A writer in *Scientific American* described industrial capitalism as producing a democracy of things, one "which invites every man to enhance his own comfort and status." [24]

Although evident as early as the mid-eighteenth century, this culture of consumption blossomed during the second half of the nineteenth century, nourished by an unprecedented increase in the consumption of goods and a distinctive new spirit of fashion-driven acquisitiveness. New five-and-dime stores and mail-order catalogues dispersed the realm of material desire beyond the city, entangling an entire nation in a web of wish fulfillment. The era's raging materialism, lamented writer Gamaliel Bradford, "has lured us all more of less from the things of the spirit." [25]

By ritualizing the display of alluring goods, the consumer culture both symbolized and accentuated the growing emphasis on tangible realities as opposed to spiritual or poetical ideals. Cluttered parlors, overflowing store shelves, and the heightened attention given to fashion and pretense bombarded people with material images and standards of value. Commodities came to represent objects of passion, to be viewed, sampled, modeled, lusted for, fought over, and purchased. "Can anything be more distractingly tempting than the dry-goods stores at this moment?" asked an exasperated reader of *Harper's Weekly*. "Alas for our republican virtues!" [26]

An Aesthetic of Things

On one level the rise of cultural realism mirrored the rise of a conspicuous materialism during the Gilded Age. The penchant for material display among affluent urbanites found tangible expression in a new style of domestic architecture designed to attract envious attention. Victorian houses overladen with ornamental scrollwork, fluted and carved windows, door frames, and lintels displaced the chaste simplicity of the Greek revival and the picturesqueness of the English Gothic. Inside these eye-catching houses, residents indulged what Henry James called their "fierce appetite" for furniture and furnishings. The moneyed classes filled their parlors and drawing-rooms with Turkish ottomans and divans, overstuffed upholstery, fringed carpets and curtains, picture albums, Currier and Ives lithographs, porcelain figurines, painted plates, ceramic bud vases, mirrors, and upright pianos. Such commodity fetishism led one journalist to claim that in "no department of our social life is the pecuniary standard more fully exhibited than in the interiors of our houses."[27]

"Houses are our fate," sighs Newland Archer in Edith Wharton's *Age of*

Lauterbach Residence Drawing Room, 1899. *(Museum of the City of New York. The Byron Collection.)*

Innocence. His comment reveals how houses and their cluttered interiors assumed heightened significance as expressions of self-worth and social status. In a short story entitled "The Ravages of a Carpet," Harriet Beecher Stowe described a middle-class mother's determination to keep up with "our fashionable neighbors" and "have a parlor fit to be seen." She bought a new carpet, only to decide that the aging furniture needed to be replaced as well. Her husband catalogued the transformation: "In a year, we had a parlor with two lounges in decorous recesses, a fashionable sofa, and six chairs and a looking glass . . . and great, heavy curtains that kept out all light." Their once simple, informal, and pleasant parlor had been turned into a stuffy, gloomy, impractical menagerie of "things."[28]

The new abundance associated with urban life became the most visible indicator of the nation's changing cultural landscape. During the 1830s, noted Henry James, the United States had been a society of "very few things, and even few persons," an environment in which, virtually by default, "the things of the mind did get themselves admirably well considered." But no more. Post-Civil War America overflowed with material desires and accoutrements. In *The Bostonians* (1886), James detailed the decline of "the age of plain living and high thinking, of pure ideals and earnest effort, of moral passion and noble experiment" and the rise of a culture of "things." What strikes southerner Basil Ransom most vividly when he visits his cousin Olive Chancellor in Boston is the contents of her parlor. He had "never before seen so many accessories." Olive's self-esteem requires that she have "the refreshment of a pretty house, a drawing-room full of flowers, . . . an imported tea-service, a Chickering piano."[29]

The representation of this spreading material ethos became a major emphasis of realistic literature and art. The Boston painter Childe Hassam spoke for many when he observed that the "tangible attractions of the physical world at hand furnish more inspirations than the spiritual glories of kingdom come." Of course, earlier writers and artists had incorporated material objects as allegorical symbols, but nineteenth-century realists achieved a new level of specificity in their meticulous renderings of urban structures and interiors. Henry James stressed that his hero Balzac was at his realistic best when conveying "his mighty passion for *things.*" In this regard it is worth recalling that the term "realism" itself derives from *res*, the Latin word for "thing," and cultural realists lavished attention on physical objects and material desires. "We think we are individual, separate, above houses and material objects generally," wrote Dreiser in *The Financier* (1912), "but there is a subtle connection which makes them reflect us quite as much as we reflect them."[30]

Commodities and fashions thus served as the key props and icons in realistic literature, art, and drama. "An interest in things," John Sloan remarked, "is and always was at the root of art." In this regard he and other realists agreed with William James that "things of sense"—clothes, books, homes, furnishings, carriages—were what most people perceived as real. In Frank Norris's novel *Vandover and the Brute*, the contents of Vandover's new apartment offer an inventory of the author's own world of value:

> The walls were hung with dull blue paper of a very rough texture set off by a narrow picture moulding of ivory white. A dark red carpet covered with rugs and skins lay on the floor. Upon the left-hand wall, reaching to the floor, hung a huge rug of sombre colours against which were fixed a fencing trophy, a pair of antlers, a little water colour sketch of a Norwegian fjord, and Vandover's banjo; underneath it was a low but very broad divan covered with corduroy. To the right and left of this divan stood breast-high bookcases with olive green curtains, their tops serving as shelves for a multitude of small ornaments.

The description continues in similar detail for two more pages and attests to the widespread assumption that individuals expressed themselves through their material surroundings.[31]

Edith Wharton was especially sensitive to the role of physical trappings in shaping self-esteem and social consciousness, and she became justly famous for her painstakingly rendered interiors. She even co-authored a book on interior decoration which claimed that most women tended "to want things because other people have them, rather than to have things because they are wanted." Wharton described Undine Spragg in *The Custom of the Country* (1913) as dedicated to the pursuit of luxury items and pride of possession: "To have things had always seemed to her the first essential of existence." At one point, as she gazes about the garish parlor, Spragg reflects that "in that one small room there were enough things of price to buy a release from her most pressing cares."[32]

In Wharton's *The House of Mirth*, Lily Bart displays the same material calculus. Beautiful and self-absorbed, she had been reared in luxury, only to be impoverished by the premature death of her parents. Her legacy was a compulsive taste for the sumptuous life and a keen awareness of the patriarchal custom that society women serve as decorative accessories and objects of spectatorial male desire. She had been fatally conditioned to see in money and things *the* key to social esteem and personal happiness. Her "whole being dilated in an atmosphere of luxury; it was the background she required, the only climate she could breathe in." Bart explains that if she "were shabby no one would have me: a woman is asked out as much for her clothes as for herself. The clothes are the background, the frame, if you like: they don't make success, but they are part of it. Who wants a dingy woman? We are expected to be pretty and well-dressed till we drop." Elegant clothes and plush furnishings testified to her commodified sense of self, and she became at once the victim and parasite of fashionable society, a muddle of emotional greed and misguided aspiration.

Having been "brought up to be ornamental" and "fashioned to adorn and delight," she has "no real intimacy with nature," preferring the "whirling surface" of society life.[33]

A thirst for things and a fashioning sense of self also pervade Dreiser's *Sister Carrie*. Self-conscious about living in a run-down Chicago flat, Carrie Meeber finds in the ritual of department-store shopping "the delight of parading here as an equal." At "The Fair" department store, she "passed along the busy aisles, much affected by the remarkable displays of trinkets, dress goods, stationery, and jewelry. . . . She realized in a dim way how much the city held—wealth, fashion, ease—every adornment for women, and she longed for dress and beauty with a whole heart." Bereft of funds, Carrie leaves with only fantasies, but, having been seduced by the merchandise and singed by the "flame of envy," she later returns with twenty dollars to purchase tangible images of grandeur. "Fine clothes were to her a vast persuasion; they spoke tenderly and Jesuitically for themselves. When she came within earshot of their pleading, desire in her bent a willing ear."[34]

The "Woman Question"

Still another new social reality associated with the emergence of an urban culture was the changing status of women. After the Civil War, thousands of women entered the work force as nurses, teachers, librarians, civil servants, seamstresses, secretaries, retail clerks, and social workers. Between 1870 and 1900, the number of women working for wages outside the home tripled, and five million women (17 percent of all women), many of them overseas immigrants or rural migrants, had full-time jobs. By the end of the century, every vocation and profession had at least a few women representatives. One male editor feared that he would be "swept away by the rising tide of femininity." There was, he added, "scarcely an occupation once confined almost exclusively to men in which women are not now conspicuous."[35]

In a letter to the editor of the *Nation* in 1867, a "working woman" described the changing nature of gender roles in American life. Most middle-class women, she acknowledged, still lived in a "world of love, of a sweet and guarded domesticity, of drawing-rooms and boudoirs, of dainty coquetries, of quiet graces, where they flourish like fair flowers in a south window." But each year thousands of women were entering a different world "of mud, carts, ledgers, packing-boxes, counting-houses, paste, oils, leather, iron, boards, committees, big boots, and men who . . . meet women on a cool, business level of dollars and cents."[36]

At the same time, women also gained access to higher eductaion. Dozens of women's colleges were founded, and many all-male colleges and universities began admitting women. By 1910, 40 percent of college students were female.

"After a struggle of many years," a New York woman boasted, "it is now pretty generally admitted that women possess the capacity to swallow intellectual food that was formerly considered the diet of men exclusively." To be sure, college women were often shunted into home economics classes and other "finishing" courses. Still, the doors of the professions—law, medicine, science, and the arts—were at least partially opened by the turn of the century.[37]

In this context, then, the much discussed "woman question" of the second half of the nineteenth century involved far more than the issue of voting rights; it also concerned the liberation of some women from the home and from long prescribed social roles and character traits. In addition to new jobs and professions, women of means took advantage of other new venues for female interaction and expression: charitable associations, women's clubs, literary societies, church work, and so on. "If there is one thing that pervades and characterizes what is called the 'woman's movement,'" E. L. Youmans remarked, "it is the spirit of revolt against the home, and the determination to escape from it into the outer spheres of activity."[38]

Yet liberation from the "cult of domesticity" often exacted a high price. Many women who desired to play a more active role in the "real world" contracted a peculiar new affliction called neurasthenia, a debilitating psychological and physiological disorder whose symptoms usually included dyspepsia, insomnia, hysteria, hypochondria, headaches, depression, and a general state of enervation.[39]

Although neurasthenia plagued both sexes, it most often afflicted college-educated middle-class women. George M. Beard, a distinguished New York neurologist who popularized the term neurasthenia, concluded that women were "more nervous, immeasurably, than men," and women neurasthenics tended to be those who had become "overly active" outside the home. This association of neurasthenia with independence led one doctor to insist that the malady provided the best "argument against higher education of women."[40]

Several women authors incorporated their own experiences with "nervousness" into their artistic works. Charlotte Perkins Gilman, for instance, intended her short story "The Yellow Wallpaper" as a piece of "pure propaganda" to expose the horrors of the "rest cure" she was subjected to at age twenty-seven. A doctor had ordered her to "live as domestic a life as possible; have your child with you all the time; lie down an hour after each meal; have but two hours intellectual life a day; and *never touch pencil, brush, or pen as long as you live.*" This regimen, Gilman explained in a letter to Howells, took her "as near lunacy as one can, and come back."[41]

Social worker Jane Addams also struggled with neurasthenia. After graduating from Rockford College in 1881, she found few opportunities to use her degree and lapsed into a state of depression during which she developed an intense "desire to live in a really living world." Women, she charged, were "so besotted with our [sentimental] novel reading that we have lost the power of seeing certain aspects of life with any sense of reality because we are continually looking for the possible romance." The desire to engage "real life" eventually led her to found Hull House, the famous settlement house in Chicago where immigrants could make the transition to American life, and young women like herself "could learn of life from life itself."[42]

The example of Addams and others helped inspire other women to enter the "real" world and adopt a more "realistic" frame of mind. By 1890 a writer in *The Arena* could announce that women were "beginning to look upon the world with their own eyes. They are examining facts and theories." He urged progressive-minded people to dismiss the traditional view of women as something "hollow, false, and unreal." More and more people did just that. Upon returning from England to America in 1904, Henry James reported that the predominance of "new" women and their struggle for autonomy had become "the sentence written largest in the American sky." Several years later he applauded the shifting emphasis "from the idea of woman's weakness to the idea of her strength." Representing and encouraging such a transformation in female consciousness and status became a central concern of realistic fiction and painting. As writer Charles Dudley Warner declared, the main "question that society is or should be interested in is whether the young woman of the future . . . is going to shape herself by a Realistic or an Ideal standard."[43]

The Media of Realism

In Edgar Allan Poe's superb short story, "The Man in the Crowd," the narrator gazes out upon the kaleidoscopic urban scene "with a delicious novelty of emotion," accumulating in the process a collection of sensuous observations, "now in poring over advertisements, now in observing the promiscuous company in the room, and now in peering through the smoky panes in the street." Poe's references to advertisements attested to the ways in which the rise of mass advertising and the modern metropolitan newspaper generated a media revolution during the nineteenth century. Along with two other new forms of representation, photography and moving pictures, they imprinted the myriad surfaces and values of the material world into public consciousness and artistic expression. These "mirrors of reality," one journalist remarked, exploited and sanctified the "concrete, commonplace, every-day, matter-of-fact way of thinking."[44]

Between 1870 and 1900 expenditures on advertising increased more than tenfold. As early as 1867 a journalist could report that advertising "is the monomania of the time. There is no relief from it in all the earth." Ads were invading "every department of life, and no privacy or exclusiveness is a bar to them."

The burgeoning advertising industry changed buying habits and crowded the cultural environment with alluring new signs and images. In 1875 a writer in the *Nation* asserted that "the variety and vivaciousness of advertisements have greatly increased in the last ten years." The reasons for this were clear: the crowded cities provided an endless array of enticing surfaces for advertisers to exploit. A spokesman for the Immigration Protection League commented that foreigners arriving in American cities first noticed "the pretentiousness of signs and advertisements." The same was true for many rural migrants to the metropolis. As Dreiser's Carrie Meeber approached Chicago on the train, she "gazed out the window" and saw an endless array of advertisements: "Signs were everywhere numerous."[45]

Advertising signs and slogans shouted for attention and accustomed people to the material imperatives of commercial capitalism. A writer in *Harper's Weekly*, for example, noted that advertisements "are now part of the humanities, a true mirror on life." Ads, of course, were intended to convince people that a particular product or service represented the "real thing." In the same way, creative writers trumpeted the "truth" or "reality" of their fictional works. Per-

Hudson Street, New York City, ca. 1865. (*Collection of the New-York Historical Society.*)

haps the most common subtitle of nineteenth-century works of fiction was "A True Story," a certifying phrase that catered to the growing public fascination with authenticity and contemporaneity. In a relentlessly commercial environment increasingly detached from spiritual priorities and the integrity of local markets, people craved the moorings of both product credibility and artistic veracity.[46]

People also desired knowledge about the immediate goings-on in their own community. This desire for topicality helped give rise to the mass-circulation metropolitan newspaper. As writer Edward Everett Hale, Jr., observed, the newspapers offered "a daily advertisement of the attractions of city life." This was especially true after the development of the so-called "penny press" during the Jacksonian era. A newspaper revolution began with the invention of the steam press in the 1830s and picked up speed with the development of the Hoe rotary press in 1846, both of which enabled publishers to mass-produce newspapers, lower the cost of a newspaper to a few pennies, and thereby greatly expand readership. Within a few decades the nation was awash in newsprint. In 1840 there were 138 daily newspapers in the country; thirty years later there were 574; by the turn of the century the total was 2600. Overall circulation during the same period increased from less than two million to over 24 million.[47]

To attract and sustain mass interest, editors called for more comprehensive reporting of the news of the day, including the activities of ordinary people. James Gordon Bennett launched the New York Morning Herald with a pledge "to record facts, on every public and proper subject, stripped of verbiage and coloring. . . ." To be sure, Bennett and other editors frequently highlighted sensational events, but most observers felt nonetheless that newspapers offered, overall, the most realistic representation of contemporary life. "Newspapers," Walt Whitman claimed, "have become the mirror of the world," reflecting and shaping the nation's "reality and life."[48]

It was no coincidence that many of the most prominent "realistic" writers and artists served apprenticeships on newspapers or weekly magazines. Whitman, Howells, Twain, Dreiser, Crane, Willa Cather, Winslow Homer, John Sloan, and others began their careers as journalists, editors, or sketch artists. According to Dreiser, newspapers exposed aspiring writers like himself to "vital, dramatic, true presentations of the life that is being lived today." In addition, newspaper work honed their skills at precise observation. It also inculcated in them a respect for the stubborn significance of facts and a desire to unravel the twisted motives of the human heart. George Ade credited his experience as a reporter for the Chicago Record with shaping his literary outlook. "My ambition," he maintained, "was to report people as they really were, as I saw them in everyday life, and as I knew them to be." Ade's friend William Dean Howells likewise referred to city journalism as "the school of reality."[49]

Increased use of illustrations reinforced journalism's claim of providing the most "realistic" accounts of contemporary life. Illustrated magazines such as *Ballou's Pictorial* in Boston and the *New York Illustrated News*, *Frank Leslie's Illustrated News*, and *Harper's Weekly* in New York all began publication during the 1850s. The editors of *Harper's Weekly* promised to "present an accurate and complete picture of the age in which we live," while Frank Leslie claimed that the liberal use of illustrations would enhance journalism's "actuality and attraction." Readers appreciated his authenticating efforts. A New York subscriber commended Leslie for providing Americans with a "realizing sense of what is going on in the world." Newspaper and magazine illustrators, Leslie wrote, provided a "grand daguerreotype of American events, American life, and American associations."[50]

Leslie's use of the daguerreotype as a metaphor is especially illuminating, for photography emerged during the 1840s as a new medium peculiarly suited to the era's preoccupation with visual reality, factual scrutiny, and precise representation. By seizing a privileged moment and providing a tangible image of untouched reality, the promiscuous camera lens conferred significance on common people and the trivial particulars of the physical world. "In the epoch of fact," wrote an art critic, "photography is the concrete representation of consummated facts." Seductive in its seeming veracity and neutrality, the camera provided a magical new means of enlarging the realm of the visible and thereby helped to define public perceptions of reality.[51]

Whitman, for example, loved to let his "peering gaze" linger in front of the daguerreotypes on display at Plumb's gallery on lower Broadway in Manhattan. Surrounded by "a great legion of human faces" from floor to ceiling, he found their intense *"realities"* overwhelming. To him the "spectacle" of mute images on display in the daguerreotype galleries revealed how the romantic imagination "would be infinitely outdone by *fact*."[52]

Photography quickly became the most democratic of the arts as people from all classes flocked to have their pictures taken. Its mass appeal reflected the public's passion for visible facts. People demanded documentary evidence of their own selves and times. Their interest in the truth of things led them to ask of a product or a work of art: Is it authentic? Is it an accurate representation? Photographers assured the public that the camera offered the closest rendering of reality possible. As a daguerreotypist proclaimed, his pictures were the "real thing itself."[53]

Philosophical idealists cringed at photography's "objective" celebration of the visual world. As a writer in the *Photographic Art-Journal* insisted, the camera's accurate standards reinforced the emerging realistic movement, "while the idealists, however vigorous, will be forgotten." Those who saw true reality as a spiritual rather than material phenomenon now were put on the defensive by a mindless mechanical instrument grounded in the sciences of chemistry and optics, an instrument that seductively demonstrated that matter *was* real. "Photography is realism; everything else is romance," novelist Maurice Thompson concluded. As a self-described "idealist," he admitted that photographs furnished palpable evidence of the "real" world of faces and buildings and "things" to be portrayed, examined, and enjoyed in and of themselves.[54]

Portrait painters especially felt the competition of the daguerreotype studios because people now could obtain a camera likeness for a much smaller fee and with much less bother. Yet several painters and art critics argued that competition with photography actually enhanced the quality of art. It encouraged greater fidelity and challenged painters to mimic its ability to record instantaneous moments of action observed from seemingly accidental points of view. A writer in *The Crayon* predicted that the camera's proliferating influence would "educate the popular eye to the keener perception of truth" and help "drive juggling out of Art, and compel painters to be more conscientious and studious."[55]

Many painters used photography as a tool to enhance the accuracy of their depictions. Mary Cassatt found it remarkable "how much our modern painters use the camera." Photographic aids increased the precision and seeming objectivity of paintings. An art critic noted that there was a growing "submissiveness to the matter-of-fact" emphasis of photography among American painters. Artists also often used photography as a metaphor for realism in art. The neo-classical painter and art critic Kenyon Cox, for instance, described William Chase as "a wonderful human camera—a seeing machine—" showing "us the beauty of the commonest and humblest subjects."[56]

The camera similarly provided creative writers with a new standard of representational accuracy. In presenting the title character in *Uncle Tom's Cabin*, Harriet Beecher Stowe pledged to "daguerreotype [him] for our readers." In the same vein, Whitman boasted in *Leaves of Grass* that "everything is literally photographed," while novelist Joseph Kirkland observed that modern fiction demanded "photographic exactitude in scene-painting—phonographic literalness in dialogue—and telegraphic realism in narration—these are the new canons for the art of fiction."[57]

Yet many writers maintained that photography was not an art itself; it was too mechanical, too superficial, and provided a false sense of perspective. Even the camera lens had its inherent blind spots and distortions. The monocular view of the camera could never match the breadth, complexity, or ambiguities of human vision. Nor did a photograph expose reality's own riddles. True art, some critics contended, included the development of the *inner* reality of the subject as well as its external description. While acknowledging that a camera provided "the external facts of whatever visage is placed before it," the editors of *The Galaxy* maintained that it "does not tell the whole truth."[58]

The Dreadful Chill of Change

Taken together, urban-industrial growth, social and ethnic turmoil, a burgeoning consumer culture and an attendant rise of secular materialism, the emergence of "new women," and the raging popularity of metropolitan journalism
and the camera transformed the very nature of American civilization. The recurring theme of cultural commentary during the second half of the nineteenth
century was an acute sense of accelerated social and intellectual change, a
feeling on the part of reflective men and women that the United States was
profoundly different in tone, tempo, values, and appearance from the country
of their childhood. Of course, life is always changing. The pressure of history,
the pneumatic push of extraordinary new social developments and ideas, never
rests. Yet the velocity and scope of change at mid-century and after seemed
especially bewildering. "When I reflect what changes I, a man of fifty, have
seen, how old-fashioned my ways of thinking have become," James Russell
Lowell confessed to a friend in 1869, "the signs of the times cease to alarm
me."[59]

In 1907 Lowell's friend Charles Eliot Norton, the magisterial Harvard professor of fine arts, marveled at "the enormous change" during the past fifty years.
The rude displacement of old faiths and artistic ideals saddened his backward
glance. Such rapid and inescapable changes had wreaked "a disastrous effect"
upon "the literature of pure imagination, upon poetry, upon romance." Now,
he lamented, secular realism was triumphant, as writers, painters, architects,
and critics concentrated their gaze on the "material things" and "daily affairs"
of contemporary life. "The whole spiritual nature of man" now found "only
feeble and unsatisfactory expression."[60]

Henry James adopted a somewhat different angle of vision in assessing modern American life. When he returned to the United States from his adopted
home in England early in the twentieth century, he reported that the "dreadful
chill of change" filled the air in the homes of the New England intellectual
elite. The romantic idealism of the antebellum era had become a "conscience
gasping in the void, panting for sensations, with something of the movement
of the gills of a landed fish." His Darwinian metaphor aptly expressed his outlook. Unlike Norton, James made no wistful plea for a return to the halcyon
days of pre-industrial idealism; the romantic era was gone, never to be retrieved.
Modernity needed to be faced and fronted: the "world as it stands is no illusion;
we can neither forget it nor deny it nor dispense with it." An unprecedented
new society demanded new aesthetic forms. To accept that world and develop
those forms was the challenge of modernity. In his notebook James reflected
that "to dig deep into the actual and get something out of *that*—this doubtless
is the right way to live." Life "*is*, in fact, a battle," he argued. People no longer

could afford to be "scared back by the grim face of reality into the world of unreason and illusion." Like Basil Ransom in *The Bostonians*, James believed that the ability "to know and yet not fear reality, to look the world in the face and take it for what it is—. . . that is what I want to preserve. . . ."[61]

Yet such rhetoric about taking reality simply "for what it is" cannot be accepted at face value. No matter how sincere their quest for truth, no matter how brassy their claims of objectivity, the realists were in fact quite diverse and complex in their representations of reality. Many of them promoted more than documentary accuracy. They wanted to use realistic representations of contemporary life for a special purpose: to rein in the runaway aspects of a turbulent new society. To them the United States seemed to have lost much of its stability and cohesion as a result of urban-industrial development and western expansion. It had become a loose aggregate of competing individuals, divided from one another by all manner of economic differences and ethnic, racial, and class prejudices.

Many of the new social scientists, for example, strove to find "realistic" ways to enhance social stability amid an increasingly turbulent society filled with labor unrest, ethnic diversity, and racial tensions. "The sociologist," Edward Ross wrote, "shows that we are bound up with one another in a kind of loose corporation, so that in many cases we have to stand or fall together." Yet he feared that traditional stabilizing forces such as the family, church, and traditional patterns of social deference were losing their potency. Religious sanctions had become "a decaying species of control."[62]

Hidden within this desire to improve human relations and social stability was a powerful element of racial and ethnic prejudice common among Gilded Age intellectuals, even the most progressive. After the Civil War, biologists and social scientists, most of them white Anglo-Saxon Protestants, developed "scientific" theories supporting the ideal of a genetically determined racial and ethnic hierarchy. Americans of "Nordic" or "Teutonic" descent topped such a scale, and below them were immigrants from southern and eastern Europe, Hispanics, Asians, and African Americans. Ross, for example, frequently expressed his disdain for Asian and East European immigrants, characterizing them as "beaten members of beaten breeds" who lacked "the ancestral foundations of American character." Yet he celebrated America's ability to assimilate them. What marvelous "change a few years of our electrifying ozone works in the dull, fat-witted immigrant."[63]

Ross argued that the new realism evident in literature and the fine arts could help stitch together a fraying social fabric of divergent ethnic groups and economic classes. Social prejudices, he claimed, resulted primarily from ignorance and provincialism. "Most of us are narrow in sympathies because our life circle is narrow. Our contacts with others are quite too few." By depicting how people of different regions, classes, races, and conditions lived, the realistic writer or artist "calls forth fellow feeling and knits anew the ever-ravelling social web." Ross therefore praised those writers and painters who were "at work cementing classes, conciliating local groups, keeping all parts of the nation *en rapport*."[64]

Many writers and artists shared Ross's assumption that they must use their pens and brushes to help defuse social tensions. William Dean Howells, for instance, observed that the old sources of social unity had lost their adhesion. "Once we were softened, if not polished by religion," Bromfield Corey tells his son in *The Rise of Silas Lapham*; "but I suspect that the pulpit counts for much less now in civilizing." Instead, "civilization comes through literature now, especially in our culture." Howells contended that writers should craft their "realistic" stories so as to encourage readers to reflect upon the complex social questions of their day rather than leave their sensibilities merely "tickled and pampered." Popular romantic writers, he charged, told sugar-coated lies that "make one forget life and all its cares and duties," but literature's fundamental purpose was "to make the race better and kinder," to help people "know one another better, that they may be all humbled and strengthened with a sense of their fraternity." The "close" study of real life would influence people to "deal justly with ourselves and one another" by eradicating prejudices and provincialism, "for if the study is close enough it is sure to be kindly; and realism is only a phase of humanity."[65]

In this sense, the realistic impulse represented more than simply an attempt to "mirror" or "document" American life: it was also an effort to ameliorate the profound social disturbances associated with unparalleled urban-industrial development. Howells declared that "realism alone has the courage to look life squarely in the face and try to report the expression of its divinely imagined lineaments." Old perspectives and illusions, he added, were bankrupt. "Romanticism belonged to a disappointed and bewildered age, which turned its face from the future and dreamed out a faery realm in the past; and we cannot have its spirit back." In a period of such flux and turmoil, realism offered coherent representation of a new social order that seemed increasingly inaccessible and fragmented. The realistic enterprise, Howells promised, offered a disoriented people a "vision of solidarity."[66]

Part III

Realism Triumphant

After its first stirrings in the 1850s, realism took on the more settled and confident air of a movement. The following three chapters survey the maturation of the realistic impulse in fiction, art, and architecture. In the process, they address several basic questions. What precisely did it mean to write realistic stories, paint realistic scenes, and design realistic structures? Who were the leading artists in these fields? How were their works received? Before dealing with such issues, a final qualification is in order. The realistic revolt constituted a potent but by no means omnipotent force in American thought and culture after mid-century. The idealistic spirit, so dominant during the antebellum years, continued to pose a combative alternative to the new realism. "The romantic and the realistic," declared the Chicago novelist Joseph Kirkland in 1886, "are engaged in a life-and-death struggle."[1]

6

Truth in Fiction

After Appomattox a new phalanx of avowedly realistic writers, many of them women, surged into prominence and developed networks of friendship and mutual support. "The idealist pure and simple hardly exists at all," announced a journalist in 1887. "The taste for the fantastic and impossible is plainly dying out." Three years later the English critic Edmund Gosse could report that in America "almost every new writer of merit seems to be a realist." For all of the ambiguity embedded in that label, most people at the time recognized what it meant. John W. De Forest, for instance, assured William Dean Howells in 1879 that Henry James "belongs to our school."[1]

In fact, Howells and James were the headmasters of the "school" of literary realism. Everyone in the literary world knew them, read their fiction, and grappled with their ideas. Through their arduous efforts, the realistic novel reached maturity in the United States and the realistic movement developed a coherent theoretical basis. "To Mr. Howells more than anyone else," Hjalmar Boyeson declared, "are we indebted for the ultimate triumph of realism in American fiction." Howells and James, like all enthusiasts, were sometimes dogmatic, occasionally inconsistent, and not always convincing in pressing their case, but they were effective campaigners for a new literature dedicated to representing contemporary morals and manners.[2]

The two writers shared more than aesthetic preferences. As boys they displayed marked "feminine" tendencies, yet they also developed compulsive ambitions to participate in the rough-and-tumble masculine arena. In the James household, William gobbled up most of the family's passion and vitality, leaving Henry to nurture a keen aesthetic sensibility. Once, when Henry asked to join his older brother and his friends on an outing, William curtly declined, saying that "I play with boys who curse and swear!" By contrast, the contemplative

Henry, alive to every sensation, spent much of his boyhood "dawdling and gaping" at life's visual panorama rather than romping with playmates. He once admitted to William that he "always looked forward . . . to the day when I . . . at least might become more active & masculine." For his part, Howells remembered that in his youth he was so intimately attached to his mother that he grew terrified at the prospect of being labeled a "girl-boy." Throughout his long life, he treasured memories of the maternal domesticity that shaped his childhood, and he simultaneously worried that he lacked the spine of true masculinity.[3]

Howells and James came to maturity during the 1860s, when women writers and readers "ruled" the literary marketplace. Their strident attacks against feminine sentimentalism resulted in part from their continuing efforts to assert their own manliness. Gender anxieties permeate their fiction and criticism. Before the Civil War, Howells observed, the male writer had been a heroic figure, but afterward he was dismissed as "a kind of mental and moral woman, to whom a real man, a business man, could have nothing to say." Howells himself was determined to break away from the "preening and prettifying" tendencies of most male writers and produce a literature rooted in the "world of men's activities." Howells and James diagnosed American culture as suffering from what James called the "emasculate twaddle" of taffeta sentimentalism, and they confidently prescribed the antidote: a robust school of realistic fiction.[4]

James was born in 1843 into a home of wealth, leisure, and learning in New York's fashionable Washington Square neighborhood. More interested in "being" rather than "doing," he moved within an elite world, strictly bounded, thinly populated, and insulated from the poor, the toiling, the vulgar, and the vicious. His philosophical father, Henry James, Sr., used his sizable inheritance to provide the children with French-speaking governesses, private tutors, and prolonged excursions to Europe. Henry Jr. thus spent much of his "rootless and accidental childhood" as a privileged traveler obtaining a "sensuous education" in hotels, museums, libraries, theaters, and drawing rooms in New York, London, and Paris. An acute observer of the buzz and boil of urban life, James described himself as someone addicted to "pedestrian gaping." He nurtured a genius for place. Few writers then or since have been so capable of describing city culture, so alive to the sensual energies of *things* and the ruthless ironies of high society.

Although his two younger brothers fought in the Civil War, Henry suffered a vague "physical mishap," a "horrid if obscure hurt" (a strained back suffered while fighting a fire in October 1861) that forced him to watch the conflict from the "almost ignobly safe stillness" of a Newport garden while an art student and from a Cambridge library while briefly enrolled at Harvard Law School in 1862. His failure to join the ranks left him with a nagging sense of having missed a crucial initiation into manhood.[5]

In 1864 James abandoned plans to become a lawyer and fastened upon literature as his chosen vocation. The next year he recommended in the *North American Review* that American writers exchange the prevailing method of "descriptive idealism" with "the famous realistic system" developing in France. Thereafter James hugged what he called "the shore of the real," and his commitment to literary realism "gathered further confidence with exercise." By 1876 he could exclaim that the "real is the most satisfactory thing in the world, and if once we fairly get into it, nothing shall frighten us back."[6]

While a friend of the New England patricians, James liberated himself from many of their brittle assumptions about life and art and became a penetrating analyst of drawing-room mores. He wrote twenty-two novels and over a hundred short stories before his death in 1916. But he was more than prolific; he had true genius. Few writers had such a sensuous eye for the savor and shimmer of words and the nuances of gesture and conversation. Even fewer wrote such scrupulously refined and elaborated prose. What James once said about the French novelist Alphonse Daudet applies equally to him: his "great characteristic" was a splendid "mixture of the sense of the real with the sense of the beautiful." As the aristocrat of realism, James mixed with the very few and wrote for the few. "The multitude," he once told his brother William, "has absolutely no taste—none at least that a thinking man is bound to defer to." Soberly fastidious in his own dress and manners and fiercely literate in outlook, James, a lifelong bachelor, grew justly famous for a fussy self-regard. "I am," he once admitted, "interminably supersubtle and analytic."[7]

Howells, on the other hand, was remarkable more for the breadth and authority of his work as editor, critic, and novelist than for his artistry. He published too much too quickly, authoring thirty-five novels and over a thousand magazine pieces. Howells was not so much interested in conveying ideas as he was preoccupied with exploring the texture of social intercourse. His novels displayed more poise than penetration, more warmth than insight, more charm than drama. As later critics delighted in pointing out, a sense of Victorian prudery and unrelenting optimism about the social condition hobbled Howells's literary imagination. He pushed realism only to its nearest limits. James had a subtler and more penetrating sensibility, a bladed mind of far greater reach and intensity. He once noted that his friend Howells lacked "a really *grasping* imagination."[8]

Howells, however, was a man more of his age: a man with sympathy to share—and genius to judge—the dominant sentiments and movements of his time. Unlike James, he was a commoner from the West who dreamed of a society where a man "is valued simply and solely for what he is in himself, and where color, wealth, family, occupation, and other vulgar and meretricious distinctions are wholly lost sight of in the consideration of individual excellence." Howells was remarkably tender and generous in his treatment of people, both real and fictional. It was not only his nature but his mission to be encouraging to others. Where James staged a worried withdrawal from mass society that eventually turned his critical detachment into mere disengagement, Howells struggled—not with complete success—to moor himself to what Whitman called the "divine average."[9]

Howells's rise to authorial fame and editorial influence was a literary rags-to-riches story. One of eight children of a small-town Ohio couple who professed fierce antislavery convictions and mystical religious beliefs, young Howells quit school at age nine to help his printer father set type. The self-conscious, introspective boy had to take charge of his own education, and the printer's shop became his schoolroom, his father's many books providing the texts. Yet this self-made and self-taught outlander would eventually be elected the first president of the American Academy of Arts and Letters and be offered distinguished professorships at Yale and Harvard.[10]

Writing a campaign biography of Lincoln earned Howells a consulship in Venice in 1861 and thereby freed him from service in the Civil War. While in Italy, Howells realized that his literary talents were best suited for prose rather than poetry. Before going abroad, he later recalled, "I was an idealist" who assumed that in fiction "persons and things should be nobler and better than they are in sordid reality; and this romantic glamour veiled the world to me, and kept me from seeing things as they are." In Venice, however, Howells developed a consuming interest in contemporary affairs. He began writing sketches of Italian scenes that confirmed his ability to evoke the "living human interest" in everyday life.[11]

Howells left Italy in 1865 with a fresh determination to pursue a career in literary journalism and fiction. After a short stint with the new *Nation* magazine in New York, he moved to Cambridge and joined the editorial staff of the *Atlantic Monthly*. In 1871, at age thirty-four, he was named editor-in-chief, leading James to label him the "monarch absolute" of the magazine.

By the 1880s Howells had become a celebrated author in his own right. His novels of contemporary life offered neither hope nor despair but simply the necessity to fashion a moral life. Mark Twain assured Howells that "only you see people & their ways & their insides & outsides as they *are*, & make them talk as they *do* talk. I think you are the greatest artist in these tremendous matters that ever lived." Such effusive appreciation suggests the height of Howells's prominence. He became in the words of one literary critic "the sole, supreme, infallible dean of American letters" during the last quarter of the nineteenth century. And he loved to use his editorial pen to puncture romantic illusions and deflate idealistic pretensions. "From the first," he wrote in recalling his efforts on behalf of literary realism, "it was a polemic, a battle. I detested the sentimental and the romantic in fiction, and I began at once to free my mind concerning the romanticists, as well dead as alive."[12]

Howells served as the keystone in the arch of the realistic movement, without whose granitic leadership the others would have tumbled. As editor of the *Atlantic Monthly* during the seventies and then of *Harper's Monthly* in the eighties, he championed the efforts of his two contrasting friends, Twain and James. He also gave earnest encouragement to younger and frequently more audacious writers of realistic bent. James appreciated Howells's uncommonly beneficent leadership: "You showed me the way and opened the door." Howells returned the compliment. American literary realism, he wrote in 1882, "finds its chief exemplar in Mr. James; it is he who is shaping and directing American fiction."[13]

ALTHOUGH MARK TWAIN is usually included in the realistic "school" along with Howells and James, he was at best a truant member. True, he burlesqued the genteel tradition, addressed controversial topics, mastered the use of authentic dialect in his fiction, and hated "sentimentality & sloppy romantics." But he wrote few critical essays, rarely articulated a concrete literary perspective, and never used the term realism. His personal preference lurched between the truth-telling restraints of realism and the inventive freedom of romanticism. Unwilling to sacrifice his imaginative independence to any single creed, Twain is perhaps best described as an idiosyncratic fellow traveler to his friends among the professing realists.[14]

Huckleberry Finn vividly illuminates Twain's inner dialectic. Living in a boy's fantasy world of medieval chivalry and buccaneering adventure, Tom Sawyer repeatedly fabricates "stretchers," or lies, to protect his illusions from the assault of mundane reality. Huck tolerates and even participates in such extravagances but eventually tires of Tom's unreal conceits. In rejecting one of Tom's "dream" escapes from civilization, Huck explains: "So then I judged that all that stuff was only just one of Tom Sawyer's lies. I reckoned he believed in the A-rabs and the elephants, but as for me, I think different." Of the two, Huck is more the realist, impassively observing the flow of life, accepting the world as he finds it, and adapting himself to its demands. Floating down the river with Jim, the runaway slave, Huck steers away from Tom's fantasies as well as those perpetuated by society to camouflage its hypocrisies. In his stoic sensitivity to the facts and nuances of common life, Huck provides the reader with illuminating insights into the human condition. "For Huck," Robert Penn Warren has remarked, "the discovery of reality, as opposed to illusion, will mean freedom." Twain, too, set out to discover reality, but he always kept Tom Sawyer's capacity for self-deception as a companion.[15]

Catalysts for Change

The writing of fiction is never simply an act of the isolated imagination. It also reflects the influences of particular social settings and historical forces. In developing their realistic literary credo, Howells, James, and others drew inspiration from the English and European writers making up what James called "the young realists": Trollope, Balzac, Flaubert, Zola, Turgenev, and Tolstoy. As a literary critic observed in 1889, "Just now we are trying to be French; yesterday we were cultivating the Russians; last week the English had us under their thumbs."[16]

Yet too much can be made of these foreign influences. Although deeply indebted to transatlantic developments, literary realism in the United States resulted largely from indigenous causes. The Civil War, for example, created a surge of national pride that promoted a new literature representing contemporary American life. A writer in *Harper's Weekly* observed in 1862 that "this great struggle, by revealing to us our own manhood, releases us nationally from our childish dependence upon European criticism, so it will emancipate our literature from foreign subservience." Four years later, the outspoken young literary critic Eugene Benson urged that a new generation of "strong and virile" authors supplant the "dainty, respectable, weak" writers who presume to "protect us from the defilements of reality" and "think they must preach [to] or instruct" the reader. He implored younger writers with the "daring to express life" to depict in a "masculine" way "the myriad life of this most complicated age."[17]

Enormous difficulties, however, confronted those seeking to represent the mosaic of contemporary American life in fiction. In an essay entitled "The Great American Novel," John W. De Forest noted the challenges facing the writer who aspired to capture the whirring diversity of "this eager and laborious people, which takes so many newspapers, builds so many railroads, does the most business on a given capital, wages the biggest war in proportion to its population, believes in the physically impossible and does some of it." The realist, De Forest acknowledged, necessarily pursued an impossible task in trying to give faithful representation to a nation whose strange medley of humanity seemed in a state of perpetual motion. "Can a society which is changing so rapidly be painted except in the daily newspaper?"[18]

Unable to compete with the immediacy or breadth of newspapers, De Forest and other self-conscious realists tried to render a fragment of social reality in words. "America," Howells observed, "is so big and the life here has so many sides that a writer can't synthesize it." The best one could hope for, according to Twain, was a vivid depiction of "the ways and speech and life of a few people grouped in a certain place." In prose seasoned with authentic details and local dialects, self-conscious realists portrayed heretofore ignored areas and aspects of American life.[19]

Realists benefited from a literacy explosion that greatly expanded and diversified the reading public. Throughout the nineteenth century, more and more states adopted compulsory school attendance laws. As a result, the number of high schools in the country soared from 800 in 1878 to 5500 in 1900. In 1870 only 16,000 students graduated from high school (56% being women); in 1900 there were 95,000 graduates (60% women).

Even more important for authors of "serious literature" was the dramatic increase in the number of college graduates who eagerly exploited new opportunities for cultural enrichment. During the 1870s, middle-class Americans flocked to take part in the new movement for adult education and artistic enrichment based in Chautauqua, New York. By the 1880s over 100,000 people participated in Chautauqua's traveling lecture series, home reading courses, and musical concerts. Thousands of others joined book clubs and literary societies or enrolled in university extension courses. At the same time, the number of libraries grew enormously. Only a handful of public libraries existed in 1850, but over 8000 had opened by 1900.[20]

These developments helped fuel a huge expansion in the market for fiction. A writer in 1881 noted that Americans had become "a nation of readers." Hundreds of new low-priced, mass-circulation magazines appeared in print, newspapers began publishing serialized fiction, and the sales of novels skyrocketed. "The almost universal ability to read and the consequent love of reading," remarked a southern critic, "have developed in this nation especially an immense middle class of ordinary readers of average intelligence," most of whom were women.[21]

Technological advances in printing and new national marketing and distribution networks transformed fiction into a mass-production industry. As the volume of books increased, prices fell, further expanding the market, and writers directed their attention to the ever-broadening pool of middle-class magazine subscribers and book buyers. Mark Twain announced that he was not interested in trying "to help cultivate the cultivated classes"; he wrote for people of ordinary comprehension. The surge in "uncultivated" readers helped recast the literary profession from being a small New England and New York coterie to a national community of diverse writers from every class and region. Their livelihood depended upon placing stories or novels in magazines or newspapers. "Literature has become a business with us," Howells confessed, and he did rather well by it, earning today's equivalent of $250,000 a year.[22]

Then, as now, the "reading public" was in fact a cluster of many publics with quite different tastes. Most middle-class readers still preferred uplifting or amusing fiction that helped them escape from everyday concerns. But an increasing number wanted to know more "about facts, about concrete things, the things of the world." Willa Cather reported that people were putting "the historical play and the historical novel on the shelf, they have seen their day.

The public demands realism and they will have it." These "realistic" readers wanted stories they could relate to, stories that would help them make sense of their own lives, stories that featured recognizable people making difficult choices. "The people of this century," Cather insisted, "have a right to demand something that is close to them, something that touches their everyday life."[23]

The Feminine Factor

Even though an enlarged and more diverse literary marketplace helped emancipate writers from conventional subjects and treatments, it did not pry loose every Victorian restraint. The need to earn a living made authors cater to the tastes and morals of the predominantly female literary market. De Forest lamented to Howells that womanhood "furnishes four-fifths of *our* novel-reading public." He harbored "a profound contempt" for those who wanted tragic love stories with happy endings. Too many women readers, he charged, suffered from "a fibreless soft-heartedness that cannot bear to have pretty women disappointed in fiction."[24]

Yet writers could not ignore prevailing moral standards and gendered taste in fiction. New Hampshire-born Constance Woolson, the grandniece of James Fenimore Cooper, testified to such constraints when she confessed to being "very fond of the 'real,' almost too much so, I fear." After Woolson's first local color stories appeared in print, female friends and critics chastised her for being "too realistic; you have gone after false gods. Give us something ideal, something purely imaginative." Woolson responded with *Castle Nowhere* (1875), a collection of fanciful tales about early French settlers near the Great Lakes. Afterward, however, she regretted this romantic escapade and assured Howells that she would go back to her own preference for "realistic" fiction.[25]

Others experienced similar tensions between their professional aspirations and realistic proclivities. On numerous occasions Rose Terry Cooke, who described herself as "an emancipated female," clashed with editors who wanted her to idealize particular incidents or characterizations in her stories. She repeatedly insisted to her publisher that the original version of a story should be retained because it "*is* true." Louisa May Alcott felt the same pinch of compromise. Financial exigency, she explained, forced her to write pseudonymous "blood and thunder sensation rubbish" and "moral pap for the young." She yearned to progress from writing about "fairies and fables to men and reality," but "publishers are very perverse & won't let authors have their own way so my little women must grow up & be married off in a very stupid style."[26]

Henry James also grew irritated with what he called the "friction of the market." The leisure class of young girl readers, he claimed, unduly influenced the content and style of his own fiction, for literary taste had become "largely *their* taste." He would have preferred "nothing better than to write stories for weary lawyers and school-masters," for hardy men who did the world's work, but he had to content himself with an audience composed primarily of affluent girls caught up "in the great romance of doing nothing," mostly sisters, daughters, and spinsters "who stay at home all day to practise listless sonatas and read the magazines."[27]

The pervasive influence of young women readers—whether real or imagined—helps explain the fastidious prudery so characteristic of the fiction of the day. No book should be written, Howells proclaimed, that would "offend the modesty of a pure woman," and this precluded any overt treatment of sexual passion. To this end he counseled his own daughter not to read most novels, and he allowed his wife to censor anything in his own writings that she deemed unpleasant or distasteful. While "very Victorian in my preference of decency," he confessed that his "devilish idea of propriety" sometimes made him feel ashamed. His ambition to be a popular success had made him hypersensitive to the expectations of "female society."[28]

Yet if Howells and others let their conception of public taste and gentlemanly decorum delimit the moral and sexual boundaries of their fiction, they also rejected many of the mawkish pieties and romantic deceptions they saw infecting both serious literature and pulp fiction during the decades after the Civil War. In this sense, literary realism was as much a rebellion against the tradition of popular feminine fiction as it was an outgrowth of profound social changes and intellectual developments.

Although males led the realistic crusade against romanticism, several women writers agreed that American literature had become overly feminine. Elizabeth Barstow Stoddard, a brassy New Englander, resolved to write fiction "with a masculine pen." In her journal, she thanked God "for suddenly feeling virile." Mary Abigail Dodge, who wrote under the pen name Gail Hamilton, likewise alerted her readers that she did not write like a "conventional" woman author. "There is about my serious style," she observed, "a vigor of thought, a comprehensiveness of view, a closeness of logic, and a terseness of diction, commonly supposed to pertain only to the stronger sex." In her effort to "fight the truth's battles," she armed herself with the "concentrativeness which is deemed the peculiar strength of man."[29]

Self-conscious realists—whether male of female—especially gagged on the jam of sentimentalism. Of all the literary sins, it most aroused their contempt. The moral platitudes and easy tears associated with sentimentalism coerced an emotional response from the reader and thereby encouraged people, as the literary critic Henry Seidel Canby charged, "to dodge the facts of life—or to pervert them."[30]

At its worst sentimentalism served as a debasing emotional straitjacket that limited women to the domestic sphere and the realm of vicarious fancy. Popular authors fed young female readers a pabulum of cheap pathos and false role models that encouraged them to accept self-sacrifice and quiet suffering as their only outlets for self-expression. "Ages of false thinking about her on the part of others," Ellen Glasgow stressed, "have bred in woman the dangerous habit of false thinking about herself, and she has denied her own humanity so long and so earnestly that she has come at last almost to believe in the truth of her denial." Glasgow called for writers to depict women not through the "colored spectacles of tradition and sentiment, but by the clear, searching light of reality."[31]

Howells, too, savaged the popular "romances of no-man's land." He was convinced that exposing people to "the facts of life seldom does any harm; it is the distempered imitations that are mischievous, with their exaggerated emotions, their false proportions, their absurd motives, their grotesque ethics." In his own fiction, he offered examples of more responsible feminine behavior and perspectives to help cloistered women find a way out of their circumscribed social roles and gauzy outlook. For example, his novel *April Hopes* (1888) centers on a young single woman who fashions a "rose-coloured idea world" from her reading of "genteel novels." Such vicarious idealism, Howells felt, perpetuated girlhood and left readers woefully unprepared to confront the realities of a rapidly changing American life. In *Annie Kilburn* (1889) he describes the title character as indulging in "reveries so vivid that they seemed to weaken and exhaust her for the grapple with realities"; she was virtually paralyzed "in the presence of facts." The following year, in *A Hazard of New Fortunes*, he crafted Miss Woodburn to represent his own ideal of the realistic woman: "She was not a sentimentalist, and there was nothing fantastic in her expectations; she was a girl of good sense and right mind."[32]

Mary E. Wilkins (she would become Mary Wilkins Freeman after her marriage in 1902) also saw in literary realism an opportunity to use pointed facts to deflate airy fictions. In her short stories about New England village life, she suggested possibilities for women to stretch the boundaries imposed by traditional notions of domesticity. A woman, she believed, had a right to define herself through what she *did* rather than submit to age-old gender stereotypes. In "Old Woman Magoun," a story about a feisty, elderly woman who dominates her little hamlet, she observes that the "weakness of the masculine element in Barry's Ford was laid low before such strenuous female assertion." Woolson echoed Wilkins's belief that too many women "go swimming through life in a mist of romantic illusion," and she sought to counter such fantasies with an "intense realism of description and dramatic action." Likewise, the "coming woman" in American literature, Louisa May Alcott promised in *An Old-Fashioned Girl*, would be "strong-minded, strong-hearted, strong-souled, and strong-bodied."[33]

Kate O'Flaherty Chopin, a Louisiana widow and mother of six, fit such a description. She wrote about restless, discontented women who aspired to freedom and trespassed across the boundaries of social and literary convention. Her short stories include a gallery of distinctive characters: Louise Mallard, a widow who in the midst of her grief at the loss of her husband experiences an exhilarating sense of freedom: "when the storm of grief had spent itself she went away to her room alone . . . .[and] said it over and over under her breath: 'free, free, free!'" ("The Story of an Hour"); Mrs. Baroda, a married woman who dreams of an affair with her husband's best friend ("A Respectable Woman"); a woman who gives another woman a "long and penetrating kiss" ("Feodora"); and Paula Von Stoltz, a young woman who decides to pursue her musical career rather than a husband because marriage "doesn't enter into the purpose of my life" ("Wiser Than a God").[34]

The Awakening (1899), Chopin's scandalous novel of female self-assertion, shocked Victorian moralists. Its heroine, Edna Pontellier, an upper-middle-class wife of an affluent Creole, engages in a steamy affair with a much younger man. Tired of being treated as a "possession" gone stale, she resolves to have her "own way" in life. Edna's sensual "awakening" leads her to reject the traditional model of "women who idolized their children, worshiped their husbands, and esteemed it a holy privilege to efface themselves as individuals and grow wings as ministering angels."[35]

Such depictions of liberated women outraged defenders of the cult of domesticity. Amelia E. Barr, a genteel novelist and literary critic, abhorred the new woman appearing in realistic fiction: "She is not a nice girl. She talks too much, and talks in a slangy, jerky way that is odiously vulgar. She is frank, too frank, on every subject and occasion. She is contemptuous of authority, even of parental authority, and behaves in a high-handed way about her love affairs. She is alas! something of a freethinker." Barr and others did not want the light of realism cast into every corner of human experience. They also exaggerated the extent of a realistic revolution. Devotional literature, didactic domestic novels, historical romances, and sensational melodramas remained the most popular literary genres throughout the Victorian era. Yet a growing number of readers embraced fiction which depicted the various new experiences and attitudes of contemporary life. By 1890 Helen Cone could announce the "almost entire disappearance" of the sentimental woman's novel. "More and more," reported another critic, "women are learning what this world is in which they live; and, as they learn, they are inclined to write and talk about it."[36]

What explains this seeming conundrum? If women dominated the reading public and were naturally inclined toward idealized fiction, why did realistic stories and novels experience growing popularity? After all, such fiction offered little exciting action, timely escapes, or emotional consolation. Often, in fact, realistic fiction could be physically and emotionally draining. One reader, for example, complained to Howells that he "felt beaten out at the end" of *The Undiscovered Country* (1880) because of its "painfully powerful" realism.[37]

The attraction of realistic literature was its appeal to a perennial human

desire: the pleasure of recognition. "It is of the life we know best," Florence
Jackson wrote in the *Overland Monthly*, "or think we do, that we like best to
hear. More readily can we put ourselves in touch with it, feel a near relation
to and part of it." After reading a book, she wanted to be able to say to herself:
"Yes, yes, that is the way life seems to us." In this regard the prevailing stereotype
of female readers was inaccurate: women were not addicted to escapist romance
simply by virtue of their gender. Many displayed a natural affinity and affection
for the world of everyday actualities. "Woman interests herself in the concrete,
the particular, the practical," observed one female writer. Women writers and
readers fronted common life with an intimacy unknown to many men. Life in
the kitchen, parlor, nursery, church, and market exposed them to the sources
of basic human experience that fed their literary perception with insight and
understanding. "Women are delicate and patient observers," Henry James
emphasized; "they hold their noses close, as it were, to the texture of life. They
feel and perceive the real with a kind of personal tact, and their observations
are recorded in a thousand delightful volumes."[38]

Intrepid writers such as Alcott, Wilkins, Chopin, and Sarah Orne Jewett,
like Rose Terry and Rebecca Harding before them, fastened upon the representation of their local life not simply as a stylistic preference or as an expression
of regional provincialism. For them a literary approach grounded in personal
experiences, common tasks, and familiar dialects exposed important truths about
the life they knew and respected. They demonstrated that quiet lives were not
necessarily uninteresting. Prior to the publication of *Little Women*, for instance,
Alcott predicted that if the novel succeeded, it would be because it was "not a
bit sensational, but simple and true, for we [she and her family] really lived
most of it." Such "real" fictions offered readers accessible emotional worlds
while enlarging their sense of humanity. Jewett, who often accompanied her
father on his daily rounds as a Maine country doctor, contended in *Deephaven*
that to "enjoy the every-day life one must take care to study life and character,
and must find pleasure in thought and observation of simple things, and have
an instinctive, delicious interest in what to other eyes is unflavored dullness."[39]

Feminine Realism

This "instinctive, delicious interest" in the commonplaces of rural and village
life nourished what has since been labeled "feminine realism." Writers such as
Jewett and Alcott recognized a basic truth: many of the most interesting and
ironic situations in life occur in the home and its immediate surroundings.
What mattered to them was the strange spectacle of everyday humanity; common life harbored a matter-of-fact magnificence. Their remarkable ability to
render familiar experiences with a pebble-like clarity led Howells to proclaim

that "the sketches and the studies by the women seem faithfuler and more realistic than those of the men."[40]

When male critics complained about Jewett's preoccupation with widows and spinsters who sewed, gossiped, brewed tea, and engaged in other "trivialities and commonplaces in life," she responded that such ordinary subject matter "gives everything weight and makes you feel the distinction and importance of it." Her fictional world, small in size but drawn exactly to scale, is saturated with the tang and scent of the Maine coastal villages and islands she knew so well. Jewett's moral aesthetic, her need to communicate humanly through homely details, prosaic beauties, ineffable truths, and quiet lives, ordered her outlook as a woman as well as directed her pen. Her description of "Aunt Cynthy Dallett" typifies her approach:

> She was the tiniest little bent old creature, her handkerchiefed head was quick and alert, and her eyes were bright with excitement and feeling, but the rest of her was much the worse for age; she could hardly move, poor soul, as if she had only a make-believe framework of a body under a shoulder-shawl and thick petticoats. She got back in her chair again, and the guests took off their bonnets in the bed-room, and returned discrete and sedate in their black woolen dresses. The lonely kitchen was blessed with society at last, to its mistress's heart's content.

There is no straining or asserting in such observant prose. Jewett's delicately granular stories about lonely old fisher-folk and decaying seaside villages develop effortlessly through the accretion of plain lived moments and the calm eagerness of emotion. Grain by grain, her fiction acquires weight and gives the reader a sense of witnessed life.[41]

Although determined to avoid being "preachy" and quite conscious of the danger of letting sentiment fall into sentimentality, Jewett believed that fiction should include moral messages in the form of "silent scripture." She prized the stability inherent in familiar places such as Green Island in *The Country of the Pointed Firs*, a community "solidly fixed into the still foundations of the world." Women in her stories fight to preserve the dignity of domestic virtues and the solidity of old values in the face of turbulent social changes. As Jewett wrote at the beginning of "By the Morning Boat," the "weather-beaten houses of that region face the sea apprehensively, like the women who live in them."[42]

Readers such as Willa Cather celebrated Jewett's "gift of sympathy" and her ability "to understand by intuition the deeper meaning of what she saw." After finishing Jewett's *A Country Doctor* (1884), another reader praised her approach: "There is a basis of real life in every character—none of them are stagy or artificial." He and his wife had found "the real thing in [her] writings," and in the process they had learned "indirect real lessons" about human affairs. A female correspondent also thanked Jewett for the "*real* men and women" in her stories, recognizing that "there is something" about the immediacy of such fiction that "appeals to women—perhaps especially to younger women." She most appreciated Jewett's fictional "encouragement to other women" to persist against stiff circumstances in creating new spheres for self-expression.[43]

An obvious question arises: What made some writers lean toward a realism of commonplace incidents and experiences and others toward a romanticism rooted in what Howells called "the extravagant, the unusual, and the bizarre"? Individual temperament certainly played a major role. Howells recalled that he "was a helplessly concrete young person, and all forms of the abstract, the air-drawn, afflicted me like physical discomforts." Jewett attributed her own commitment to realism to her father's repeated command: "Tell things *just as they are!*"[44]

Others testified that their reverence for literary truth mimicked the era's pervasive scientific spirit. In this "scientific, rationalistic age," a literary critic wrote in *Cosmopolitan*, the "works of the imagination reflect the spirit of the times by becoming realistic instead of symbolic." This meant that a realistic novel should deal with characters whose behavior was psychologically explicable and with events that were familiar rather than exotic. Thomas Sergeant Perry, a Boston literary critic of supple intelligence, reported in 1883 that new scientific standards of accuracy and objectivity were helping purge literature of its "melo-dramatic element, which can certainly be well spared." The best American writers, Perry contended, were those who showed "us what we really are."[45]

The evolutionary theories of Darwin and Spencer furnished writers of real-istic fiction with compelling analogies and a new vocabulary. As a literary critic asserted at the end of the century, writers had borrowed "such fundamental phrases of the biologist as 'natural selection,' 'the survival of the fittest,' and 'the struggle for existence.'" Kate Chopin, for instance, avidly read modern scientific works and concluded that "it's impossible to ever come to a true knowledge of life as it is—which should be everyone's aim—without studying certain fun-damental truths" of natural science. Scientific study, she felt, would especially benefit those women writers who continued to produce "hysterical, morbid, and false pictures of life." Perhaps what most distressed literary traditionalists was the tendency of many realistic authors to incorporate the Darwinian assumption that chance rather than providence determined the rewards and punishments of life.[46]

Democratic Realism

Most realistic writers promoted a social agenda as well as a literary aesthetic. They shared Whitman's vision of literature as an expression of America's dem-ocratic spirit and as a means of promoting greater social toleration. "The great word Solidarity has arisen," Whitman announced in 1871. The time had come for "a native expression-spirit" intent upon "democratizing society." To Whitman and others, ethnic, racial, and class tensions posed the greatest danger to the reunited Republic. "The arts must become democratic," Howells insisted in 1888. His western disciple Hamlin Garland agreed that the realistic impulse represented the "democratization of literature." Garland credited Whitman, the "genius of the present" and "master of the real," for opening the eyes of creative Americans to the glories of common activities and common folk.[47]

Of course, Emerson and other romantics had also assaulted the barriers of artistic convention and celebrated a spreading egalitarianism. The post-Civil War realists, however, wanted to see this democratic spirit manifested in social cohesion rather than in the celebration of an egocentric freedom. They preferred to examine how people related to other people rather than plumb the depths of the individual soul or celebrate the sovereign self. "As the world is now built," Howells remarked, "a man can no more live to himself than he can live to others exclusively. . . ." People living in the modern city could not, like their rural ancestors, practice self-reliant isolation. The very success of urban culture depended upon creating and sustaining a web of mutually dependent relationships.[48]

Literary realists thus saw themselves as tour guides on behalf of American democratic ideals. They sought to expose affluent readers to the character and virtues of America's different regional communities and racial and ethnic groups. "One of the uses of realism," Howells declared, "is to make us know people." Although opposed to explicit preaching in literature, he hoped that his fiction would "teach a lenient, generous, and liberal life." In *The Rise of Silas Lapham*, he used the marriage of the patrician Tom Corey to the bourgeois Penelope Lapham to suggest a symbolic fusing of two antagonistic social classes. Howells envisioned society as a *community* of persons, a community admittedly held together by tenuous bonds, but one that accorded fundamental respect to individual worth and mutual responsibility. As a character observed in A *Modern Instance*, "We're all bound together."[49]

Jewett echoed the sentiment, proclaiming that "the best thing that can be done for the people of a state is to make them acquainted with one another." She hoped that her fiction might "help people to look at 'commonplace' lives from the inside instead of the outside." For all of her regret at the transformation of the cohesive village culture of her youth, Jewett recognized the need to deal with the harsh new fact of social conflict, and in her fiction she tried to weave together the threads of seemingly insignificant aspects of life into a resilient social fabric. "Possessed by a dark fear that townspeople and countrypeople would never understand one another, or learn to profit by their new relationship," she introduced the two types to each other in her stories. In the process Jewett hoped to broaden the horizon of her readers as well. She especially prized fiction that "makes you think of the people afterward." Such intimate exposure to others was especially needed, said another writer, by "those of us who, leading sheltered lives, are inclined to sit too comfortably in our bodies, feeling that the world is not so bad after all."[50]

This desire to introduce diverse Americans to one another and cultivate fraternal bonds helps explain why realists often focused their stories on family reunions, friendly "visiting," or travelers who encounter strangers in hotels, summer resorts, railroad stations, train cars, around dinner tables, or along the streets and byways of the teeming city. In *Deephaven*, for example, Jewett describes two genteel, "but not sentimental girls" from Boston, Helen Denis and Kate Lancaster, who spend a long summer vacation together in the house of Kate's late aunt. While in Deephaven, they develop an excited curiosity about the picturesque locals. One day they come upon a destitute family scratching out a hardscrabble existence on the Maine coast: "We succeeded in making friends with the children, and gave them some candy and the rest of our lunch."[51]

A woman reader praised Jewett's "wonderful gift of understanding lonely lives, especially those of elderly women." After finishing *The Country of the Pointed Firs*, she told Jewett that "my heart filled with a loving pity for those whose limitations were greater and harder than my own." She assured the author that many others shared her reaction, especially those "whose lives are so out-wardly blessed with everything." Another reader declared that it was impossible for people of means to read Jewett's stories and not be "bettered by the inspi-rations of kindliness which they afford or consoled by the companionship of those beings that live and breathe along your genial page."[52]

Jewett and other "democratic realists" were thus anything but objective observers: they wanted to help readers *see* the essential humanness of their fellows and provoke them to examine the motives behind their own social prej-udices. Howells intended the realistic novel "for the benefit of people who have no true use of their eyes." He knew, however, that it would not be easy to break through the thick crust of ignorance and prejudice. "Nothing is so hard," Basil March confesses in *Their Wedding Journey*, "as to understand that there are human beings in this world besides one's self and one's set."[53]

Yet what Howells saw as bourgeois sympathy for the unrepresented could at times represent uninformed condescension. In *Their Wedding Journey* (1873), the two honeymooners, Basil and Isabel March, take a train from Albany to Niagara Falls. Along the way they enjoy the vistas of the Mohawk Valley. Eventually they tire of sentimentalizing over the landscape and turn their atten-tion to the people sharing their railway car, "an ordinary carful of human beings" experiencing the "habitual moods of vacancy and tiresomeness." The narrator explains how such an enclosed setting enables one to imagine the other's "shallow and feeble thoughts, to be moved by his dumb, stupid desires, to be dimly illumined by his stinted inspirations, to share his foolish prejudices, to practice his obtuse selfishness." Such contemptuous parlor-car compassion led Howells and other writers to fashion fictional works that were often unfair to those they were seeking to know and help.[54]

The Art of Realistic Fiction

In a letter to Robert Louis Stevenson, Henry James confessed that he embraced literary realism because he wanted to "leave a multitude of pictures of my time." This documentary objective helped explain the growing emphasis on verisimilitude. But *how* did writers construct recognizable representations of contemporary reality? James, for whom the aesthetic question remained far more compelling than the social question, declared that the "supreme virtue of the novel" was its "air of reality." In creating a credible "illusion of life," the novelist "competes with his brother the painter in *his* attempt to render the look of things, the look that conveys their meaning, to catch the color, the relief, the expression, the surface, the substance of the human spectacle." What gave a novel such authenticity was its "solidity of specification," its plausible plot and dialogue and its familiar characters and meticulously described settings, all of which enabled it to "compete with life" itself, enticing from the reader a nod of recognition.[55]

In this regard realists felt that the besetting sin of romantic literature was imaginative or emotional excess. Howells, for instance, praised Jewett for consistently refraining "from every trick of exaggeration." Fiction, he and others believed, lost its credibility and dignity as soon as melodrama, didacticism, or emotionalism took control. People in realistic fiction must be authentically human, possessed of mixed motives, confused consciences, and changing natures. "Let fiction cease to lie about life," Howells urged; "let it portray men and women as they are, actuated by the motives and the passions in the measure we all know; let it leave off painting dolls and working them by springs and wires." Realists believed that characters should be capable of making genuine choices. Instead of being governed by implacable fate, people in realistic novels confront viable options and retain a sense of agency. Their own decisions and reactions to events give shape to their lives.[56]

Graphic scene-painting, recognizable characters, and plausible dialogue and narration formed the technical foundation of literary realism. But Pilate's famous question remains stubbornly insistent: What is "truth"? As one critic noted, "the whole issue between realism and idealism depends upon this point." Realists embraced the commonsensical assumption that both "reality" and "truth" could be readily transcribed into literature. Few of them doubted the self-evident "reality" of the visual world. In the preface to *The American*, Henry James wrote that realism "represents to my perception the things we cannot possibly *not* know, sooner or later, in one way or another."[57]

But did this intuitive sense of reality penetrate below surface appearances? Some writers claimed that a truthful literary work functioned as a mirror or a photograph of the present-day world of fact. Harriet Beecher Stowe used such similes to express her own evolving realistic credo in the preface to *Oldtown Folks* (1869): "I have tried to make my mind as still and passive as a looking-glass, or a mountain lake, and then give you merely the images reflected there." She promised that her stories had been taken from "real characters, real scenes, real incidents." They represented "simple renderings and applications of facts," and she described her role as that of "a sympathetic spectator. I propose neither to teach nor preach."[58]

Other advocates of literary realism, however, argued that great fiction must do more than merely imitate appearances. When realism "heaps up facts merely, and maps life, instead of picturing it," Howells insisted, "realism will perish too." In other words, the literary realist must employ the tools of the creative imagination to elevate fiction above the merely documentary. A superb rendering was a necessary but not sufficient element of the realistic enterprise.[59]

Here is where realistic fiction departed from journalism. If the reporter was a connoisseur of facts, the realistic novelist was a surgeon who probed beyond fact into motive. Henry James maintained that a mature realistic perspective enabled readers to "see beneath the surface of things." Writers must extract revelations of meaning from the tissue of bald facts. As James once told a friend, fiction should be "bravely and richly, and continuously psychological." It should focus on "moral and intellectual and spiritual life" rather than a series of "vulgar chapters of accidents . . . which rise from the mere surface of things." The sophisticated literary realist, in other words, should strive to *represent* life in all its mental and emotional complexity, not simply *reproduce* its visible appearances.[60]

James and others knew that a work of literary realism could never approach the objective fidelity of a photograph or the documentary authenticity of a newspaper. Nor should it. Whatever its degree of descriptive precision, however crowded its weave of fact and event, literature remained a product of the probing imagination. The realistic writer started with facts, but the details must somehow be endowed with life—and with art. In this regard Whitman opposed any author who sought "merely to copy and reflect existing surfaces" by "daguerreotyping the exact likeness. . . ." Insight must inform sight. "In short," as the writer and editor George Parsons Lathrop concluded, "realism reveals."[61]

Genteel Realism

Not surprisingly, realists did not always practice the virtues they preached. For all of their bold rhetoric about exposing the facts of life, most Victorian realists operated according to strict guidelines in the selection and treatment of their literary subjects. "Let only the truth be told," trumpeted Joseph Kirkland, "and not all the truth." While cheering the retreat of historical romance, he urged that writers erect sturdy barriers on behalf of good taste: "Much that is true is not worth telling: more is not proper to be told." In America, he stressed, prose fiction "must be written for men, women, and children." What is "not good for every member of a household is excluded from its library table."[62]

Jewett likewise called on American writers to seize "the middle ground" when selecting their subjects. They should highlight the typical experiences of ordinary life and avoid dealing with the exceptional, the morbid, or the violent. Woolson agreed that "the mid-world is best," as did Garland, who proclaimed that American realism "will not deal with crime and abnormalities, nor with diseased persons. It will deal, I believe, with . . . average types of character, infinitely varied, but always characteristic."[63]

Terms such as "abnormal" and "diseased" were veiled references to the controversial writings of Emile Zola and other French "naturalists." Zola's coarse fictional portraits of the French underclass awed and disgusted American readers. Zola peopled his fictional world with debased drunkards and prostitutes, slum dwellers and smugglers, homicidal maniacs and unwed teen-aged mothers. He also asserted that heredity and environmental forces rather than reason and free will determined individual fate.[64]

While respecting Zola's audacity as well as the tragic grandeur of his many novels, most Americans winced at his brutish tendencies and his "vulgar" subjects. When Whitman referred in *Democratic Vistas* to the "growing excess and arrogance of realism," he had Zola in mind. On this point he found himself in rare alliance with Henry James. "On what authority," James demanded in a review of *Nana*, "does M. Zola represent nature to us as a combination of the cesspool and the house of prostitution?" Many areas of life, James argued, did not warrant inclusion in fiction: "It would be difficult for what is called *realism* to go further than in the adoption of a heroine stained with the vice of intemperance." True, in "leaving so many corners unvisited, so many topics untouched," writers in the United States lacked the candor and breadth of the French naturalists, but James nevertheless preferred the "good-natured, temperate, conciliatory" realism practiced by American writers. It "tells us, on the whole, more about life, for it is more at home in the moral world."[65]

James's reference to the "moral world" raises an issue central to the realistic sensibility: the role of morality in literature. If realism meant anything, James contended, it meant abandoning the notion that fiction must primarily edify and instruct the reader. To him, literary sermonizing constituted the fume and spume of a misdirected moralism. The true realist did not preach but instead encouraged readers to reflect upon issues of significance. "Morality is hot," he told a friend, "—but art is icy!" In representing "things as they are," authors should induce readers to *feel* responsible emotions. They should never dictate right conduct.[66]

James applied such a standard to Rebecca Harding Davis and found her wanting. In 1867 she published *Waiting for the Verdict*, a novel addressing the compelling question of the day: What was to be the fate of the freed slaves? Were they to be treated as equals? given the vote? provided education? Set during the Civil War, *Waiting for the Verdict* revolves around Margaret Conrad, a tall, statuesque Kentucky belle who falls in love with a mysterious Philadelphia surgeon, Dr. Broderip. The doctor, however, harbors a terrible secret: he is a mulatto. When Broderip discloses his mixed racial ancestry to Margaret, she recoils in horror, for the "negro blood is abhorrent." After she reveals Broderip's genetic stain to the white community, his hospital fires him and "polite society" shuns him. With no place else to turn, Broderip embraces the black community he had earlier ostracized. He organizes a regiment of freed slaves, only to be shot and mortally wounded while leading his victorious troops through Richmond.[67]

In an unsigned review of *Waiting for the Verdict*, James scolded Davis for the novel's manipulative emotionalism. He approved of its central idea and applauded its authentic detail, but he found her execution "monstrous." Throughout the story, he claimed, Davis injected mini-sermons about racial and class attitudes. Her didactic perspective resembled one of the characters she described: "After the habit of women," she "could not leave any one with their trouble in quiet, but must peer curiously into it, to cry over it afterward, and fill her own heart with aching and pity." According to James, Davis and other writers of overheated virtue defied their own realistic credo. They drenched characters in "a flood of lachrymose sentimentalism," splashing readers with "tears and sighs and sad-colored imagery." In the process they washed "life's honest lineaments out of all recognition."[68]

The rasp of hostility and gendered language in James's assessment of Davis illuminates the distinction between his own "analytical" realism and the kind of "reformist" realism she and others practiced. After reading James's critique of *Waiting for the Verdict*, Harriet Beecher Stowe sent Davis a consoling letter in which she noted that the unidentified reviewer obviously had "no sympathy with any deep & high moral movement—no pity for human infirmity." James was not *that* callous, but he did prefer aesthetics over homiletics. Deference to morality constituted an "essential perfume" in literature, but "moral fiction" was not the same as "moralizing fiction." Too many authors tried to "idealize the victims of society, to paint them impossibly virtuous and beautiful." Authors should instead present realistic situations and recognizable people without prescribing solutions or dictating emotional responses. He found nothing "more refreshing" than a writer's "disinterested sense of the real."[69]

In this debate between representation and didacticism, most literary realists occupied a middle ground between James's detached aestheticism and Davis's preachy moralism. Edith Wharton, for instance, stressed that no "novel worth anything can be anything but a 'novel with a purpose,' & if anyone who cared for the moral issue did not see in my work that *I* care for it, I should have no one to blame but myself." Every realistic "picture of life contains a thesis," she added. What differentiated the "literary artist from the professed moralist" was leaving it to the reader to "draw his own conclusions from the facts presented."[70]

Like Wharton and James, Howells despised writers who bluntly instructed the reader in correct moral behavior, who clumsily flavored their fiction with "sentimentality or religiosity," and who wrote in a "gasping and shuddering" style. Howells once told Stephen Crane that a novel "should never preach and berate and storm" because it "does no good." Yet a novel should "have an intention," a social "conscience and purpose," and through his own fiction he aspired "to move" the reader. Writers should help sharpen the reader's responsibility to make conscious distinctions between vice and virtue: "If the sermon cannot any longer serve this end, let the novel do it."[71]

Howells and other democratic realists found James's breadth of sympathy too narrow, his tone too "patronizing," and his critical detachment too disengaged. In a review of James's first book, *The Passionate Pilgrim and Other Tales*, Howells criticized his friend's snobbish preference for "a certain kind of cultivated people" who "are often a little narrow in their sympathies and poverty-stricken in the simple emotions." Howells expressed his own position in a letter to E. C. Stedman: "I am devotedly a realist, but I hope I keep always a heart of ideality in my realism. Nothing is worth doing without that."[72]

The Smiling Aspects of Life

The complex "ideality" expressed in the fiction of Howells and most other literary realists focused on the need for greater social "sympathy" rather than any concrete economic or political reforms. This reflected their widely shared assumption that the quality of life during the 1870s and 1880s was steadily improving. Howells provided what became the most quoted—and criticized— explanation for the lack of explicit social criticism in post-Civil War American fiction. After praising Dostoevsky's *Crime and Punishment* (1886) for provoking "the deepest sympathy and interest," he cautioned that the powerful Russian novel could be appreciated only within the context of the degrading "social and political circumstances in which it was conceived." In other words, what was appropriate subject matter in Russian literature was not suitable in American fiction. Dostoevsky's somber descriptions of ghastly social conditions and criminal violence gained their power from the distinctive circumstances of Russia's despotic political system and poverty-stricken economy. In the United States, however, "very few American novelists have been led out to be shot, or finally exiled to the rigors of a winter at Duluth." America had relatively few paupers, Howells claimed, and class conflict was "almost inappreciable." He admitted that "we are still far from justice in our social conditions, but we are infinitely nearer it than Russia."[73]

Few would have disagreed with this comparative assessment of social conditions, but Howells went on to draw a conclusion he would later recant. Because America enjoyed more social equality and political liberty than Russia or France, he argued, aspiring writers should "concern themselves with the more smiling aspects of life, which are the more American" and should "seek the universal in the individual rather than in the social interests." The little suffering evident in the United States, he believed, resulted from personal rather than systemic causes. It "is mainly from one to another one and oftener still from one to one's self."[74]

Howells's ever-shrewd assessment of the literary marketplace flavored his social outlook. "The American public," he claimed, "does not like to read about the life of toil." If writers were to dwell at length on "how mill hands, or miners, or farmers, or iron-puddlers really live," then bourgeois readers would quickly let them know that they "did not care to meet such vulgar and commonplace people." Readers did not want their noses rubbed in the sordid realities of the underclass.[75]

Howells's bland emphasis on "the smiling aspects of life" and on the moral responsibility of individuals rather than on the structural inequalities of capitalism or the perennial evils of racism and ethnic prejudice reveals the moral blind spots in his field of vision during the two decades after the Civil War. At this stage in his career he lacked a full sense of social complexity. Walt Whitman emphasized that Howells was "not revolutionary: he goes a certain distance— then hauls himself in with a shock: that's enough—quite enough, he is saying to himself." On the whole, Whitman concluded, Howells had "so little virility" that he was "unable to follow up radically the lead of his rather remarkable intellect." The same could be said for most of the other literary realists among Howells's generation. In focusing their attention on the family life of the middle class, they frequently ignored the raging inequalities outside their parlors. Many of the truth-tellers told half-truths.[76]

Yet for all of its timidities Howellsian realism was quite controversial for its own time. "Realism," scowled Hamilton Wright Mabie, the literary critic for the *Christian Union* and *The Independent*, seemed determined to crowd "the world of fiction with commonplace people, whom one could positively avoid coming into contact with in real life." He dismissed Howells's *Rise of Silas Lapham* because it "throws no spell over us; creates no illusion for us, leaves us indifferent spectators of an entertaining drama of social life." The realism of Howells and James, Mabie charged, was "practical atheism applied to art." It emptied "the world of the Ideal" and denied "the good God."[77]

Those who sneer at the limitations of the "refined realists" ignore how far they had moved from the furred and gowned literary world guarded by Mabie, Lowell, and their patrician colleagues. As Whitman recognized, people should appreciate how progressive Howells and others were: "Rather than complaining that it is not more, we ought to be glad it is not less." The efforts of literary realists to represent the diversity of American society during the Age of Enterprise, however tentative and discreet, prepared the way for even bolder writers at the end of the century. Mary Austin, a young California author, testified that Howells gave aspiring writers like herself "the assurance that in American life is to be found the material of an adequate American art."[78]

Realists during the 1870s and 1880s also helped create among at least some readers a reservoir of social conscience that more aggressive writers and more tenacious reformers would tap at the turn of the century. Theodore Roosevelt, for example, acknowledged that his youthful reading of Jewett's and Wilkins's short stories helped "make me feel the need of arousing the public interest and the public conscience as regards conditions of life in the country." Five years later he added that he belonged to "the generation whose youth was profoundly influenced by Mr. Howells's books. They not only gave us pleasure as works of literature, but they helped us toward a spirit of kindliness and justice in dealing with our fellows, and they stirred our souls to the strife for national ideals."[79]

7

Realism on Canvas

The debate between literary realists and idealists found a ready parallel in the art community after the Civil War. In 1880 the prestigious critic S.G.W. Benjamin lamented that the new vogue in painting "is realistic, satisfied with the surfaces of the objects it represents, and not aspiring after a conception of the spiritual and the ideal." Benjamin implored artists to disavow such literal representation of common things and common folk and instead portray beautiful subjects and explore transcendent themes. Their objective must be to "search after the ideal good, to live in an ideal world, to yearn after and try to create the harmony of the ideal."[1]

Many painters and art patrons wholeheartedly agreed with Benjamin's standard of excellence, and such "ideal" painting remained popular throughout the late nineteenth century. George Inness, for example, continued to produce poetic landscapes suffused with an atmosphere of contemplative solitude. Albert Pinkham Ryder, on the other hand, composed mystical works featuring spectral riders under moonlit skies, rudderless boats on treacherous seas, enchanted lands of water sprites and Norse heroes bathed in an aura of ghostly mystery. The artist, he believed, "should strive to express his thought and not the surface of it. . . . His eyes must see naught but the vision beyond." Still other avowed idealists such as George de Forest Brush, Thomas W. Dewing, Abbott Thayer, and Kenyon Cox glorified the feminine form, painting angelic, languid young women draped in classical garb, playing lutes and harps or sheltering cherubic children, serving as chaste embodiments of all that was good and noble in the cult of domesticity.[2]

Although Benjamin championed such otherworldly and idealized offerings, he grudgingly admitted that "the tide also sets strongly toward realism." The realistic painters of the post-Civil War years varied considerably in style and emphasis, but they shared a common desire to provide accurate depictions of

Abbott Thayer, *Caritas*, 1894–1895. *(Courtesy, Museum of Fine Arts, Boston.)*

familiar scenes. The sublime landscapes, historical subjects, and allegorical themes so popular during the antebellum era now seemed bankrupt. The New York artist John Ferguson Weir stressed in 1878 that "art, in common with literature, is now seeking to get nearer the reality, to 'see the thing as it really is,' to grasp the object with palpable truth in its expressive values and characteristic facts." Two years later, art critic William C. Brownell singled out East-man Johnson and Winslow Homer among the more established "realists" and listed William Harnett, William Merritt Chase, and Thomas Eakins as the most talented of "the new men" devoted to the representation of "real life."[3]

Weir claimed that these painters shared a common objective "to see and describe" the "palpable truth" of their immediate surroundings rather than historical or foreign subject matter. They relished observing the world, exploring its textured surfaces, and rendering its mundane activities on canvas. Just as writers felt a surge of cultural nationalism following the Civil War, realistic painters and art critics sensed "the increased interest in things American." Instead of viewing common life as trivial and ugly, they savored the look and feel of ordinary objects and people. The realists, an art critic observed, were eagerly seizing upon "the sights and scenes of our own times."[4]

Realistic art was rooted in unmistakably American values and experiences. "What I mean by truth in a painting," explained the Philadelphia artist Thomas Anshutz, "is as follows: Get up an outfit for outdoor work, go out to some woe-begotten, turkey chawed, bottle-nosed, henpecked country and set myself down, get out my materials and make as accurate a painting of what I see in front of me as I can." Realistic art, he added, was "based on knowledge and knowledge on fact."[5]

Such a forthright mimetic approach exasperates many art historians who believe that the surface appearance or evident "facts" of a painting are less important than its concealed meanings, unconscious motives, symbolic insinuations, and higher purposes. For many nineteenth-century artists, however, seeing *was* believing. Self-described literalists sought to render in paint what was perceptible, tangible, and readily accessible to the senses. Their "resistless curiosity to peer into the world of things" led them to prize an artist's attentive eyes and reliable hand more than an excited imagination.[6]

The Reality of Appearances

The rising popularity of still-life painting after mid-century testified to the growing appeal of literal representation and the fascination with material facts. In their *trompe l'oeil* (that which deceives the eye) still lifes, for instance, Philadelphians William Harnett and John Peto debunked the traditional assumption that art must deal with elevated subjects and noble themes. They demonstrated instead that everyday objects and prosaic activities provided appropriate subject matter for serious painting. In catering to the middle-class rage for photographic likenesses, they painted hyper-realistic assemblages of common things—beer mugs, pipes, newspapers, advertisements, musical instruments, bank notes, knives, guns, dead game, books, flowers, fruit, and letters. Such still-life montages provided a visual inventory of familiar possessions and a documentary record of popular recreations. By asserting that meaning inheres in particulars and by making prosaic things fit subjects for art, they contributed to the widespread "fetishizing" of commodities that characterized the age. They also revealed how much beauty the talented can wrest from the mundane.[7]

The son of an Irish immigrant shoemaker, Harnett was a virtuoso of appearances whose passion for the density of the actual, for the indisputable life of what "is there," captured the attention and respect of observers. "Everything is delineated with truth in the minutest detail," reported an art critic after viewing a Harnett exhibition in 1890. "This realistic fidelity in the representation of such familiar objects appeals to very many." Such popular appreciation of Harnett's realizing abilities and tactile qualities was exactly what he aspired to achieve. Disavowing any transcendent purpose, he wanted his works to be appreciated on their own terms without loading them with symbolic significance. His goal was to mimic—and celebrate—the appearance of things and to create a painting that was less a representation than a piece of reality itself.[8]

Harnett and other *trompe l'oeil* painters developed a reliable bag of tricks—sharp edges, dark tonality, varied textures, and true-to-life scale—to narrow the gap between image and object and bamboozle the viewing public. Harnett's *After the Hunt* (1883), for instance, is a cunningly rendered collection of masculine hunting paraphernalia hanging from a door. For years the painting adorned a wall at New York's famous Stewart saloon on Warren Street, and customers frequently passed their fingers over the canvas to confirm that it was indeed paint and not the "real thing."

Observers delighted in such optical trumpery. The images of things in *trompe l'oeil* paintings have specificity and are verifiable. In this sense, they fulfill a large element of the realistic enterprise: to substantiate the observable data of the physical world. In reviewing a Peto exhibition, a New Orleans critic concluded that the "imagination plays no part here. It is all realism, and realism in the extreme; but we insist that the artist has produced delightful results with his experiment." An Ohio art critic used similar terms to describe the visual artifice of Harnett's *Toledo Blade* (1886): it was "so like the real—that in looking upon it you are likely to forget it is a picture and feel that you are looking at a true violin, a genuine candle, etc." In celebrating Harnett's accomplishment, the critic provided a tribute to mimetic art: "The highest triumph of artistic genius is in approaching the actual—in the perfect reproduction of the subject presented."[9]

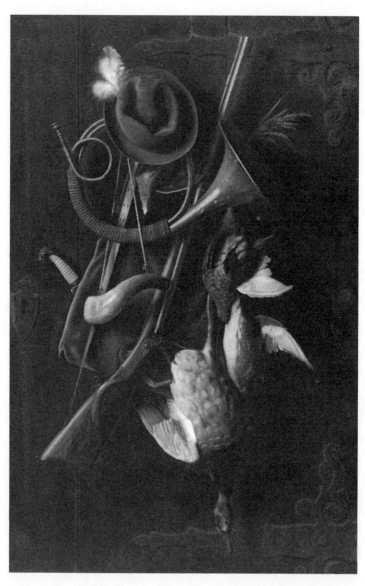

Michael William Harnett, *After the Hunt*, 1883. *(Columbus Museum of Art.)*

Conservatives balked at such brazen materialism, charging that it lacked the nobility of feeling or stature which genuine "art" required. Realists countered that a painting's grandeur derived from how well it depicted the evident truth of actual things. In this regard, observers frequently credited the pervasive scientific spirit with helping to legitimize the new interest in the exact representation of everyday subjects. "In this modern day," a writer in the *New Path* asserted, "when all things are brought under the reign of science," art's primary function must be the accurate "representation of things." Brownell echoed the point when he argued that realism had become science's counterpart in the arts. Idealism was being driven from the field. "So thoroughly has the spirit of realism fastened upon the artistic effort of the present," Brownell concluded, "that temperaments least inclined toward interest in the actual feel its influences and show their effects."[10]

The Genre Realism of Winslow Homer

In addition to still-life paintings, the growing middle-class art public also relished uplifting scenes of everyday life. In fact, genre painting displaced landscape paintings in popularity after the Civil War. "Nature pure and simple, the wild, untamed wilderness," Charles Moore acknowledged, "does not so much attract or so long hold our admiration" as it once did. Instead, "the realistic idea which animates the artistic activities of our time" was leading people to prefer "scenes of rural and pleasant industrial life."[11]

The focus of genre painters on "pleasant" scenes of contemporary life reveals how people placed strict limits on how literal and inclusive realistic art should be. Few artists—even those most open to innovation and committed to realism—painted ugly or disquieting subjects. Most of them offered sanitized versions of the "smiling aspects of life." They preferred what Henry James called a *"delightful realism"* depicting cohesive families, distinguished men, attractive women, joyous children, pleasurable outdoor recreations, and picturesque city scenes. Only a few bold rebels such as Thomas Eakins depicted a sitter's physical blemishes or personality quirks. Even fewer accented rural distress, factory labor, or urban poverty. The lives of the low and the humble would receive little candid attention until the end of the century.[12]

Of all the painters of everyday life during the post-Civil War era, Winslow Homer best epitomized the appeal and complexity of the realistic outlook. One critic referred to him as "the leader of the invading school" of realists, while another labeled him a "flaming realist—a burning devotee of the actual." Having attained a national reputation during the Civil War, Homer thereafter became one of the nation's most revered illustrators and painters. Art patrons found his renderings of common scenes deliciously "American."[13]

However different in subject or approach, however varying in mood or strength, Homer's paintings carry the signature of his fierce feeling for the color and drama embedded in actualities. He was determined to do more than produce look-alike imitations of the apparent. With a grasping eye for the essential fact, an almost austere economy of means, an abiding respect for humanity and a raging contempt for gentility, Homer illuminated the poetry embedded in familiar scenes and activities, revealing their plain beauty and disclosing their simple dignity. He was, as a critic wrote, "the Walt Whitman among our painters." Like Whitman, Homer promoted a self-consciously virile cultural outlook. In 1865 a New York critic asserted that Homer's sturdy "manner of painting is healthy and manly," a welcome contrast to the "timidity of weakness" displayed by the "effeminate" works dominating exhibitions. "No American artist," announced the editor of The Art Amateur, "is more interesting and original than Mr. Winslow Homer."[14]

After the Civil War, Homer, a short, raw-boned, and erect man with a handlebar moustache, continued to work as an illustrator for Harper's Weekly and other middle-class magazines ("a treadmill existence," he labeled it, "a form of bondage"). Although a city dweller, he relished the backwoods and made frequent forays into upstate New York and New England to canoe, hike, fish, observe, and paint out-of-doors. Homer adopted an "open-air" outlook long before the term became a commonplace. In discussing one of his paintings, he noted that the "picture is painted fifteen minutes after sunset—not one minute before—You can see that it took many days of careful observation to get this." Yet Homer was not simply an undiscriminating copyist. "You must not paint everything you see," he told a fellow artist, "you must wait, and wait patiently, until the exceptional, the wonderful effect or aspect comes."[15]

Homer frequently painted scenes of country life during the 1870s, and his broad-planed works convey warm images of simpler and happier times. Although rich in sentiment, Homer's rustics are rarely sentimentalized. The difference is subtle but substantial. His renderings of farm children are simultaneously documentary and picturesque, a combination of qualities that evoked empathy as well as interest and served as a defining attribute of his work. Homer was, an art critic announced, the "realist of the realists, with a kind of haughty and half-grudged poetry breaking out from him against his will."[16]

Homer's genre paintings of common folk engaged in common activities offered a robust expression of America's democratic ideal. His bucolic scenes are so beautifully trivial and prosaic that they refuse to mean less than everything. He took ordinary people seriously, treated them with respect, and steadfastly asserted their right to be represented in art. For example, Homer was one of the first painters to provide straightforward depictions of working women and to highlight African Americans as real people rather than subjects of racist satire. Women in his works are not delicate moral mothers confined to the parlor and kitchen, but independent, self-reliant, enigmatic souls performing strenuous tasks. Similarly, his African-American subjects display a refreshing dignity and autonomy. Art critic George Sheldon observed in 1878 that Homer's numerous "negro studies" were "the most successful things of the kind that this country has yet produced."[17]

Earlier artists who had featured blacks had typically produced what the *New York Times* called "nerveless and inane" caricatures "which do not contain the most evident characteristics of the race." Homer was so determined to reveal the essential humanity of African Americans that he made several excursions to Virginia to paint newly freed slaves after the Civil War. *Near Andersonville* (1865–66) offers a grave, emphatic image of a lone black woman standing in a cabin doorway in south Georgia, watching captured Union troops being marched away in the distance. Their impending confinement in the infamous Andersonville prison seems to bind their fate to her own.

Near Andersonville reveals one of Homer's favorite devices for puncturing surface complacency: the distant gaze or reflective glance. To capture the attention of the viewer, he often gave his characters quirky angles of vision, portraying them aslant or in profile. Frozen in rapt concentration, their wan, distant countenances suggest the passage of time and depth of feeling. Homer's self-absorbed subjects look outward but seem to peer inward as well. Many of his sun-bonneted and straw-hatted rustics project a lonely, deliberative sadness that raises them above the sentimental.

Homer's singular style both impressed and perplexed Henry James. In 1875 he reviewed a show at the National Academy of Design in New York and announced that Homer displayed "the most striking pictures in the exhibition." James, however, did not know what to make of Homer's "little barefoot urchins and little girls in calico sun-bonnets. . . ." There was vast confidence in James's malice and ironic laughter in his disdain for such hayseed subjects. Despite his cosmopolitan jaundice, however, James perceptively diagnosed Homer's distinctive virtues:

> Mr. Homer goes in, as the phrase is, for perfect realism, and cares not a jot for such fantastic hair-splitting as the distinction between beauty and ugliness. He is a genuine painter; that is, to see, and to reproduce what he sees, is his only care; to think, to imagine, to select, to refine, to compose, to drop into any of the intellectual tricks with which other people sometimes try to eke out the dull pictorial vision—all this Mr. Homer triumphantly avoids.

Homer's unpretentious paintings, for all of their freshness and spontaneity, were too crude and angular for James's exquisite taste; they lacked finish and grace. "He is almost barbarously simple, and . . . he is horribly ugly; but there is nevertheless something one likes about him." James could not help but admire

Homer's "vigorous way of looking and seeing" and respect his "honest, vivid, and manly" work.[18]

James mistakenly assumed that Homer painted only rustic scenes. In fact, he often depicted bourgeois women at work and at leisure—playing croquet, promenading at seaside resorts, riding in the mountains, reading novels or letters, and gathering flowers. True, Homer painted few scenes of urban or industrial life, but he did produce dozens of magazine illustrations of city and factory scenes that warrant attention. *New England Factory Life—Bell Time*, which appeared in *Harper's Weekly* in July 1868, illustrated an article describing Charles Dickens's tour of the textile mills at Lowell, Massachusetts. Compared with conditions to his native England, Dickens claimed, Lowell's factory life "has less of horror" and less "disease and misery." None of the American workers "gave me a painful impression." Yet Homer's accompanying drawing of a dense mass of workers headed home from a bleak mill in nearby Lawrence belies such a rosy assessment. The pallid faces and hunched shoulders suggest the hardships of work at the looms.[19]

In the spring of 1881 Homer went abroad to Cullercoats, England, a picturesque Yorkshire fishing village on the North Sea. His twenty months living amid the harsh climate and almost primitive conditions marked a crucial turning point in his style. In representing stalwart fisherfolk wrenching a crude subsistence from an unforgiving sea, Homer's canvases grew more atmospheric and dramatic, his forms became more sculpted and weighty, and his colors deepened into grays and blues as he depicted the tremendous mystery, force, and peril associated with the sea. Critics applauded the simpler composition, stronger masses, graver themes, and more rugged tones of these English works. Yet for all of their raw power, they seem contrived and theatrical; they lose in authenticity what they gain in grandeur.[20]

Upon his return from England, Homer settled hermit-like in a red cottage on the rocky coast of Maine, at Prout's Neck, a small village on a peninsula south of Portland, where Homer had occasionally vacationed with his family. He lived in splendid isolation, content to hug his happiness and focus his labors. His nearest neighbor was half a mile away, and his only companions were a wirehaired terrier named Sam and a well-stocked liquor cabinet. Homer's solitude suited his briny personality. A friend remarked that Homer's four favorite words became: "Mind your own business."[21]

Having crossed the threshold of middle age, Homer grew more metaphysical in his concerns, preoccupied with humankind's elemental relationship to nature, and he resolved to focus his energies on that subject. Between 1883 and his death in 1910, he drew most of his inspiration from the sea or the Adirondacks woods, where he and his brother camped for one or two months each summer. Homer mixed comfortably with the guides and fishermen among the region's white pines, hemlocks, and crystal rivers. When not angling or canoe-

Winslow Homer, *Near Andersonville*, 1865–66. *(Collection of the Newark Museum.
Gift of Mrs. Hannah Corbin Carter, Horace K. Corbin, Jr., Robert S. Corbin,
William D. Corbin, Mrs. Clementine Corbin Day in memory of their parents Hannah
Stockton and Horace Kellogg Corbin, 1966.)*

ing, he produced a series of watercolors based on incidents he had observed. His sunny, almost bucolic country themes of the 1870s now gave way to scenes of Darwinian struggle. Fusing savage action and balsam tranquility, he borrowed the impersonal perspective of the camera in depicting deer being pursued and killed, hooked trout thrashing on the line, and ravenous crows stalking a red fox struggling through deep snow. Homer viewed nature through a secular lens that isolated moments of violent confrontation and exuberant masculinity.

The subject to which Homer regularly returned during the eighties and nineties was the awesome majesty of the sea. His passion for the ocean enabled his true genius to flourish. Homer's celebrated marine paintings reveal with almost brutal force the sea's unrelenting power and treacherous beauty. Many of his seascapes feature the primeval drama of white-tipped waves sullenly gnawing upon slaty rocks. Others focus on rocklike men doing battle with the sea. Works such as *The Life Line, The Undertow,* and *Fog Warning* deserve the label masterpieces, yet several academic critics shuddered at Homer's lack of finish. One lamented that he included not "the slightest trace of mere decorative beauty either in composition or coloring." Yet a few commentators acknowledged that even though his canvases were "often frankly ugly," they offered an eager new audience "the realism of stern fact . . . all unadorned, uncheapened, direct, truthful, grand."[22]

Winslow Homer, *Gloucester Farm*, 1874. *(Philadelphia Museum of Art: John H. McFadden Collection.)*

Winslow Homer, *The Life Line*, 1884. *(Philadelphia Museum of Art: George W. Elkins Collection.)*

Homer may have fastened upon seascapes and hunting scenes because they enabled him to create dynamic contrasts between clashing elements. So, too, did he remain a powerful hybrid of warring values. For all of his widely recognized realistic tendencies, he retained potent strands of romantic idealism. His preference for the country over the city, his self-imposed isolation and his fascination with the essential loneliness of the human condition, his brooding preoccupation with nature's dramas rather than those of the human community, his intoxicated awe in the face of the crashing sea, his fertile coupling of robust masculinity and lyrical delicacy, all attest to his complex character and mixed motives. Homer's tangy blend of romantic realism gave his work its invigorating originality. Like Whitman he displayed the virtues of a synthetic vision; he respected the dignity of facts but probed beneath them to explore deeper levels of feeling and meaning, rendering the transitory eternal.[23]

Thomas Eakins

When a journalist asked Thomas Eakins to name the country's greatest painter, he quickly replied, "Winslow Homer." The artistic hermit at Prout's Neck would have especially appreciated the assessment, for he and Eakins were kindred spirits in their rejection of the brittle gentility regulating American taste. Both men scorned academic formalism and Victorian prudery. Beauty in the decorous sense of the term never concerned them; they revered the truth as it appeared to their eyes and minds. Both were rugged, vital spirits who delighted in portraying energetic people engaged in outdoor recreation and sporting events. As older men they both developed a prickly individuality and growing sense of alienation from a professional art community they deemed stifling. And they both contemptuously dismissed academic painters who added "finish" to their canvases as being "lady-like" in their approach to their art. [24]

Yet where Homer enjoyed considerable popularity, Eakins was relatively unknown among the general public and deemed a scoundrel among official art circles. His unconventional ways provoked controversy and scandal. "My honors," he observed late in life, "are misunderstanding, persecution and neglect." The differing degrees of acclaim accorded Homer and Eakins reflect an obvious fact: Eakins was the more thoroughgoing realist, more willing to shock and displease prevailing tastes in his relentless scrutiny of physical fact. His discomforting candor posed a direct challenge to popular notions of the true, beautiful, and discreet. While Homer had a shrewd sense of the limits of public taste and chose his picturesque subjects and decorative values accordingly, Eakins became a painter of stupefying force and disconcerting objectivity, almost clinical in his devotion to facts and in his refusal to hide blemishes or coat unpalatable subjects in cloying sentiment. [25]

Eakins shared Homer's "masculine" prejudices, but his outlook was more cool and cerebral, his vision more analytical, his purposes more profound. He strove "to peer deeper into the heart of American life." To do so he employed the scientific rationalism of an engineer and the anatomical knowledge of a surgeon. Eakins was not in the least sentimental: his eye was sharp and his perspective was angular. Many of his most famous works grew out of detailed preliminary drawings using photographs and sophisticated geometrical studies of perspective and proportion. As a result his canvases are convincing three-dimensional representations: well rounded, massive, and substantial in their presentation. The price Eakins willingly paid for such meticulous design was the loss of the charm and spontaneity characteristic of Homer's genre studies. [26]

Eakins (pronounced Akins) was born in 1844, the first child of a celebrated Philadelphia writing master, Benjamin Eakins, and his Quaker wife, Caroline Cowperthwait. The elder Eakins taught penmanship at a Quaker school and also inscribed deeds, diplomas, and other official documents. Over the years, his astute speculations in stocks and real estate brought him a comfortable income that enabled him to subsidize his son's art endeavors. Thus, unlike most of his peers, Thomas Eakins enjoyed the luxury of financial independence, and this enabled him to be more daring in his choice of subjects and techniques. [27]

By his late teens Eakins had developed an intense interest in art and a dogged determination to master its tools. After graduating with honors from high school in 1861, he entered the Pennsylvania Academy of the Fine Arts. His experiences there confirmed his artistic inclination, but did little to refine his natural skills. At that time the Academy offered no full-time instructors, and the conventional curriculum required students to spend months drawing from plaster casts of antique figures before they could even observe a live model. Eager to learn the intricacies of the human figure, Eakins took classes at Jefferson Medical College, where he dissected cadavers and briefly considered becoming a surgeon. By age twenty he had gained a thorough understanding of anatomy that would give his later work its distinctive signature.

In 1866 Eakins joined the throng of aspiring American painters sailing for France. Benjamin Eakins bestowed his blessing as well as financial support: "You learn to paint as best you can, Tom; you'll never have to earn your own living." In Paris Eakins enrolled in the Ecole des Beaux-Arts, the most renowned art school in the world, and began study under Jean-Léon Gérôme, a celebrated genre painter famous for his hard-edged, polished technique and historical subject matter. (Mary Cassatt, a fellow Philadelphian, had earlier arranged to take private lessons under the great Gérôme.)

Gérôme ordered his students to copy nature, not interpret it; he abhorred "poetical" painting and prized a detached, almost scientific approach to the subject. Initially, Eakins adopted such traits as his own. In a long letter to his sister, he castigated painters who injected "unnatural or sickish sentiment" in their work. Gérôme, by contrast, excelled at painting "living, thinking, acting men, whose faces tell their life long story." He "makes people as they are— their virtues, their vices & the strongest characteristics of those he represents."[28]

Eakins especially appreciated Gérôme's emphasis on painstaking drawing and flawless perspective. By 1868, however, he had come to question his teacher's preoccupation with historical subjects. Too many of the instructors, he wrote his father, "read Greek poetry for inspiration & talk classic & give out classic subjects & make a fellow draw antique." For his own part he had no interest in ideal beauty; he wanted to paint those subjects he knew, "as I see them."[29]

Eakins also decided that representing visual facts alone, no matter how accurately observed and faithfully rendered, would mean little unless animated by the unseen forces that endow real things with meaning. As a consequence, he resolved to develop a visually accurate *and* emotionally expressive artistic perspective that would reveal the charged vitality embedded in nature. The "big artist," Eakins wrote, "does not sit down monkey-like & copy a coal-scuttle or an ugly old woman . . . but he keeps a sharp eye on Nature & steals her tools"— light, shade, and color. In a good painting, "you can see what o'clock it is afternoon or morning, if it's hot or cold, winter or summer, & what kind of people are there & what they are doing & why they are doing it."[30]

An increasingly self-confident Eakins reported in October 1868 that he had "made great progress" under Gérôme's tutelage. Soon he would be able "to

paint what I can see before me better than the namby pamby fashion painters."
A trip to Spain in the fall of 1869 confirmed the trajectory of Eakins's ambitions.
At the Prado in Madrid he fell in love with the seventeenth-century works of
Diego Velásquez and José de Ribera. He relished their emphasis on character
rather than idealized beauty and preferred their freer use of the brush to the
tightly controlled technique taught at the Ecole. The Spaniards displayed a
weighty brilliance, a solidity and seriousness achieved through accurate anatomy
placed against a background of rich, almost opaque, earth tones. In an epiphany
of self-confirmation, Eakins realized while viewing such paintings that true art
required fusing delicacy with force, energy with insight.[31]

Upon his return to Philadelphia in 1870, Eakins converted a third-floor
room in his family home into a studio and immediately threw himself into his
work. His initial American paintings depicted his sisters engaged in their daily
activities—playing the piano, attending to children, skating, playing with a cat.
The sober tone and subdued feeling evident in these family portraits anticipated
the distinguishing traits of his future works. Eakins refused to follow conventional practice and "prettify" his subjects, even those he loved. He captured one
sister's big nose and dumpy figure and another's flat, round face and oversize
head. S.G.W. Benjamin grumbled that Eakins "cares little for what the world
of taste considers beautiful."[32]

Instead of worshiping the beautiful or expressing the poetic, Eakins revered
what one scholar has called "the heroism of modern life." Unimpressed by a
person's wealth, cosmetic beauty, or claim to social standing, he admired people
who achieved success in their chosen field. His respect for vigorous accomplishment appears in an ambitious series of early paintings depicting a new recreational sport: rowing. Eakins was himself an avid rower, as were many men and
women of all classes in cities across America. In Philadelphia, students, doctors,
lawyers, clerks, mechanics, and artisans organized rowing clubs during the 1860s
and began to hold races on the Schuylkill River which attracted thousands of
spectators.

Among the most celebrated of Philadelphia's amateur rowing champions
was Max Schmitt. In *The Champion Single Sculls* (1871), now labeled *Max
Schmitt in a Single Scull*, Eakins depicted Schmitt, his high-school friend,
pausing in between strokes during a late afternoon workout, as if captured in a
snapshot. The lean, muscular, mustachioed Schmitt looks back over his right
shoulder at the viewer, his inviting eye contact suggesting an elemental kinship
among strangers. Beyond him in another shell is Eakins himself, rowing off into
the middle distance, his identity confirmed by his signature on the bow. Eakins
frequently inserted himself in his paintings to authenticate a scene, and in *Max
Schmitt in a Single Scull* he is both admiring spectator and expert witness. Such
self-inscription enabled Eakins to transform his paintings into documents. The
striking particulars, tight linear control, and simple eloquence of *Max Schmitt*

in a Single Scull led an art critic to conclude that Eakins "shows promise of a conspicuous future."[33]

Over the next three years Eakins painted more than a dozen rowing studies. For most of them he prepared detailed perspective drawings to enhance the three-dimensional illusion and to fix the facts of a scene. Eakins firmly believed that a painting must first convince the viewer of its subject's reality. Yet some people found his scrupulous engineering and cool gaze unnerving. After acknowledging Eakins's "exact, uncompromising, hard, analytic style," one critic confessed that "the spectator's approval is not solicited, but extorted." Others better appreciated his arduous efforts. Earl Shinn, Eakins's Philadelphia friend who had also studied under Gérôme before becoming an art critic, alerted the public to Eakins's distinctive new style. In the *Nation* in 1874, Shinn highlighted Eakins's "remarkably original and studious boating scenes" and revealed that the artist wanted to be known as a "realist, an anatomist and mathematician."[34]

Shinn especially encouraged readers to view Eakins's "curious and learned" portrait of Dr. Samuel David Gross, the eminent professor of surgery at Jefferson Medical College. *The Gross Clinic*, as it came to be known, features a silver-haired Gross pausing to address his students while directing an operation in the

Thomas Eakins, *Max Schmitt in a Single Scull*, 1871. *The Metropolitan Museum of Art, purchase, 1934, Alfred N. Punnett Fund & Gift of George D. Pratt.)*

College's teaching amphitheatre. Five surgical assistants aid him in removing a
piece of dead bone from the thigh of a young male charity patient, clad only
in socks. Among the students sits a sketching Eakins, intent upon the dignified
doctor's instruction. The background of dark-suited students and gloomy tiers
includes a curious melodramatic touch: seated behind Gross is the patient's
mother, her clenched hands shielding her anguished face (relatives were
required by law to be present during such operations). The light catches her
tensed, extended knuckles and gives to the act its peculiar force. Above her sits
a young clerk, pen in hand, carefully recording Gross's words and actions. His
dispassionate countenance and that of Dr. Gross provide a powerful antithesis
to the horror-stricken woman unable to face the harsh facts.

Although *The Gross Clinic* presents a collective drama in which every figure
is a portrait, every face highly individualized, the god-like Gross is the focal
point. Eakins once told his sister that he especially admired "living, thinking,
active men," and Gross exemplified all those virtues. As a high-school student,
Eakins had attended Gross's classes and had come to respect him as both a
dexterous surgeon and a splendid teacher. His reverence for this man of science
shines through the canvas. *The Gross Clinic* reveals Eakins's uncanny ability to
arouse not simply interest but an engagement that is deep, complex, and lingering. He displayed in *The Gross Clinic* the difference between painting objects
and merely rendering surfaces.[35]

A pyramidal composition and ray of light concentrate attention on Gross's
leonine head, intellectual self-possession, and august grandeur. Dressed in a
dark suit and holding a blood-tipped scalpel in bloodied fingers, he turns from
the patient on the operating table to accent a point while an assistant probes
the incision. As with *Max Schmitt in a Single Scull*, Eakins freezes Gross in a
moment of contemplation, his face directly confronting—and engaging—the
viewer.

A monumental canvas, nearly eight feet by six feet, *The Gross Clinic* con-
stitutes an epic tribute to a venerated figure and offers a splendid example of
the epistemology of common-sense realism. "Gentlemen," the doctor seems to
be saying, "simply by observing the real thing closely, by peering intently at this
physical evidence, by facing the facts of the matter and using rational analysis,
you will learn proper techniques and procedures." The heroicizing immediacy
and intensity of the canvas provoked a Philadelphia critic to exclaim that "nothing greater" had "been executed in America."[36]

Yet members of the selection committee of the 1876 Centennial Exposition
in Philadelphia so objected to *The Gross Clinic*'s "brutal" theme that they
refused to hang it. Eakins's casual display of blood and the patient's gashed thigh
horrified those committed to art as a serene sanctuary from life-and-death realities. Brownell complained that "the sense of actuality about it was more than

Thomas Eakins, *The Gross Clinic*, 1875. Jefferson Medical College, Philadelphia.

impressive, it was oppressive." Although the committee members selected five
of Eakins's works for the exhibit, their veto of *The Gross Clinic* confirmed his
belief that prudes still controlled the art establishment. Only through Dr. Gross's
intervention did Exposition officials agree to display the painting in the U.S.
Army Post Hospital building, which had been set up to demonstrate how Civil
War injuries were treated.[37]

The incident reveals how conservative the official art world remained in the 1870s. Writing in the *New York Daily Tribune*, an art critic acknowledged that *The Gross Clinic* was "one of the most powerful, horrible and yet fascinating pictures that has been painted anywhere in this century," and he condemned authorities for exhibiting it "where men and women of weak nerves must be compelled to look at it. For not to look at it is impossible." Others testified to the painting's mesmerizing power but insisted that it had no claim as art. The *New York Times* reviewer harrumphed that Eakins, in his relentless quest for blood and bone authenticity, "had no conception of where to stop, or how to hint a horrible thing if it must be said. . . ."[38]

Eakins brought his mulish personality and rebellious candor to the Pennsylvania Academy of Fine Arts when he was hired to teach anatomy and the male life class during the mid-1870s. By the early 1880s Eakins had become the academy's director of instruction. Great art, he stressed to the students, entailed painstaking preparation which necessarily included drawing from live models, sculpting, and anatomical study, even to the point of dissecting human cadavers. "It quite takes one's breath away, does it not?" reported Brownell. "Exhaustive is a faint word by which to characterize such a course of instruction."[39]

During his years at the academy, Eakins developed a keen interest in photography. What drew him to the camera was its ability to freeze objects in motion and thereby reveal the unexpected distortions caused by movement. While conducting his own camera studies of horses and people in motion, Eakins was commissioned to paint the elegant four-in-hand coach owned by Fairman Rogers, a member of the academy's board of directors and himself an amateur photographer and horseman. *A May Morning in the Park* (1879) displayed all of Eakins's scientific skills and photographic expertise. He repeatedly had the coach driven past him; he built wax models of the four horses, their legs positioned in accord with photographs; and he made numerous preliminary studies of parts of the horses.

Eakins's labors produced a pathbreaking canvas. *A May Morning in the Park* was one of the first paintings in the world to capture the correct gaits of trotting horses. But in fact it is too accurate: the horses seem suspended rather than in collective motion. Their static quality contradicts the spinning motion of the wheels. Eakins's effort to break down the blurred effect of moving horses violated his own principle that the painter should not be so meticulous as to depict what the unaided eye cannot see. *A May Morning in the Park* was an optical experiment gone awry. Realism in art, Eakins discovered, had practical limits.[40]

Throughout these years of experimentation and teaching, Eakins saw his relations with the academy's board of directors grow increasingly strained. The directors tolerated his bringing a live horse into the studio for a lecture on anatomy, but they balked at his preoccupation with nudity. In 1882 a disgruntled mother wrote a letter to the president of the academy in which she expressed the priggishness common to the era:

> I acknowledge that every effort should be made and sustained with enthusiasm that promotes true Art. By *true art* I mean, the Art that purifies and ennobles the mind, elevates the whole intellect, increases the love of the beautiful. . . . Now I appeal to you as a Christian gentleman, educated amidst the pure and holy teaching of our beloved Church . . . to consider for a moment the effect of the teaching of the Academy, on the young and sensitive minds of both the male and female students. I allude to the Life Class studies, and I know where of I speak.

She then lamented the degrading effect of "horrid nakedness" on a young woman whose parents had responsibly shielded her from the human body and "every thought that might lead her young mind from the most rigid chastity."[41]

Eakins's loyalists have savagely caricatured such Victorian skittishness, but the painter's mythic image as a dedicated artist unfairly assaulted by pinched prudery seems both overdone and overly convenient. His interest in nudity bordered on the compulsive, and his utter lack of discretion invited trouble. Once, he himself disrobed in order to show a female student the movement of the pelvis, and he repeatedly badgered students and models—male and female—to pose nude for his camera or his brush. Weda Cook Addicks, the subject of *The Concert Singer*, reported that nudity was an "obsession" with Eakins. Another said that he was "hipped on nudes."[42]

In early 1886 Eakins provoked a showdown. While conducting an anatomy class for women, he removed the loincloth from a male model to show how muscles were attached to the pelvis. The directors, already nervous about the academy's declining enrollment, demanded Eakins's resignation. Students, both male and female, protested the decision, but to no avail. The dismissal of the forty-two-year-old Eakins led many others among the established art community to deem him a rogue. Embittered by what he considered a campaign of falsehoods against him, this stout, sturdy man suffered an emotional and physical collapse.

During the summer of 1887, Eakins took a therapeutic excursion to the Dakota Territory. In the Badlands he lived and worked alongside cowboys, hunting antelope, roping steers, and riding the line. Ten weeks of such rugged life restored his spirits and his health, and he returned to Philadelphia with two horses, a buckskin outfit, and a revived commitment to an authentic art. Walt Whitman, whom Eakins had recently befriended, reported that the artist was "sick, rundown, out of sorts: he went right out among the cowboys: herded: built up miraculously."[43]

Perhaps because of their shared reputation as rebels living under a cloud of public disapproval, Whitman and Eakins developed an intense comradeship in 1887–88. Although separated by over twenty years in age, they had much in common in addition to their scandalous reputations. Both were sprung from Quaker stock, and both spurned religious orthodoxy and genteel decorum. They revered scientific study and admired the aesthetic qualities of American technology. Each exuded a rugged sincerity and enjoyed outdoor life and strenuous recreations. Both professed a shameless reverence for the naked human figure. And they each refused to compromise their devotion to "truth" in order to win critical acclaim or popular success. Perhaps most important, both strove to bring to life the people they knew and respected without resorting to distortion or false sentiment. The aspect of Whitman's writings that Eakins most admired, he told a friend, was "the realistic; the observation, the truth, the sense of coming direct out of life."[44]

In late 1887, Eakins brought his painting gear to Whitman's home and began working on his portrait. The poet liked what he saw as the canvas took form: "it is going to be realistic & severe I think." Although he had sat for many other portraits, Whitman quickly realized that this one would be "different from any preceding—plain, materialistic, very strong & powerful." By April 1888 the canvas was finished. Whitman's sycophants regretted its "fleshy" quality, claiming that it made their hero look fat and self-indulgent, but the author of *Leaves of Grass* pronounced it a masterpiece: "I like Eakins' picture (it is like sharp cold cutting sea brine)." He especially admired its "strong, rugged, even daring" quality. Eakins's honest rendering of Whitman's eyes and facial features illustrates his uncommon ability to connect between body and spirit, to construct a portrait from the inside out, using bone and muscle structure to build strength of character. Eakins offered viewers insights into the spirit of his sitters. "Eakins' picture grows on you," Whitman reported. "It is not all seen at once—it only dawns on you gradually."[45]

As he himself crossed the threshold of middle age, Eakins highlighted the passage of time in his subjects. Having experienced the deaths of his fiancée in 1879 and two of his sisters in 1882 and 1884, he had grown intimate with mortality, and a profound moral gravity ballasted his many late portraits. The singers, musicians, educators, lawyers, and churchmen in Eakins's portraits are scarred, introspective individuals, precarious survivors of an arduous age, people stripped of innocence. Their life-battered faces, lined with anxiety and frozen in thought, expressed both Eakins's own brooding introspection and sense of estrangement as well as the ravaging impact of time and stress upon sensitive minds. The thought-burdened intensity evident in his diverse portrait gallery encourages viewers to identify with the sitters' emotional states, to share their existential predicament and fragility.[46]

Two-thirds of Eakins's portraits were of women, and, like Homer, he refused to depict them as idealized maidens or as delicate creatures of fashion and leisure. Such confectionary women were not those of his own experience. He had lived most of his life in a household full of talented, spirited, self-reliant women, and in 1884 he married a former student, Susan MacDowell, a gifted painter in her own right who made great sacrifices to promote his career. Most of Eakins's female subjects are strong figures, their characters intensely individualized. Usually caught performing a skilled task or engrossed in furrowed thought, they project an autonomous dignity and introspection rarely seen in more popular representations of women. Often they stare blankly into space and seem alone and self-absorbed.

All of these qualities unite in *Miss Amelia C. Van Buren* (1891), a portrait that encapsulates the virtues of Eakins's best works: precise detail, anatomical modeling, scrutinizing vision, and an evocative mood—even *mystery*—built upon the tension between delicacy and power. Van Buren, a former student of Eakins who later became an accomplished photographer, sits in a ponderous upholstered armchair with huge round knobs and red plush covering, her finely sculpted head tilted away from the viewer and staring off toward the source of light, presumably a window. The portrait is both mimetic and expressive. It summons insights that lay beneath superficial appearances. Eakins exhausted every outer semblance so as to give the viewer the illusion of seeing the soul behind the mask. Van Buren's face is full of truthful and terrible implications; time seems to burden her thoughts. Her forlorn eyes reveal a hidden reality, a deep, solemn grief or ennui more powerful and noble than any photograph could convey. The juxtaposition of her melancholy, tight-lipped expression and her pink-and-white flowered dress creates a subtle friction that entices extended scrutiny and empathy.

How different such probing portraits were from those produced by the more celebrated society painters of the era. Cosmopolitan artists such as Cecilia Beaux, William Merritt Chase, and John Singer Sargent used their considerable skills to flatter the vanity of opulent businessmen and silken ladies. Such "decorative" painters produced marvelous likenesses and glistening surfaces, but they lacked Eakins's subcutaneous scrutiny and moral gravity. Beaux, Chase, and Sargent showed little interest in burrowing beneath exterior appearances to explore the sitter's personality. Sargent's stunning canvases, painted with gestural panache and bursting with brilliant colors, capture the eye but rarely provoke sustained reflection or empathetic identification. Sargent himself confessed his superficial emphasis. "I paint what I see," he claimed. "I don't dig beneath the surface of things that don't appear before my eyes."[47]

By contrast, Eakins peered into his sitters. He refused to indulge their vanity or the expectations of the art public. Indeed, he often stripped people of their conceits, leading several of his subjects to flee the studio. Others refused to accept his portraits or destroyed the canvases. Edward Abbey, the celebrated painter of saccharine murals, recoiled at the prospect of posing for Eakins because his devasting candor "would bring out all the traits of my character that I have been trying to hide from the public for years."[48]

Thomas Eakins, *Miss Van Buren*, ca. 1886–1890. *(The Phillips Collection, Washington, D.C.)*

An uneasy, complex, serious soul, Eakins revealed that realism could be more incisive than generally believed. He did not merely imitate what he saw. At the same time that he used scientific techniques to heighten the illusion of reality, he injected a powerful expressive quality into his art that gives his paintings their perennial attraction. There is something formidable and enduring about Eakins's ability to mine nuances of feeling from his subjects and explore their inner reality. No artist's style was ever more personal, more autographic, more wholly his own creation or more evident in the painting itself. He shared

Whitman's recognition that true realism required the artist to "force the surfaces and the depths also." Whitman spoke for many, then and since, when he concluded that "Eakins is not a painter, he is a force."[49]

Impressionistic Realism

Eakins and Homer demonstrated that realistic art was neither as objective nor as straightforward as commonly assumed. Subjective notions of order and significance repeatedly intervened in their efforts to represent visual reality. This was even more true of the colorful new art associated with the impressionist movement. Beginning in the 1860s a group of young French insurgents— Edouard Manet, Claude Monet, Pierre Auguste Renoir, Edgar Degas, and others—challenged the static formalism of academic art standards.[50]

Like the purposeful urban promenaders and shoppers who frequented city streets, the impressionists took voluptuous delight in the ephemeral spectacles of bourgeois life in town and country. Degas and the other French impressionists, along with their American recruit Mary Cassatt, who lived permanently in France after 1874, populated their canvases with fashionable people in shops, cafés, theaters, ballet and opera houses, at city parks and gardens. They self-consciously fostered a strong, vigorous, "manly" treatment. When Degas saw Cassatt's early canvases, he declared: "I will not believe that a woman can draw so well!" Cassatt herself bridled at the thought of being considered a "lady painter." She wanted "to achieve force, not sweetness; truth, not sentimentality or romance."[51]

The impressionists' disdain for genteel sentiment and their preference for bourgeois subject matter aggravated Academicians, but even more disconcerting was their rejection of conventional mimetic technique. Impressionists painted things as they *appeared* to the roving eye rather than as they *were* in fact. Reality to them was not so much a static condition as it was a fluid process. Their spectatorial mode of perception led them to replace the precise description of people and objects with the amorphous delineation of the mood and atmosphere enveloping their subjects.

Impressionists realized that people do not see objects with equal clarity; things on the periphery and in the distance tend to be vague or blurry. In this sense the roving gaze of the impressionists was unapologetically photographic: it captured snapshot, blurred images of the picturesque life around them. For the impressionists, unlike Eakins, immediacy was superior to veracity. The painter's hand should serve as an extension of the eye rather than the calculating mind. Henry James commented in 1876 that the impressionists were "partisans of unadorned reality and absolute foes to arrangement, embellishment, selection."[52]

By vividly rendering their fleeting perceptions, the impressionists brought to art the nervous vivacity of modern city life, a crowded field of vision where reality was ephemeral and fragmented, contingent upon time and place for meaning. The seemingly haphazard brushstrokes and "unfinished" appearance of impressionist paintings reflected the overstrained nerves of the metropolitan personality and resembled the throngs of people in random motion on city streets, separate individuals merging to form an undifferentiated whole. For impressionists, flux rather than fixity governed modern experience; they saw the real world as an ever-changing chimera of object and surfaces.[53]

Childe Hassam

The first major exhibition of the European impressionists opened at the American Art Association in New York in 1886. It provoked one critic to call the 300 canvases on display "magnificent insanities." Others, however, were more sympathetic to "this ultra-modern group," and large crowds testified to the favorable reaction among the broadening art public. The Chicago art critic Henry Stephens assured Americans that "impressionism is the apogee of realism." The "older realists," he explained, "were false to fact in their treatment of light" whereas the new rebels "give the impression of the actual as we see it in the world of daily experience." Stephens noted that the best American impressionist was Childe Hassam.[54]

Born in Dorchester, Massachusetts, in 1859, Hassam was the eldest son of middle-class parents descended from New England farmers and sailors. When his merchant father suffered sharp financial reverses, Childe dropped out of high school and apprenticed himself to an engraver. Within a few years Hassam had become a freelance illustrator, producing drawings of contemporary American scenes for *Harper's Monthly*, *Century*, and *Scribner's*. Yet he yearned to become a professional artist, and in 1878 he enrolled in the evening life class at the Boston Art Club. By 1883 he had gained enough stature to begin teaching art classes himself. That year Hassam embarked on an extensive tour of Europe, traveling the Continent with a Boston friend, painting scenes he encountered along the way. Upon his return to Boston, he offered a one-man exhibition of his European paintings, and they sold well enough for him to establish his own studio at 282 Columbus Avenue across from Boston Common.[55]

Between 1884 and 1886 Hassam engaged in a kind of urban espionage, capturing on canvas the familiar scenes he witnessed. His shimmering views of Bostonians at twilight or on rainy or snowy days reveal his interest in the evocative effects of light, air, and color. These Boston scenes capture the tone and texture of middle-class urban life during the Gilded Age, and several have since been used as cover illustrations for the novels of Howells and James.[56]

Childe Hassam, *Rainy Day, Boston*, 1885. *(The Toledo Museum of Art.)*

Rainy Day, Boston (1883), for instance, shows the prospect of Columbus Avenue to the left looking east to the Public Gardens, with Appleton Street veering off to the southeast, a location Hassam passed often on the way to his studio. In the foreground a mother lifts her skirt to avoid puddles as her clinging child carries a small book. On either side, rain-soaked cabbies attend solemnly to their tasks. The broad expanse of damp pavement, steely in its wetness, glistens in contrast to the dark figures crouching under umbrellas or raincoats.[57]

Where Hassam and other impressionists often concentrated on evanescent scenes and the ballet of suffused light, *Rainy Day, Boston* conveys a haunting permanence. On the surface it represents an ordinary moment in Hassam's actual experience, yet it can also be read as an early account of the urban alienation that has pervaded twentieth-century cultural expression. Is it only the rain that keeps people aloof from one another or is it also the terrifying indifference of their urban environment? The brownstone houses receding into the distance form a seemingly continuous wall that encloses with looming disinterest the people on the street. Huddled figures suggest the vulnerability of the urban self and the hostility of the city. They are reminiscent of what Mrs. Horn says in Howells's *A Hazard of New Fortunes*: "New York is full of people who don't know anybody." Shelterless and alone, Hassam's beleaguered city dwellers seem to have staked out invisible boundaries to shield their privacy from an intrusive public arena. Umbrellas and hats protect them as much from each other as from the rain. In other early paintings of lonely crowds, Hassam similarly showed people rushing past one another as though they had nothing in common and would prefer to have nothing to do with one another. Such scenes illuminate the paradox of modern urban experience: nothing is more alienating than the close proximity of too many people; nowhere does one feel more alone and forsaken than among a throng of strangers.[58]

In 1886 Hassam traveled again to Europe, this time to study at the Académie Julian in Paris. He reveled in Baudelaire's "ant-swarming city," with its incessant traffic, its plunging perspectives, and its panoramic activity. The flickering brush strokes and lighter palette evident in his paintings of Parisian street scenes betray the influence of the French impressionists. Still, Hassam never embraced a pure impressionism; he refused to sacrifice fidelity to physical form for the pure glories of light and color. Instead he struggled to synthesize figurative realism with spontaneous plastic effects. Like Eakins, he had no respect for "the man who would emancipate art from discipline and reason."[59]

Hassam returned to the United States in 1889 and settled in New York, where he and his wife resided for the rest of their lives. An open-minded and warm-hearted urbanite, Hassam cast a loving eye on Manhattan's diverse peoples, incomparable skyline, material splendor, and throbbing energies. "There is nothing so interesting to me as people," he told an interviewer. "I am never tired of observing them in every-day life. Humanity in motion is a continual study to me."[60]

In his paintings of Central Park, Fifth Avenue, Madison Square, Union Square, and Washington Square, Hassam captured the restless change and movement as well as the somberness and allure of thronged thoroughfares, fashionable avenues, and pleasant parks. Whether seen through the filter of snarling blizzards or radiant spring days, his New York works portray the ceaseless ebb and flow of people and vehicles, giving the urban landscape its unique sense of palpitating life and random motion. Rather than simply draw figures and objects, he enveloped them in space and atmosphere, giving them a blurred definition that encourages the viewer to sense the vibrating effects of light and color in motion.

A robust man, plump, bull-necked, pink-skinned, and tweedy in attire and demeanor, Hassam brimmed with lusty tastes and strong opinions about art. New York, he exclaimed, was "the most beautiful city in the world," and he resolved to reveal its picturesque charms without sacrificing veracity. "Good art is, first of all, true," he declared. In practice this meant that figures seen in the distance were to be represented as the indistinct images they created in the eye rather than in close-up detail. "Art, to me, is the interpretation of the impression which nature makes upon the eye and brain." This required painting what the artist saw rather than invented in a flight of fancy: "The true impressionism is realism. . . . I believe the man who will go down to posterity is the man who paints his own time and the scenes of every day life around him."[61]

In selectively representing the peopled spectacles of urban life, Hassam avoided the city's darker moods and deeds. He did not, as one critic noticed, "indulge in the sentimental aspect of squalor, nor look with sympathetic eye upon the picturesque life of the humble." Hassam, like Howells, chose to tell gentle truths palatable to bourgeois tastes. His untroubled and unquestioning paintings of middle-class city life evoked the poetry rather than the prose of the pavement. Art for him was neither a means of raising social consciousness nor a passage to profound truths. It was simply a warm expression of life: to be loved and cherished for what it was.[62]

A master of atmospheric effects, Hassam was a better colorist than figure painter, and his art at times suggests the airiness of spun confection, lovely to observe and consume, but lacking in solidity. No one accused Hassam of harboring a penetrating vision. He reveled in surfaces. Yet by engulfing the whirl and lurch of contemporary life in an aura of respect and charm, he helped legitimize the city as a fit subject for art. "Look at some of our great buildings, our bridges and viaducts," he stressed. "They in themselves are works of art." He was, as a journalist concluded, one of the very few artists who "found beauty in the teeming busy streets of New York."[63]

Of course, the same could be said for Eakins and Homer. Despite their considerable differences in temperament and technique, they enticed people who had little or no academic preparation to appreciate the aesthetic potential of their everyday surroundings. Like their literary counterparts, most realistic painters and critics discovered that reality was not so matter-of-fact as it seemed; artistic truth required more than a simple mirroring of the external world; it also entailed projecting the artist's impressions and emotions onto the canvas. This distinction between duplicating life *as is* and representing life *as perceived* both complicated and enhanced the realistic enterprise. "All good art," insisted the editor of *The Art Amateur* in 1886, "is impressionistic in character."[64]

8

Form Follows Function

Realistic tendencies in fiction and art are relatively easy to document and describe. Architecture, however, is another matter. A building design is of necessity a compromise among utilitarian needs, engineering imperatives, financial constraints, and aesthetic values. What makes a building "realistic"? Piles of steel, stone, glass, and wood do not easily serve a mimetic function. What can they possibly represent outside of themselves? Yet however difficult it is to assess buildings as works of art, the fact remains that people during the nineteenth century frequently discussed architecture in relation to larger cultural trends, including the rise of realism. In 1886, for instance, architect Henry Van Brunt admitted that "architecture has not kept pace with the general advance of the other arts of civilization."[1]

Despite Horatio Greenough's mid-century crusade for an indigenous American form of design related to contemporary needs and "real" uses, most architects after the Civil War continued to borrow from earlier periods and foreign lands. The result was a jumble of warring styles: castles with colonial façades, Greek public buildings adjacent to Gothic churches, and French Second Empire brownstones and Elizabethan mansions across the street from one another. Such eclecticism did not satisfy those thirsting for a distinctively American style of design. Art critic James Jackson Jarves lamented that architecture in the United States formed an "incongruous medley as a whole, developing no system or harmonious principle of adaptation, but [remaining] chaotic, incomplete, and arbitrary."[2]

The imitative historicism prevalent in American design appeared most vividly in 1893 at the World's Columbian Exposition, also called the Chicago World's Fair. The colossal Exposition paid an imposing physical tribute to American ambition and enterprise. An army of workers converted the mile-and-a-half stretch of sand dunes and marshes between Lake Michigan and Michigan

Avenue into a series of low, broad terraces on which they erected some 400 buildings intersected by canals, lagoons, and wide promenades.[3]

Chicago architect Daniel Burnham, the director of works who assembled a team of the nation's most celebrated architects to design the buildings, prized a monumental classicism. "Make no little plans," he advised the group; "Make big plans; aim high in hope and work. . . . Let your watchwords be order and your beacon beauty." The architects adopted a neo-classical theme for the Court of Honor, site of the main exhibition buildings and the centerpiece of the fair. Grouped around a central basin, these stately buildings, all the same height and "perfectly white," were meant to embody the order and nobility prized by genteel tradition. Burnham hoped that the Exposition would "inspire a reversion toward the pure ideal of the ancients."[4]

For all of its imposing grandeur, the White City betrayed a nation still uncertain of its own aesthetic merits and self-identity. The neo-classical motif of the World's Fair symbolized how most architects during the post-Civil War era continued to look to the past for models and inspiration. In part this was a self-conscious effort to be cosmopolitan, to demonstrate that American designers could recreate the best that Western civilization had to offer.

World's Columbian Exposition, Chicago, Illinois, 1893. The Court of Honor, looking west to the administration building. Photo by C.D. Arnold. *(Art Institute of Chicago.)*

There were, however, a few cultural nationalists who called for a more "realistic" approach to design that would reflect both engineering innovations and the massive technological and social changes transforming the pace and nature of everyday life. Van Brunt, for instance, observed that the art of building was much more dependent on scientific invention than any other art form. This meant that it could least afford to cling to "the principles of conservation and conventionalism." Instead of plagiarizing motifs from previous eras and foreign countries, American architecture "should be a living and growing reality." Echoing Greenough, he encouraged designers to reflect "our own iron age" and its "eminently mechanical spirit." Modern urban-industrial America demanded not "instruction or emotional impulse" from a building, but "fitness for its peculiar purposes."[5]

This emphasis on a structure's "fitness" rather than its decorative frills or historical echoes served as the axis of a new architectural program that crept into prominence during the 1880s. People began to talk of a "realistic" architecture attuned to functional needs and contemporary values. By 1891 William P. Longfellow, the editor of the *American Architect and Building News*, could announce that the debate between idealism and realism that had long preoccupied writers and painters had finally surfaced among architects. The two "things which attract our people to art," he wrote, "are the ornamental element and the realistic." Longfellow charged that most designers had overemphasized decorative effects at the expense of utility. As a result, architecture had grown effeminate and enfeebled, characterized by "softness, languor, extreme delicacy, [and] the appearance of fragility." He found this particularly disconcerting because architecture represented "the most virile of the arts." In his view, commercial design afforded opportunities for the boldest expression of the materialistic and "masculine" concerns pervading the age. A writer in the *Western Architect* echoed this equation of architectural realism with a virile functionalism when he claimed that modern design "tries to derive its beauty from the purely realistic, practical solution of the problem, as much as possible; from the manner in which one shapes and groups and not in which one ornaments and decorates."[6]

Yet even the most "realistic" architects acknowledged that a *purely* utilitarian aesthetic was neither desirable nor possible. Like their literary and artistic counterparts who balked at being labeled merely "photographic," architects who championed a more "realistic" style still preserved a role for artistic expression and social statement in their designs. Most of them promoted an "organic" approach in which art and utility, spiritual ideals and practical needs, united to form the whole. Montgomery Schuyler, the distinguished architectural critic, observed that "the art of architecture is not to produce illusions or imitations, but realities, organisms like nature."[7]

The Chicago School

Chicago served as the seedbed for a new "commercial school" of architecture after the Civil War. The brawny rail center became the hub for the entire midwestern economy and the main artery for the region's bustling commodities markets. Chicago's feverish materialism and entrepreneurial energy attracted swarms of new residents and drew international attention. Novelist Henry Blake Fuller described the metropolis in *With the Procession* (1895) as "the only great city in the world to which all its citizens have come for the one common, avowed object of making money."[8]

In the midst of flat prairies and out of the ashes of the Great Fire of 1871 that razed 18,000 buildings, a new skyline of stone and steel arose that dazzled observers with its originality. Architect Peter B. Wight claimed in 1880 that young Chicago designers were working from principles "uncontrolled by the traditions of previous styles." This "Chicago school," reported Van Brunt,

Henry Hobson Richardson. Marshall Field & Co. Wholesale Store, Chicago, Illinois, 1885–1887. *(Chicago Historical Society.)*

"seems to have fairly won the distinction of being the fountainhead of architectural reform in the West." In their commercial designs, the leading Chicago
architects—William Le Baron Jenney, John Wellborn Root, Dankmar Adler,
Louis Sullivan, William Holabird, and Martin Roche—expressed both the city's
vibrant materialism and a keen interest in modern needs rather than historical
traditions. Their warehouses, office towers, department stores, and banks served
as gesturing symbols of a new commercial order and influenced city-builders
across the world.[9]

Although quite profitable, multi-story structures posed immense engineering
challenges and required new interior designs as well as new construction methods and technological innovations—the hydraulic passenger elevator, terracotta
fireproofing, riveted steel-frame construction, the electric light, and new foundation systems. They also demanded courageous architects willing to risk innovation and to embrace new principles of design appropriate to the new commercial and industrial culture dominating American enterprise.

Henry Hobson Richardson, the magisterial Boston designer, led the way. A
Louisiana native, he graduated from Harvard in 1858 and the following year
went to Paris, where he enrolled in the prestigious Ecole des Beaux Arts. Upon
his return to the United States in 1865, Richardson settled in New York and
began specializing in Gothic designs. After relocating to Brookline, Massachusetts, a suburb of Boston, he discovered the simplicity and repose embodied in
the Romanesque and Moorish structures built in southern France and Spain
during the Middle Ages. The "quiet and massive" Romanesque style resembled
Richardson's own personal qualities. An enormous man (weighing well over
300 pounds) with a vigorous, colorful personality, he designed libraries,
churches, and government buildings with massive courses of multicolored,
rough-finished stone, low-browed, rounded arches, and soaring towers with
pointed roofs.[10]

Richardson's Romanesque affinities fixed his initial style and made his reputation, but late in his career he developed an original vocabulary of form
appropriate to the modern commercial society in which he lived. Richardson's
new approach appeared most boldly in his celebrated design of the wholesale
warehouse-store for Chicago's Marshall Field and Company (1885–87). He
explained to a Chicago newspaper that he had "long had certain ideas he wished
to embody in such a building," most notably an emphasis on "the beauty of
material and symmetry rather than of mere superficial ornamentation." Based
on a simple design remarkable for its unity and proportion, the massive seven-
story structure and its huge, rough-faced foundation stones epitomized the gigantic ambitions associated with commercial capitalism.[11]

The Marshall Field store displayed Richardson's heightened concern for
functional values. He applied fundamental design principles—an essentially
square shape, repeating window patterns, a flat, heavy cornice, semicircular

arches, and rusticated masonry—to suit the building's prosaic purposes and embody its commercial mission. By minimizing detailed ornamentation and disdaining striking façades and roofscapes in favor of a straightforward treatment of the wall surfaces, Richardson pried himself loose from the grip of romantic historicism and proved that utilitarian structures need not be ugly. As Louis Sullivan remarked, the Field store represented "a monument to trade, to the organized commercial spirit, to the power and progress of the age," providing a refreshing oasis "amidst a host of stage-struck wobbling mockeries."[12]

Richardson's focus on a building's elemental purposes inspired younger Chicago architects to discipline their own picturesque tendencies and develop a distinctive new style that synthesized practical and artistic elements. In the process, many of them began to use the term "realism" to describe their efforts. The German-born architect Frederick Baumann, for instance, declared that he and others among the "Chicago school" of builders promoted a "coalescence of the natural with the ideal, and this I conceive as realism." The best buildings, he added, displayed "a happy poise between the two elements." Baumann and others steadfastly believed that modern structures should express spiritual needs while at the same time serving practical uses. A building must embody "integral verities," another Chicago builder asserted. The best designed buildings, he believed, had a "realistic" structure and a spiritual center. "The plan, the purpose, the inner soul of the building, determine the exterior, its forms and features."[13]

John Wellborn Root

In 1908 Frank Lloyd Wright recalled that when he entered the architectural profession in the 1880s the freshest approaches were evident "in the work of Richardson and of Root." Wright later termed John Wellborn Root a preempted "genius," for his death in 1891 at age forty-one robbed the architectural community of one of its most innovative leaders. The editor of the *Inland Architect* wrote that "we have thought him our greatest architect, both in achievement and promise."[14]

Born in rural Georgia in 1850, Root grew up in Atlanta during the roiling debates over slavery and sectional conflict. When Union troops commandeered the Root home in the fall of 1864, John and his brother fled to the coast and took passage on a blockade runner headed for England. Three years later, Root returned to America and settled in New York City, where his parents had moved. At New York University he earned a degree in science and civil engineering, graduating fifth in his class. Thereafter he worked as a draftsman for several New York design firms. In 1871, sensing the potential for a dramatic building boom in Chicago after the Great Fire, he relocated to the midwestern metropolis.[15]

Eager to be his own boss, Root and another talented young draftsman, Daniel H. Burnham, formed their own firm in 1873. Unfortunately for the two aspiring architects, however, their partnership began just as a sharp financial panic reverberated through the nation's credit markets. Throughout the rest of the decade, in the face of the prolonged economic depression, they struggled to sustain a foothold in the Chicago market, eking out a living by designing personal residences. In the early 1880s, however, Burnham and Root received their first commissions for tall office buildings, and their reputations soared alongside their commercial structures.

In the midst of his newfound success, Root discarded his enthusiasm for Gothic forms and resolved to design an "American" structure that would "express the conditions of life about and within it." Architecture, he insisted, "is, like every other art, born of its age and environment"; designers must therefore accept the commercial and scientific dictates of their own age. Root dismissed contemporary American architecture as a carnival of derivative styles: "Victorian Cathartic, the Tubercular, the Cataleptic, and the Dropsical," none of which addressed modern needs. A new style of commercial design must emerge "through the frankest possible acceptance of every requirement of modern life in all its conditions, without regret for the past or idle longing for a future and more fortunate day."[16]

Root's brash new outlook reflected his excited interest in modern science. During the 1870s he had grown increasingly enamored of evolutionary theory and the scientific spirit. People "no longer believe profoundly," he told the Architectural Sketch Club; "the subtler ether of pure imagination is displaced by the heavier air of demonstrated fact." Yet out of the heightened respect for facts, he predicted, would "come a new and stronger art." Old architectural forms would give way to new designs that better related to their contemporary environment. For Root the object of both scientific investigation and architectural design was identical: "the search for truth—truth in its purest form." Modern American architecture, he pleaded, should not war against scientific and commercial actualities, "it must sympathize with them."[17]

The Chicago skyline, Root decided, suffered from "the blight of mid-century heaviness"—ostentatious moldings and cornices, French mansard roofs, Romanesque turrets, Queen Anne gables, and jigsaw ornamentation. "The random application of ornament of all sorts, without thought or purpose," he charged, "is the crying evil of modern architecture." Like other young architects associated with the "Chicago School," he frequently characterized such anachronistic forms and superfluous ornament as the structural equivalent of sentimentalism in literature and painting. Root's sister-in-law, the poet Harriet Monroe, noted that he "took always the rational point of view—even to hardness, rather than the sentimental." A "realistic" philosophy of design, Root argued, required that architects leave their "petticoat fortress and fight our battles in open field."[18]

Like Richardson, Root was a vigorous, hearty, sensual man who mandated that "feminine ornamentation" be subordinated to the building's more essential features and its "masculine" presence. While disdaining the "servile imitation" of Old World styles, Root celebrated designs that expressed the nation's virile material spirit and served the client's practical needs. He saw in Chicago's new office towers the most appropriate expression of America's business ethos and its "great middle class." No type of building "is more expressive of modern life in its complexity, its luxury, its intense vitality" than the skyscraper. Such buildings should reflect "the ideas of modern business life—simplicity, stability, breadth, dignity." Worried that a commercial building might be "emasculated by ornaments too delicately wrought," Root detested theatrical effects. "The value of plain surfaces in every building is not to be underestimated," he asserted.[19]

Root's rhetorical tribute to functionalist realism should not be taken at face value. He in fact designed several buildings that contradicted his expressed commitment to a "modern" architecture reflective of contemporary life and masculine virtues. The Women's Temple and the Masonic Temple, for instance, revealed his willingness to apply both the Gothic and Romanesque traditions to tall buildings. Of course, such historicist elements may not have been of his own choosing; the owners may have dictated such imitative forms. Root's role in such decisions remains unclear, and this makes any generalizations about his architectural philosophy problematic.

Two of Root's most important Chicago buildings, the Montauk Block (1881–82) and the Monadnock Block (1889–91), were commissioned by Peter Brooks, a Boston shipping magnate and real-estate investor who took a keen interest in their design. Brooks stressed to his Chicago agent that the Montauk building must convey the "idea of strength." It "is to be for use and not for ornament. Its beauty will be in its all-adaptation to its use." Such a starkly utilitarian philosophy shocked even Root, who struggled to find ways to accommodate Brooks's severely practical requirements while maintaining a role for his own expressive talent. Sheathed in fireproof red brick, its sheer mass accented by sloping basement walls and an imposing arched entrance, the Montauk Block satisfied Brooks's desire to minimize ornamentation, highlight function, and project an image of "masculine" strength and stability. At ten stories, the Montauk Block was the tallest building in Chicago, the first to incorporate an iron frame, and the first to be called a skyscraper.[20]

In designing the Monadnock building, Root again received precise instructions from Peter Brooks. He wanted it to "have no projecting surfaces or indentations, but to have everything flush, or flat and smooth with the walls. . . ."

Decorative projections, he explained, merely gathered dirt and attracted pigeons. Brooks later suggested that Root model his plans after "Richardson's exteriors." Indeed, Root came to appreciate Richardson's tendency "to make architecture more simple and direct."[21]

The Monadnock Block (1891), the tallest building ever constructed with masonry walls, rose sixteen stories above the corner of Jackson and Dearborn streets in Chicago. With a flat roof and shallow, undulating four-window bays, it was a triumph of unified design—a huge, vertical slab of dark brick virtually devoid of ornament. Barr Ferree, a professor of architecture at the University of Pennsylvania and editor of *Engineering Magazine*, remarked that the Monadnock building "is intended solely for commercial purposes; it is not a monument devoted to artistic ends." He then provided a succinct statement of what people at the time meant by architectural realism: "In this building, as in all high buildings, it is not what should be that is to be considered, but what is." The president of the Boston Architectural Club applauded the "dignity and worth" of the Monadnock's "simple statement of fact." In such a towering testimony to modern commercial "facts," observers saw the poetry of practicality. Root, however, did not live to see the building's inaugural opening; in early 1891 he died of pneumonia.[22]

Louis Sullivan

Like Root, Louis Sullivan celebrated the virile accomplishments of corporate capitalism. In the process of implementing his famous dictum that "form follows function," he became, according to one critic, "a great master of realism in architecture." Labeling the mystical Sullivan a "realist," however, stretches the elasticity of the term to its breaking point. True, he disdained historical conventions, advocated an art of and for the present, and often asserted that true architecture "finds its goal in the real." In his autobiography, he claimed that his goal as a designer had been to develop "a realistic architecture based on well defined utilitarian needs."[23]

In practice, however, Sullivan, like his hero Walt Whitman, was an unstable hybrid of romantic and realistic tendencies. He fancied himself a lyric nature-poet and democratic philosopher who strove to bridge the gulf between transcendent ideals and practical needs by designing buildings that embodied both rational ideas and soaring emotions. Unlike twentieth-century functionalists, Sullivan never believed that beauty was the *inevitable* product of a utilitarian form. True, a form must first embody actual needs, but the designer must then make the form beautiful.

Born in Boston in 1856, the son of an Irish immigrant father and a Swiss-French mother, Louis Henry Sullivan grew up in a household devoted to the arts. His father, Patrick Sullivan, a taciturn dancing master, and his mother, Andrienne List, a warm-hearted pianist, imbued young Louis with an artistic sensibility. As a boy he attended the English High School in Boston but spent much of his time at his maternal grandmother's farm outside the city, where he loved to roam the fields and beaches, cultivating a transcendental outlook toward nature. The grandeur of soaring, solitary trees especially inspired him, and he later would see a similar majesty in his tall buildings. Young Sullivan, like young Whitman, also displayed homoerotic tendencies, glorying in the achievements of "big strong men" who built great things. When he saw his first suspension bridge in Newburyport, Massachusetts, he exulted "to think it was made by man! How great men must be, how wonderful; how powerful, that they could make such a bridge."[24]

Sullivan's bedazzled appreciation of "muscular" accomplishments led to his precocious decision at age twelve to pursue his "heart's desire": he would become an architect. So strong was his resolve that he remained with family friends near Boston (where educational opportunities were greater) rather than accompany his parents when they relocated to Chicago. At age sixteen Sullivan entered the Massachusetts Institute of Technology, only to discover that the professors depressed rather than inspired him. Preoccupied with the achievements of the past rather than the needs of the present, the faculty taught a "misch-masch of architectural theology" and paid homage to a "cemetery of orders and of styles." Formal schooling ultimately disappointed him because it lacked "the vibrant values of immediate reality."[25]

Sullivan dropped out of MIT. After serving a brief apprenticeship with the Philadelphia architectural firm of Furness & Hewitt, the seventeen-year-old rejoined his family in Chicago and found work in the office of William Le Baron Jenney, a Union army veteran and pioneer designer of "tall buildings." Sullivan saw in Chicago a "magnificent and wild" city displaying "a crude extravaganza, an intoxicating rawness, a sense of big things to be done." The city's business leaders impressed him as the "most savagely ambitious dreamers and would-be doers in the world," yet they "had vision. What they saw was real." Sullivan decided that he now had found a permanent home. Chicago, he exclaimed, was *"the Place for me!"*[26]

During his leisure hours Sullivan pillaged books about design, science, and intellectual history. "In Darwin," he later recalled, he found "much food." The "Theory of Evolution seemed stupendous," largely because it confirmed his assumption that a new architectural style must evolve to meet changing social needs. He also relished the emphasis of modern science on "exact observation" and inductive reasoning, as well as "its honest search for stability in truth." Yet Sullivan insisted that "truth" ultimately resided in the spiritual "Idea" underlying physical realities. His developing "realism" entailed a decidedly transcendental element.[27]

In 1881 Sullivan formed a partnership with Dankmar Adler, the German-born structural engineer who had become a prominent Chicago architect, noted for his theater designs and the amount of natural light he afforded the interiors of his commercial buildings. During their fourteen-year collaboration, which ended abruptly in 1895 at Adler's request, they produced almost 200 structures and in the process gave the tall office building its logical and definitive form. Adler, twelve years older than Sullivan, possessed a problem-solving mind and an array of specialized skills, most notably acoustics. He was also a good diplomat, adept at dealing with clients, stroking their egos and assuaging their concerns. Sullivan, on the other hand, was a tempestuous dreamer, a self-professed poet *"who uses not words but building materials as a medium of expression."* Propelled by this desire to moor building design to the concrete world of human needs, he displayed an artist's demonic energy and disdain for compromise.[28]

Such qualities explain Sullivan's instant affection for Walt Whitman. In February 1887, he wrote Whitman an adulatory letter, confessing that while reading *Leaves of Grass*, "You then and there entered my soul, and will never depart." Sullivan applauded Whitman's efforts to "blend the soul harmoniously with materials," and he professed his own affinities with the "good, grey poet," quoting Whitman for effect. "I, too," have been "reaching for the basis of a virile and indigenous art." Sullivan, like Whitman and Eakins, rebelled against the constricting prudery of the Victorian era, arguing that American writers and artists had been forced to accept "a tacit fiction as to the passions" and to express only "the well-behaved and docile emotions." He, too, wanted less refinement and more strength in American culture. The true architect, he believed, "must be no coward, no denier, no bookworm, no dilettante."[29]

Yet Sullivan lacked Whitman's joyous, affable outlook; he suffered the torments common to isolated geniuses. A perfectionist whose exceptional talents bred unintentional arrogance, he argued frequently with clients, employees, friends, and family members. Frank Lloyd Wright, who worked for Adler & Sullivan as a draftsman, once described "the Master" as he appeared in the office: "About 10:30 the door opened. Mr. Sullivan walked slowly in with a haughty air, handkerchief to his nose. Paid no attention to anyone. No good morning? No. No words of greeting as he went from desk to desk. . . . The Master's very walk at this time bore dangerous resemblance to a strut." Wright, who would develop quite a strut of his own, recognized that Sullivan, the temperamental transcendental realist, was a sheaf of contradictions. The immense variety of his gifts and talents formed a temperament not unlike his famous skyscrapers: simple in its large outline, but infinitely complex in detail.[30]

Sullivan's first major commission was the Chicago Auditorium (1886–89), a complicated, multi-purpose facility that encompassed half a block and became the city's most monumental edifice. The clients called for a 4200-seat concert hall, surrounded by a 500-seat recital hall, ten stories of offices and stores, and a 400-room hotel. It constituted an imposing tribute to American commerce and high culture, as well as to Adler and Sullivan's architectural ingenuity.

Sullivan's initial design for the auditorium, however, was still-born. In a misguided effort to camouflage the structure's mammoth size, he broke up the wall surfaces with arched recesses, protruding oriels, and a pitched roof tower complete with chimneys, dormers, and Queen Anne turrets. Yet after visiting Richardson's new Marshall Field wholesale store, just a few blocks west of the auditorium site, he discovered the beauty of sheer mass. Compared with the Old World decadence and daintiness of most commercial buildings, the Field store, according to Sullivan, resembled "a man that lives and breathes, that has red blood; a real man, a manly man; a virile force—broad, vigorous and with a whelm of energy—an entire male." Eager to express the same sturdy masculinity in the auditorium façade, Sullivan stripped his design of many of its original romantic excesses. "The architecture we *seek*," he later wrote in prose as bombastic as it was typical, "shall be as a man, active, alert, supple, strong, sane."[31]

The huge auditorium building, now the home of Roosevelt University, is a bold cube embodying a simple neo-Romanesque theme modified by Sullivan's penchant for the picturesque. He modeled the stout piers, rusticated stone, narrow windows, and enormous lintels after the Field store, but where Richardson let his building's square lines and essentially flat walls express its spirit, Sullivan added to the auditorium a dramatic seventeen-story tower. Three giant arches integrated at the base of the tower welcomed patrons into the main vestibule and theater lobby. As a testament to his exceptional decorative talent and his desire to yoke "feminine" beauty to "masculine" strength, Sullivan lavished the interior with lush ornamentation.

The auditorium represented a truly breathtaking achievement, but Sullivan's later office buildings have most influenced American architectural development. Throughout the 1880s, he was "feeling his way toward a basic process, a grammar of his own," determined to strip his art of all dependence upon previous styles. "Our real, live, American problems," he declared in language echoing Root, "concern neither the Classic nor the Gothic, they concern *us* here and now." Sullivan contemptuously dismissed designers who plagiarized historical styles, asserting that "*reality* is of, in, by and for the present, and the present only."[32]

Sullivan expressed his own representation of "reality" in several office towers that reshaped the mold for commercial buildings. His formula usually called for a tripartite combination of two-story base for an entry lobby and retail stores, followed by a series of uniform office floors tenoned by slender piers, culminating in an attic housing the mechanical utilities and capped by a frieze with cornice. Although appearing cubic from the front, most of his commercial

Louis H. Sullivan. Auditorium Building, Chicago, Illinois, 1897. *(Chicago Historical Society.)*

towers actually had a U-shaped footprint that provided all the offices with access to sunlight. Sullivan understood that office workers needed a maximum of light to perform their routine tasks: reading, writing, typing, and filing. To allow even more natural light into interior offices, Sullivan narrowed piers and widened window bays. By recognizing the need for well-lit offices, he demonstrated that architectural design could grow out of a building's planned uses and structural requirements. "This was a kind of hard realism, of architectural common-sense," historian Hugh Morrison noted, "that was rare indeed in the decade of the eighties."[33]

Yet Sullivan the "hard realist" often fell into conflict with Sullivan the lyrical poet. The nation of steel, he once contended, needed "men of force and intellect and character" capable of expressing "the realities of the land and the time." His concept of "manly" architecture, however, included liberal application of "feminine" wrought-iron and terracotta ornamentation. In fact, his passion for intricate decoration has embarrassed many admirers who otherwise trumpet his devotion to "functionalism." Some have even labeled him a hypocrite for violating his famous dictum that "form ever follows function."[34]

Sullivan, however, never advocated the stripped-bare façades characteristic of much twentieth-century urban architecture. Instead he believed that a "real" building synthesized masculine and feminine characteristics. The former he associated with rational, structural, geometric, and foundational qualities; the latter he identified with emotional, decorative, efflorescent, and aspiring traits. Fascinated by the "subjective possibilities of objective things," Sullivan demanded that a well-designed building yoke both "ideal thought and effective action." Most American architecture, he charged, remained both excessively derivative and predominantly feminine. Only *manliness*, he wrote, could provide "an answer for us." Sullivan thus strove to harness the masculine force of Chicago's business barons to the artistic expression of feminine beauty and emotion. A building's brute material imperatives instilled "power" in a structure, but such commercial needs must be "subtilized [sic], flushed with emotion, and guided by clear insight."[35]

In fashioning ruggedly elegant buildings, Sullivan employed ornament as an integral design element. Too many builders, he felt, simply "stuck on" decorative treatments taken from historical designs. By contrast, he derived his own intricate forms from nature, and his striking ornamental arabesques came to serve as his special signature. Garlands of flowers, nests of leaves, and other motifs drawn from nature helped endow a "sterile pile" of stone and steel with spiritual eloquence, lifting its function beyond the mundane and integrating natural elements with commercial activities. A skyscraper, he asserted, "must be every inch a proud and soaring thing, rising in sheer exultation from bottom to top it is a unit without a single dissenting line."[36]

Although not nearly as "soaring" as some of his later designs, the Wainwright Building (1890–91) illustrated many of Sullivan's emerging principles. Ellis Wainwright, a wealthy St. Louis brewer and art collector, commissioned the building and asked that Sullivan incorporate the latest technologies within it. By using a riveted steel frame to reduce and carry the weight of the building, Sullivan increased his design options. To accentuate the building's ten-story height, he supported its weight with corner piers that ran from street level to roof. The piers framed slender, smooth-surfaced, vertical columns faced with red brick. Banks of recessed windows gave the building texture, attracted the eye to the vertical piers and columns, and formed a tight rectilinear pattern.

Other aspects of the design reined in the façade's verticality and helped create a coherent composition. The rough-faced, Missouri-red granite stones of the first two floors, laid parallel to the street, called attention to the horizontal axis, as did the overhanging slab cornice. These elements compressed the "soaring" piers and columns and forced them to serve rather than outreach the rest of the building. "The result," a recent critic has concluded, "is a nearly perfect composition, in which no element could be removed without seriously damaging the whole." Sullivan took justifiable pride in the Wainwright's originality and

Adler & Sullivan, Wainwright Building, St. Louis, Missouri, 1890–1891. *(Hedrich-Blessing photograph courtesy Chicago Historical Society.)*

claimed that all of his commercial buildings were conceived in the same general spirit."[37]

The success of the Chicago Auditorium and the Wainwright Building catapulted Sullivan into the front ranks of American architects. But his prominence was short-lived. Sullivan saw his career begin a downward spiral after his partnership with Adler dissolved in 1895. Fewer and fewer clients sought him out, while critics noticed that his efforts to weave together realism and idealism, utility and decoration, masculinity and femininity, had splayed. Inconsistency and incoherence stymied Sullivan's ambitions, and his original synthetic vision became lost in the labyrinth of his increasingly intricate ornamentation. The same man who began his career determined to "produce a realistic architecture based on well defined utilitarian needs" claimed in 1887 that "I value spiritial results only." His most recent biographer argues that the "female in Sullivan had become the dominant force."[38]

Even more perplexing were Sullivan's mounting rhetorical appeals for a truly "democratic" commercial architecture. "With me," he told a friend, "architecture is not an art but a religion, and that religion but a part of democracy." Sullivan interpreted America's democratic ethos as a reflection of the nation's tradition of self-reliant individualism and creative enterprise. In his view a "democratic spirit" shunned the outdated "elitist" styles of the past and gloried in present needs and concerns. Unfortunately, he contended, American architecture "stands aloof" from the people, isolated in a "musty school" of archaic forms and principles. Sullivan wanted a gusty wind of innovation to liberate architectural thought from the rigid formulas of Europe's academic historicism and initiate a genuinely American school of architecture "of the people, for the people, and by the people."[39]

What does this mean? For a man who specialized in precise drawing, Sullivan wrote vague prose punctuated with lyrical excesses. He never adequately explained how tall commercial buildings could express the nation's democratic commitment to the realization of individual freedom. On one level he insisted that honesty of expression was the ethical basis of a democratic architecture. Well-designed and well-integrated ornament could spiritualize commercial towers by softening the hard, angular lines of the façade, relieving the intimidating scale of the interiors, and thereby elevating the very behavior of the owners and tenants. On another level he saw tall buildings as symbols of the rising aspirations of the common folk.

Yet how could an office building housing a few dozen executives, lawyers, accountants, shop owners, and insurance agents assisted by hundreds or thousands of clerical workers performing monotonous tasks serve as a monumental exhibition of America's "democratic" ideals? How could palatial lobbies adorned with opulent ornament express egalitarian values? To many of the wage laborers moored to desks in a cellular hive of offices, the skyscraper offered not poetic grandeur but daily banality and bureaucratic anomie. As Lewis Mumford pointed out, Sullivan's effort to transform the skyscraper into a towering symbol of democracy was utterly unrealistic. It gave the skyscraper a "spiritual function to perform: whereas, in actuality, height in skyscrapers meant either a desire for

centralized administration, a desire to increase ground rents, a desire for advertisement, or all three of these together—and none of these functions determines
a 'proud and soaring thing.'"[40]

Sullivan anticipated such skepticism when he claimed that he never viewed
an office building as "what we so stupidly call reality, but, on the contrary, [it]
is a most complex, a glowing and gloriously wrought metaphor, embodying . . .
the pure, clean and deep inspiration of the race flowing as a stream of living
water from its well-spring to the sea." Wright noted that such rhapsodic passages
illustrated "the Master's" propensity for "baying at the moon." An office building
may have been a primal cultural metaphor to an artist-poet such as Sullivan,
but to investors it usually symbolized immense wealth, plutocratic power, and
social elitism. A promotional pamphlet touting a new commercial skyscraper,
for instance, promised prospective tenants that "particular care will be taken in
renting this building, and all objectionable occupations and persons will be
rigidly excluded."[41]

The Chicago writer Henry Blake Fuller appreciated the realities of life inside
skyscrapers better than did Sullivan. He chose a fictional skyscraper, the Clifton,
for the setting of his novel, *The Cliff-Dwellers* (1893). To Fuller, the tall buildings dominating Chicago's business district were not "proud and soaring" things
but "modern monsters." Their height represented less a tribute to democratic
aspirations than an index of social power and capitalist exploitation. Most of
Fuller's "cliff-dwellers" are rapacious social climbers who work in a self-con-
tained world of commerce and conspicuous display. Atwater, the building's
architect, sees little art in his creation. He considers a skyscraper to be "one
mass of pipes, pulleys, wires, tubes, shafts, chutes, and what not, running
through an iron cage of from fourteen to twenty stories." The purpose of the
architect, he explained, is to embellish this utilitarian pile with "a lot of tile,
brick, and terra-cotta." Profits, not aesthetics, dominate his labors: "Over the
whole thing hovers incessantly the demon of Nine-per-cent" return on invest-
ment. Commercial architecture, a disillusioned Atwater concludes, has "no
place for the artist."[42]

Sullivan acknowledged these facts only late in life. In his autobiography, he
looked back with embarrassment upon his lyrical excesses and confessed how
grossly ignorant he had been of the social realities governing American life. "Of
politics," Sullivan admitted, "he knew nothing and suspected nothing, all
seemed fair on the surface. Of man's betrayal by man he knew nothing and
suspected nothing." He was, in short, an "absurdly, grotesquely credulous" man
who "saw the world upside down" for most of his career. By ignoring the darker
dimensions of corporate capitalism, by failing to appreciate the new organiza-
tional hierarchies residing in his tall buildings, and by enshrouding corporate
America in majestic towers of unparalleled opulence, Sullivan called into question his democratic sympathies.[43]

Late in his career, however, Sullivan fastened upon a more effective way to express his democratic idealism. Between 1906 and 1918 he designed eight rural banks in five midwestern states. Although most scholars have dismissed these banks as the last gasp of a failed genius, they in fact embody and resolve many of the dichotomies inherent in his artistic sensibility—masculine-feminine, functional-decorative, commercial-democratic, idealistic-realistic. These banks catered to farm families. They were intended, as one board of directors observed, to ensure that "the man who has a dollar to deposit is given the same polite attention as the man who has a hundred or a hundred thousand. . . ." Such an egalitarian ethos inspired Sullivan to develop what he called his "democratic plan." It abandoned the tradition of modeling banks after classical temples and eliminated the practice of compartmentalized interiors divided into a honey-comb of office cubicles. The exteriors of Sullivan's banks were "designed from within, outward, the prime governing considerations being utilitarian."[44]

By opening up a bank's interior and organizing it along parallel and sym-metrical axes, placing tellers' cages, administrative offices, and meeting rooms all within view and easy public access, Sullivan encouraged "a feeling of ease, confidence, and friendship between officers, employees, and customers." Abun-dant natural and electrical lighting added to the building's statement that "bank-ing is a function of society and not a secluded mystery apart from the people." In these rural banks Sullivan gave concrete expression to his amorphous dem-ocratic ideal. He had finally fastened upon "an indigenous American style," one that enabled him to "grasp and deify the commonplaces of our life." Having said "goodbye" to his youth and its "charming idealism," he appealed to his peers to "express the life of your own day and generation."[45]

Residential Realism—in a Romantic Vein

While Sullivan focused his attention on commercial buildings, his extraordinary protégé Frank Lloyd Wright concentrated on designing houses that would strengthen family values and serve practical needs. In fact, Wright spent much of his long career warring against the social fragmentation and artificial qualities he associated with urban life. Such concerns grew out of his own unstable family background. Wright's father, William Carey Wright, was an itinerant preacher whose passion for music was almost as great as his love of God. His wife, Anna Lloyd Jones, was the headstrong daughter of Welsh immigrant farmers and religious freethinkers who settled in Wisconsin. Inheriting three children from her husband's first marriage and thereafter giving birth to three of her own, she early on found herself overburdened with domestic duties. Her husband seemed oblivious to her plight and envious of the "extraordinary devotion" she bestowed upon her son, Frank. Over time she saw her love for her husband leached away.

By 1883 she was sleeping in a separate bedroom. Eighteen months later, in the summer of 1884, William Wright filed for divorce and left his family to its own deserts.[46]

Soon after his father's departure, Frank found a part-time job with an architect in Madison. In the mornings he took civil engineering courses at the University of Wisconsin. Feverishly ambitious, he left for Chicago early in 1887 and landed a job with J. L. Silsbee, an architect popular among the city's social elite. A year later Wright joined Adler & Sullivan and was immediately assigned to help Sullivan with the drawings for the Chicago Auditorium. What most impressed the young apprentice about "the Master" was his emphasis on "natural" ornament fully integrated into a building's design. Wright, too, viewed nature as the basis for a truly "organic" theory of design that would disavow all contrived and "sham" elements. Sullivan early on recognized that this untutored country boy had immense natural gifts. After only two years, he made the twenty-one-year-old Wright his principal assistant.[47]

Adler and Sullivan were so preoccupied with large commercial and civic projects that they showed little interest in the design of private residences. Occasionally, however, their Chicago business associates would ask for assistance in designing a house, and Sullivan usually put Wright in charge of such residential projects. His obvious talents and relatively low fees attracted other clients, and Wright soon developed a modest business on the side in what he called "bootlegged" houses. Sullivan, however, exploded when he learned of his assistant's "moonlighting," for Wright's contract prohibited such activity. The two men had a violent argument which culminated in Wright leaving Adler & Sullivan in the summer of 1893. After more than five years together, the two tempestuous designers would not speak to each other again until the end of the century.

Now on his own, Wright, a self-proclaimed and authentic genius, focused his efforts on residential work. There was much to be done. Too many houses, he decided, were intended primarily as showpieces for the conspicuous display of the owner's affluence. He resolved to help people see that genuine architecture "is a destroyer of vulgarity, sham, and pretense." The typical "General Grant Gothic" house, he charged, was a claustrophobic hive rather than a home, a dark, dingy, and "bedeviled box with a fussy lid." Crowded with gingerbread scrollwork and tasteless ornament, stuffed with tasteless furniture, furnishings, and gimcrackery, and chopped up into box-like cells, Victorian houses lacked integrity, proportion, and "any such sense of space as should belong to a free people." Most Chicago homes were "wickedly extravagant" structures that "*lied* about everything," and their vulgar affectations made him "sick of hypocrisy and hungry for reality."[48]

Perhaps Wright's greatest insight was his insistence that a structure's essential "reality" resided not in its four walls but in the space within, the space to be lived in. The architect must see and feel the ways in which material, structure, and light shape a house's inner and outer forms and spaces. A truly "organic" home would thus be designed from the inside out. All unnecessary encumbrances and barriers would be eliminated so as to allow space to "flow" from room to room, thereby opening up new possibilities for multi-purpose uses and creating an inviting sense of intimacy and harmony of parts. The Japanese exhibit at the World's Fair reinforced Wright's belief that the interior and exterior of a house should represent a harmonious whole which in turn should blend naturally into its environment. All superfluities would be banished, and function and beauty would be intertwined with spirit and style. Form would not so much *follow* function as it would *embody* function. In essence Wright wanted each house to be a living reality rather than a mere enclosure. An organic design, he explained, ensured that a building "faces and is reality and serves while it releases life, makes daily life better worth living and makes all the necessities happier because of useful living in it."[49]

Wright centered his famous "prairie house" plans on a large fireplace whose hearth faced the main living area. Around this solid, almost primeval, core he arranged the other rooms, broadly open to each other. By using new structural techniques, he kept interior partitions and doors to a minimum, enabling space to flow naturally. In his desire for "organic" unity, Wright believed that much of the furniture and lighting should be built-in rather than free-standing. Earth-tone colors and native materials would provide a sense of warmth and humanity while the liberal use of glass would ensure "that the sunlit space as a reality becomes the most useful servant of a higher order of the human spirit."[50]

The outside of a Wright "prairie house" reflected "what happened *inside*" and at the same time helped anchor the structure in the environment so as to provide a sense of stability and security for its residents. Where Sullivan accented an office building's verticality so as to make it a "proud and soaring thing," Wright emphasized the horizontal axis of his homes. He lowered the roof line and provided extensive overhanging eaves, wide bands of casement windows, and connecting terraces, all of which helped nestle the house into the landscape and provide residents with a sense of rootedness.[51]

Although most of Wright's clients were affluent businessmen, he was eager to make his "prairie house" designs available to the larger public. To do so, he published two "middle class" house plans in the *Ladies' Home Journal* in 1901 and a third in 1907. At that time the magazine had over a million subscribers, giving it the largest circulation of any in the world. Its editor, Edward Bok, was an ardent moral reformer who believed that Americans needed a simpler, more rational, and "more realistic" way of living.[52]

Like Wright, Bok found most Victorian homes to be "repellently ornate." People wasted their money "on useless turrets, filigree work or machine-made ornamentation." To remedy the situation, Bok published plans for plain, functional houses that could be built for $1500 to $5000. He offered readers blueprints for five dollars. Soon, thousands of "*Journal* houses" began going up across the country. Architect Stanford White observed that "Edward Bok has more completely influenced American domestic architecture for the better than any man of his generation." These "*Journal* houses" eliminated the Victorian parlor, made the kitchen more compact, substituted built-in cupboards for pantries, and discarded all "senseless ornament." Wright noted that the designs he submitted to *Journal* readers were intended to enhance the "real needs" of home dwellers. "Radical though it may be," he wrote, his work was "dedicated to a cause conservative." His primary motive was "a yearning for simplicity. A new sense of simplicity as organic."[53]

The Craftsman Ideal

Similar motives inspired what came to be known as the Arts and Crafts movement in the United States and England during the second half of the nineteenth century. This crusade on behalf of "real" rather than "sham" values sought not only to purify the taste of the bourgeoisie but also to improve the living and laboring conditions of working people by encouraging "simple" living and reviving the handicraft tradition. The English writer and designer William Morris was the movement's most important prophet. He vigorously promoted a simple art and architecture for all rather than for the advantaged few. Morris designed and made furniture, tapestry, wallpaper, books, and other things with an eye for their actual use and inherent beauty rather than for gaudy display.

The Arts and Crafts ideal captured the imagination of thousands of Americans at the turn of the century. As Bok reported in the *Ladies' Home Journal* in 1900, a "William Morris craze has been developing, and it is a fad that we cannot push with too much vigor." By 1904 there were twenty-five organized Arts and Crafts societies in the United States. A major element of their appeal was the use of handicrafts to bring frazzled members of the urban bourgeoisie back into contact with "real life." Rebelling against "feminized" home interiors, reformers complained that "the masculine interest in the home has declined, almost vanished."[54]

The leading organ for the Arts and Crafts movement in the United States was *The Craftsman*, a magazine founded by Gustav Stickley, a Wisconsin-born stonemason turned woodworker. During a visit to the 1876 Centennial exhibition at Philadelphia, Stickley fell in love with the functional austerity of Shaker furniture. Soon thereafter, he and his two brothers established a furniture shop in New York in which they produced what came to be known as "mission" furniture — massive, plain oak pieces whose sturdy rusticity, homey honesty, and handcrafted quality proved to be quite popular.[55]

In 1901 Stickley began publishing *The Craftsman* to broaden the appeal of the Arts and Crafts aesthetic beyond professional craftsmen. Over the next fifteen years the magazine offered articles on art, architecture, poetry, drama, politics, economics, city planning, and education. Eager to make progressive architecture accessible to the masses, Stickley began providing readers with blueprints for what he termed "craftsman homes." His house plans offered "realistic" residential designs "suitable to the lives of the people," designs having "the simplest form" and capable of being constructed with native materials. Like Wright's "prairie houses," Stickley's craftsman homes were "well-built, democratic, well-planned houses . . . able to satisfy the needs of all the members of the family." Although not as sleek, refined, or proportioned as the "prairie houses," the craftsman homes featured beamed ceilings, simple built-in furniture, native wood and stone materials, expansive staircases, and a central fireplace. Boasting "not of elaboration, but of elimination," they were "straightforward and honest" attempts "to cut away ornament." Time and again Stickley emphasized a functionalist imperative: a home should be so planned that "it meets the actual requirements of those who live in it, and so furnished that the work of keeping it in order is reduced to a minimum."[56]

Homes designed according to the principles of the Arts and Crafts movement were especially popular in California, where a dynamic group of like-minded architects—Joseph Worcester, Charles Keeler, Irving J. Gill, Charles and Henry Greene, Julia Morgan, and others—sought to build houses that Keeler said would be "rooted in the soil of the real, the practical, the utilitarian." People on the West Coast, he added, "are growing weary of shams and are longing for reality." What Keeler meant by "reality" was the use of native building materials and straightforward design principles intended to enhance "simple living." Gill, who had worked with Wright in the Chicago office of Adler & Sullivan before setting up his own firm in San Diego, said that "progressive" designers must no longer clutter houses with "foolish ornaments and useless lines." Modern life required modern principles of order and efficiency. The architects of the West "must have the courage to fling aside every device that distracts the eye from structural beauty, must break through convention and get down to fundamental truths."[57]

These California architects cultivated a severe simplicity as much for its beauty as for its utility. Charles Greene and his younger brother Henry, for example, drew upon Japanese models in fashioning striking new homes of redwood and stone. Born in Cincinnati in 1868 and 1870, they first learned about the Arts and Crafts movement while students at Washington University in St. Louis, where they read John Ruskin and William Morris. They then studied architecture at the Massachusetts Institute of Technology. In the early 1890s the brothers headed west to Pasadena, where they formed an immensely suc-

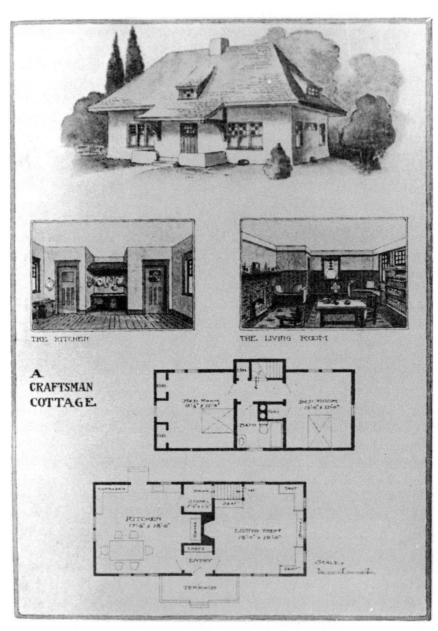

THE KITCHEN

THE LIVING ROOM

A
CRAFTSMAN
COTTAGE.

Craftsman Cottage. *The Craftsman*, April, 1905

cessful partnership. The Greene brothers wanted to produce homes that "eliminated everything unnecessary, [making] the whole as direct and simple as possible, but always with the beautiful in mind as the final goal." Like Wright,
they featured broad roof eaves, heavily windowed walls, and flowing living space.
The Greene brothers also insisted that ornament must serve a practical purpose
rather than merely aesthetic decoration.

Connoisseurs of detail, the Greenes viewed a house as a unified whole. This
meant designing the interior furnishings, picture frames, curtain rods, and even
light-switch plates. When asked to explain his "realistic" philosophy of design,
Charles Greene responded that he was a self-consciously American architect
who wanted to know "the American people of today and the things of today."
He could not begin to plan a house until he understood the personalities and
needs of the residents. Once "I find what is truly useful . . . I try to make it
beautiful," but "this cannot be done by copying old works, no matter how
beautiful they may seem to us now. When confronted with actual facts, I have
not found the man or woman who would choose to live in the architectural
junk of ages gone. The Romans made Rome and the Americans—well!—they
are making America."[58]

Of course, the Greenes, Stickley, Wright, and the other architects associated
with the Arts and Crafts movement coupled their concern for "actual facts" and
"real" rather than "sham" values with a romantic idealism resonant of Whitman.
Their emphasis on simplicity and utility and their reverence for nature and the
natural reflected a strident anti-urbanism as well as an abiding faith in the ability
of well-designed homes to foster family cohesion and social uplift. "Most of us,"
wrote Keeler, "are shaped and sized by the walls which we build about us.
When we enter a room and see tawdry furniture, sham ornaments and vulgar
daubs of pictures displayed, do we not feel convinced that the occupants of the
home have a tawdry and vulgar streak in their natures?" Over time, he added,
the residents of a "simple home" would come to realize that "the ideal is the
real, that beauty is truth, and that love is the inspiration of beauty."[59]

Although often dismissed as a fashionable fad catering to the privileged few,
the Arts and Crafts movement did, in fact, exercise a pronounced influence
over the aesthetic taste of the larger society. Early in the new century, at the
same time that Bok and Stickley were providing their subscribers with plans for
"simple" houses, mail-order giants such as Sears and Montgomery Ward began
offering craftsman home kits and "mission" furniture in their catalogs. By 1905
Sears could boast that "enough houses have been built according to our plans
and with our materials to shelter a city of 25,000 people." The landscape of
residential design had indeed changed over the previous fifteen years. In 1903
a writer in *The Brickbuilder* noted that the most dynamic tendency in residential
design was "an absence of ornament and the use of simple, almost cheap mate-

rials." The ornately decorated and gewgaw-filled Victorian home was no longer fashionable. Herbert Croly, the editor of the *Architectural Record*, announced that the decade of the 1890s represented a turning point in American residential design during which the "freakish," "out of place," and *"meaningless* eccentricities" had given way to a more sane and sensible emphasis on tasteful utility and elegant simplicity.[60]

Part IV

Extraordinary Realities

In the waning years of the nineteenth century, American society displayed a festival of contradictions. On the one hand, people talked of the "Gay Nineties," an era during which many affluent Americans displayed a self-indulgent gaiety and insouciance. On the other hand, it was a time of recurring economic calamities and unprecedented class conflicts that challenged the common assumption that the United States enjoyed a fundamentally healthy and self-correcting social order. Mounting labor violence and rising racial and ethnic tensions disclosed ugly realities and terrible injustices at the core of American society. Such revelations convinced many writers and artists that they could no longer limit their efforts at cultural representation to the "smiling aspects of life," and they began to give voice to unheard and unwanted truths.

9

Realism and the Social Question

Signs of social strain pervaded the end of the nineteenth century. Popular theories of racial superiority and fears of foreign radicalism and social degeneration gave rise to a virulent strain of Anglo-Saxon nativism. In the South a vicious new tide of racism spilled over the region as states passed Jim Crow laws and white mobs lynched blacks in record numbers. Meanwhile, in New England and along the West coast, waves of "new" immigrants from eastern Europe and Asia broke against the cliffs of nativist anxiety. The flood of new "aliens," Senator Henry Cabot Lodge predicted, threatened "a great and perilous change in the very fabric of our race."[1]

Economic tensions were even more prevalent than racial prejudices. Widespread poverty flourished alongside garish affluence, and the disparities between wealth and want helped ignite widespread social explosions. A chastened William Dean Howells confessed in 1893 that "the old American maxim that it will all come out right in the end, has less and less acceptance." Fault lines appeared throughout the social order, and they unleashed tremors that exerted what one writer called "a seismic shock, a cyclonic violence" upon the body politic.[2]

In the hinterlands, farmers grew increasingly resentful and rebellious as crop prices plummeted and foreclosures soared. Millions of rural folk, especially young people, gave up on husbandry altogether and formed a ceaseless migratory stream from the farm belt to the cities. Those who remained behind struggled to eke out a living. Farm life in the South and Great Plains bore little resemblance to the yeoman ideal so often depicted in popular literature and political rhetoric. Sodbusting in the face of droughts, blizzards, and locusts was anything but idyllic. Prolonged hard times led many small farmers to see themselves as victims of a plutocratic conspiracy. Struggling agrarians lashed out against mercurial markets, high interest rates, and discriminatory freight rates and storage fees. Millions of hard-pressed farmers, both blacks and whites, plunged headlong into national politics through the vehicle of the People's party. Their vision was apocalyptic: "We meet in the midst of a nation brought to the verge of ruin," read the 1892 Populist party platform. "A vast conspiracy against mankind has been organized," and if "not met and overthrown at once it forebodes terrible convulsions, the destruction of civilization, or the establishment of an absolute despotism."[3]

On the surface, factory workers fared better than farmers, as wages in man-ufacturing rose about 50 percent between 1860 and 1900. Such aggregate fig-ures, however, ignored the great disparity between skilled and unskilled pay scales and the sporadic nature of employment. During the recessions and depres-sions endemic to the industrial economy, layoffs and wage cuts created untold misery and imposed severe strains on municipal services and charitable resources. Working conditions also left much to be desired. The average work day was over 10 hours long, and most people labored six days a week in dan-gerous factories, mills, sweatshops, and mines. Each year thousands of workers lost their lives or limbs in on-the-job accidents. Perhaps most distressing was that almost two million children under the age of fourteen worked outside the home to supplement family income and fend off destitution.[4]

Living conditions for the poor were even more deadly than working condi-tions. The millions of immigrants and rural folk streaming into the cities crowded into filthy, disease-ridden, and crime-infested ghettos. Such enclaves, the Reverend Josiah Strong warned in his best-seller, *Our Country: Its Possible Future and Present Crisis* (1885), were filled with "social dynamite." Strong's hysterical description contained a core of truth: the gulf between rich and poor seemed unbridgeable. In 1890 one-eighth of the population owned seven-eighths of the national wealth; the numbers of poor, many of them immigrants who spoke little English, reached staggering levels.[5]

Perhaps at no time before or since have class tensions—both social and cultural—been as bitter as they were during the last quarter of the nineteenth century. The Massachusetts reformer Lydia Maria Child noticed with alarm the "strong *demarcation of classes* in this country." The "genteel classes," she reported, "do not intermarry with the middle classes; the middle classes do not intermarry with the laboring class; nothing is *said* about it, but there is a sys-tematic avoidance of it. Moreover, they don't mix socially." Resentment among the working poor exploded into open conflict during the Gilded Age, as strikes grew more numerous and violent. In 1877 railroad workers from West Virginia to California shut down the nation's train system and destroyed cars, engines, and tracks. State militia and federal troops fought pitched battles with the strik-ers, foreshadowing even bloodier confrontations to come.[6]

In 1886 a massive strike against the McCormick Harvester Company in

As the Haymarket bomb explodes in their ranks, Chicago policemen sent to break up the anarchist rally start firing wildly into the crowd. *(Library of Congress.)*

Chicago helped precipitate the Haymarket "riot." The event began peacefully enough when leaders of a minuscule anarchist movement in Chicago scheduled an open meeting at Haymarket Square to protest the killing of a striking worker. After listening under a light drizzle to turgid speeches advocating socialism and anarchism, the crowd was beginning to disperse when a phalanx of policemen arrived to hurry things along. Someone threw a dynamite bomb at the police, killing one immediately, and seven others died later. The police then fired into the crowd, killing several people, and some of the demonstrators fired back, sending shock waves across the country. In a highly prejudicial trial, seven anarchist leaders, most of them German immigrants, were sentenced to death despite no tangible evidence linking them to the bomb-thrower, whose identity was never determined. Of these, two were reprieved and later pardoned, one committed suicide in prison, but four were hanged. In *The Titan* Theodore Dreiser wrote that the Haymarket incident "had brought to the fore, once and for all, as by a flash of lightning, the whole problem of mass against class."[7]

The Haymarket riot exploded the comforting assumption that a natural harmony of interests prevailed among America's social classes. Thereafter, tensions between workers and management, farmers and creditors, reached a fever pitch.

In 1886 alone there were 1400 strikes across the country involving over 700,000 laborers. "The Eighties dripped with blood," journalist Ida Tarbell recalled in her autobiography. The sharp financial panic of 1893, which resulted in the collapse of 8000 businesses and 360 banks, spawned a new round of bloody clashes between workers and management. During the ensuing depression, the deepest and most protracted in history to that point, unemployment reached 20 percent. Every city presented scenes of unprecedented poverty and unrest. The same Chicago that constructed the beautiful "White City" at the World's Fair harbored over 160,000 slum dwellers.[8]

The Social Question

The mounting social turbulence evoked a variety of responses ranging from apathy to indignation. Widespread labor unrest and the vulgar excesses of the new rich convinced many Boston Brahmins that the pre-urban society of their youth had slipped from their grasp and was on the verge of distintegration. Believing that the poor were victims of their own moral and mental failings, patricians had little sympathy for either the agrarian or labor movements. Nor did they support the reform legislation put forward by "mistaken humanitarians." For the most part, the elderly members of the genteel elite lapsed into bitter disillusionment and nostalgic chagrin. Barrett Wendell, the crusty Harvard literature professor who labeled himself the "last of the tories," lamented that "America has swept from our grasp. The future is beyond us."[9]

While social conservatives sought private sanctuary from the horrors and banalities of modern life, others imagined brighter futures: interest in utopian literature soared during the last two decades of the nineteenth century. Then, as now, many found in such fictional havens an escape from their daily routine and chronic anxieties; for others, however, utopianism provided an enticing new vehicle for social satire intended to reform present ills. "Utopias are the noblest work of man," William James assured Howells, who wrote two utopian novels himself. "Out of them all some effect comes, for if we *want* enough, we get it." Although usually considered pure fantasies, most of the era's utopian novels reflected the impact of literary realism and the reform impulse. In their efforts to use an ideal future to shed light on the evils and excesses of the present, utopian authors, most of whom were practicing journalists, included meticulously detailed descriptions of current social conditions.[10]

By far the most popular of these utopian novels was Edward Bellamy's *Looking Backward* (1888). Within a year of publication it sold 200,000 copies. Its wide circulation, explained one economist, "is due to the fact that the earnest feeling with which it is written coincides with a very deep and widespread discontent with existing social conditions." Bellamy combined a vivid description of a technological utopia with an equally graphic account of contemporary urban squalor.[11]

Looking Backward opens on a night in 1887 as a wealthy young Bostonian named Julian West, nervous about the effect of local labor unrest on his upcoming wedding, solicits the aid of a mesmerist in his battle with insomnia. The hypnotist, however, does his job too well; West's delicious slumber in a basement sleeping chamber lasts 113 years. Awakening in the year 2000, he finds himself in a technocratic paradise, a new "fraternal civilization" with no poverty or even deprivation. Dr. Leete, his genial guide, reports that corporate property has been abolished and all industries nationalized as the result of a bloodless revolution.

In this new "nationalist" state, all men and women between the ages of twenty-one and forty-five serve in a highly disciplined industrial army. With production tied directly to consumption, and most tasks handled collectively rather than individually, everyone receives an adequate, equal income and enjoys the latest labor-saving appliances and technologies. People retire at age forty-five and devote themselves fully to "the higher exercise of our faculties, the intellectual and spiritual enjoyments which alone mean life." An incredulous West tells Dr. Leete that "Human nature . . . must have changed very much" to make such a utopia possible. "Not at all," Leete replies, ". . . but the conditions of human life have changed, and with them the motives of human action." Toward the end of *Looking Backward*, West returns to the Boston of 1887 and looks upon his native city with enlightened eyes: "The squalor and malodorousness of the town struck me," he remarks, "from the moment I stood upon the street, as facts I had never before observed."[12]

Bellamy wanted readers to see the same odious facts. Their doing so, he hoped, would lead them to endorse his imagined form of democratic socialism, which he labeled "Nationalism." His plan worked far better than he expected. Only a few months after *Looking Backward* appeared, hundreds of Nationalist Clubs sprouted across the country, determined to implement Bellamy's social ideal in the here and now (Howells was a charter member of the Boston club). "Instead of a mere fairy tale of social perfection," Bellamy observed, *Looking Backward* "became the vehicle of a definite scheme of industrial reorganization." He directed much of its anticipatory message to the working classes, those who "had quite suddenly and very generally become infected with a profound discontent with their condition" and had decided "that it could be greatly bettered if they only knew how to go about it." Bellamy offered them a blueprint because he, like "true men and humane women of every degree," shared "a mood of exasperation, verging on absolute revolt, against social conditions that reduce life to a brutal struggle for existence."[13]

Agents of Reform

The classless society of abundance and leisure that Bellamy envisioned struck many veteran reformers as utterly unrealistic and even dangerous. Walter Harte, a Boston social activist, charged that *Looking Backward* and the spate of look-alike utopian novels it inspired bred more apathy than indignation. "These pictures of life a century or so hence have a tendency to make people think our social ills will settle themselves out of court; but these moral questions are *real* questions, facts and not fictions, and they will not, and cannot, be settled by apathetic folding of hands and philosophic signs of comprehensive sympathy with all misery, and acquiescence in all wrong." Harte stressed that American reformers did not have the luxury of imagining millennial utopias; they had to confront the "concrete material fact" of "gross outrages" in their own social system in their own time.[14]

As Harte suggested, the "real spirit of the reform forces" focused on ameliorating the immediate problems of the present. Some reformers proposed legislative remedies for social injustices; others stressed private philanthropy or local charity. A few militants promoted socialism or anarchism, while a handful of colorful individuals put forward various original schemes and panaceas.

Whatever the method or approach, social reformers mobilized in great numbers at the end of the century, and their activities gave a new urgency and energy to American life. They spoke with a vigor, even a violence, that sent an electrical charge through society. Sociologist Albion Small defined the new "social movement" as "an effort for concrete, specific, definable goods" such as higher wages, shorter working hours, better working conditions and improved housing opportunities, progressive taxation, and greater government regulation of industry. "The emphasis to-day," Small concluded, "is on change of conditions rather than upon adjustment to conditions."[15]

The turn-of-the-century "progressive" reformers promoted not only a new policy of governmental intervention but also a changed notion of the nature and function of compassion. Many progressives rooted their fervor in the evangelical reform tradition, yet their methods were utterly modern and even "scientific." What Gertrude Himmelfarb has said about the emergence of an "unsentimental" compassion among British reformers applies equally well to their American counterparts:

> In its sentimental mode, compassion is an exercise in moral indignation, in feeling good rather than doing good; this mode recognizes no principle of proportion, because feeling, unlike reason, knows no proportion, no limits, no respect for the constraints of policy or prudence. In its unsentimental mode, compassion seeks above all to *do* good, and this requires a stern sense of proportion, of reason and self-control.

By merging moral fervor with both hard-headed goals and methods and a vigorous commitment to uncovering and facing the "facts" of social ills, "unsentimental" humanitarians transformed the nature of social welfare activity.[16]

The reformers involved in the settlement house movement typified the efforts
of middle-class, college-educated activists to make direct contact with "real" life
problems and in the process strip compassion of condescension. Ellen Starr,
who founded the Hull House settlement in Chicago with Jane Addams, wrote
a friend that "people are coming to the conclusion that if anything is to be done
towards tearing down these walls . . . between classes . . . it must be done by
actual contact and done voluntarily from the top."[17]

At Hull House, an enclave of beauty and concern set within the harsh
ugliness of the Chicago ghetto, Starr and Addams sought to create a sense of
community among all classes and ethnic groups. They assumed that bringing
the immigrant poor together with middle-class staff members would benefit
everyone involved. Immigrants would develop a sense of community spirit and
learn how to get along in urban America by taking classes in English, sewing,
cooking, and child care. Settlement house work would also improve the outlook
of the middle-class volunteers. Staffers told young women who applied to join
Hull House that interacting with the "other half" would "beget a broader philanthropy and a tenderer sympathy, and leave less time and inclination for
introspection, for selfish ambition, or for real or fancied invalidism."[18]

"Invalidism" referred to the rash of neurasthenia among women college
graduates who were no longer content with the corseted comforts of Victorian
domesticity yet found few opportunities to express their learning or ideals. Over-
civilized and underutilized, cut off from productive labor and social intercourse,
they often lapsed into a paralyzing state of melancholy and lassitude. Addams,
for example, guiltily recalled that she had been "smothered and sickened with
advantages" as the child of a wealthy Illinois businessman. Only months after
graduating from Rockford Female Seminary in 1881, she suffered a prolonged
bout of depression and aimless inactivity which she attributed to her difficulty
in "making real connection with life."[19]

For several years thereafter Addams continued to suffer from "nervous
depression and a sense of maladjustment." What finally liberated her from the
slough of despondency was a visit to Toynbee Hall, the pioneering settlement
house in England. The example of young people doing mission work in the
city slums exhilarated Addams. She returned to the United States with "high
expectations" that vigorous contact with the "reality of life" would provide her
with a fresh sense of purpose. Not long after Hull House was founded in 1889,
she touted to others the benefits of such vivifying experience. By ministering
directly to the working poor, people of means and leisure could make contact
with the "fundamental facts" of life and "give tangible expression to the democratic ideal." While acknowledging "the human pleasure and interest that life

among real people of many different kinds involves," she stressed that the success of the settlement house depended upon the volunteers controlling their emotions. Any trace of maudlin sentimentalism in dealing with the poor would only undermine their efforts. They must instead develop a "scientific patience in the accumulation of facts and the steady holding of their sympathies." Statistics demonstrated the appeal of such a program: in 1891 there were six settlement houses in the nation; a decade later there were well over 100.[20]

How the Other Half Lived

Progressive social reformers such as Addams knew that the first step in recruiting an army of activists involved exposing the affluent to the realities of social injustice. As the rural and immigrant poor flooded into the central cities, many among the well-to-do fled to the new "streetcar" suburbs on the urban periphery, retreating to what Theodore Dreiser called "little islands of propriety" protected from the turmoil and pain of inner city poverty. To penetrate the bourgeoisie's cocoon of ignorance and disrupt their tepid indifference, scores of investigative reporters fanned out across the cityscape, determined to reveal hidden social ills. In exposing the public to disturbing facts about their new urban-industrial society, these "muckrakers," as Theodore Roosevelt dubbed them, tended to be stronger on diagnosis than remedy, thereby reflecting their faith in the power of democratic social change. Let the people know the truth, disclose corruption and injustice, and bring government closer to the people, they assumed, and the correction of abuses and amelioration of suffering would follow automatically. The cure for the ills of democracy, in other words, was more democracy.

Among the first and most influential of these "muckrakers" was Jacob Riis, a New York reporter and earnest reformer who fashioned a powerful new mode of communication—photojournalism—in his efforts to excite public concern for the hidden poor. His book, How the Other Half Lives (1890), remains a classic example of the genre, and his career epitomizes the fact-worshiping strand of reformist realism. "Ours is an age of facts," Riis once remarked. "It wants facts, not theories, and facts I have endeavored to put down on these pages."[21]

Born in Denmark in 1849, Riis (pronounced Reese) migrated to America at the age of twenty-one. The son of a schoolmaster, he brought with him a solid education, a skilled craft (carpentry), a passable fluency in English, and the innocent hope that a free country would enable him to live out his dream of the good life. After working as a steelworker and brickmaker, however, an unemployed Riis found himself conscripted into the army of tramps wandering the streets of Manhattan, sleeping in police station lodging houses and rummaging for scraps outside restaurants. Filthy and unkempt, he "was too shabby to get work, even if there had been any to get." Riis finally made his way to Philadelphia, where the Danish consul helped him find work as a lumberjack, hunter, and trapper in upstate New York. But his "desire to roam" kept him moving, and in 1873 Riis took a job as a reporter for the New York News Association. Four years later he became a police reporter for the *New York Tribune*.[22]

Riis developed a passion for human interest stories drawn from the back alleys and tenements of Manhattan's Lower East Side. Some 334,000 people lived within a single square mile of the Tenth Ward, making it the most densely populated neighborhood in the world. In the Bowery district, Riis saw an all-too-familiar scene: throngs of poor Irish, Italians, Bohemians, African-Americans, and Jews crammed into unsanitary hovels that bred disease, ignorance, and crime. His own encounters with the pinch of poverty and the sting of discrimination seasoned his observations and ignited his fervor for reform. Convinced that a bad environment was the primary cause of poverty, Riis began writing graphic exposés of the miserable conditions in the slums. "The power of fact," he decided "is the mightiest lever of this or of any day."[23]

Yet merely printing the "facts" about urban squalor failed to summon much public outcry. "I wrote, but it seemed to make no impression," Riis recalled in his autobiography. To heighten the impact of his written testimony, he began to use photographs of urban blight and slum dwellers. The camera, he realized, had a special evidentiary power to lay bare hidden truths, and few facts were as visually compelling as poverty. So in 1887 Riis hired amateur photographers to accompany him on late-night excursions through the Bowery, and their exhausting efforts provided the first graphic images of the seamy underside of New York life. A reporter in the *New York Sun* said that Riis's illustrated articles showed, "as no mere description could, the misery and vice that he had noticed in his ten years of experience."[24]

Eager to go beyond mere reportage and promote concrete reforms, Riis began giving slide lectures to church and civic groups throughout the city and across the nation. "Neighbors," he would say at the end of his presentation, go out and "find your neighbors." Many found this a compelling challenge. A midwestern journalist reported that to those of "us who are unfamiliar with life in a large city," Riis's slide presentation "was a revelation." His photographs "were certainly more realistic than any words could be." Realizing that his middle-class audiences did not want their faces rubbed in the grime and stench of the tenements, Riis spared them from "the vulgar sounds and odious scents and repulsive exhibitions attendant upon such a personal examination." Apparently he did not worry that the price of such a distancing vision and spectatorial conscience is often superficiality.[25]

On the lecture circuit Riis met people equally committed to a realism of exposure and reform. His graphic account of bitter poverty so impressed the editor of *Scribner's Monthly* that he invited Riis to write an illustrated article

Jacob A. Riis, "Five Cents a Spot," lodgers in a Bayard Street tenement, ca. 1889. *(Jacob A. Riis Collection, Museum of the City of New York.)*

for the magazine. "How the Other Half Lives" appeared in the Christmas 1889 issue accompanied by nineteen photographs transformed into line drawings. The article described sordid tenements "nurtured in the greed and avarice" of rapacious slumlords, disease-ridden ghettos where people were shoehorned into windowless warrens. In one thirteen-square-feet Bayard Street tenement room, Riis found twelve men and women lodgers. Another two-room apartment on Essex Street hosted a family of fourteen plus six boarders.[26]

Riis's graphic revelations excited so much interest that he agreed to expand his article into a book. Published in 1890, *How the Other Half Lives: Studies Among the Tenements of New York* created an immediate sensation and went through eleven editions in the next five years. "No book of the year," proclaimed the *Dial*, "has aroused a deeper interest or wider discussion than Mr. Riis's earnest study of the poor and outcast." In spirited prose, Riis detailed the crowded life of the "unventilated and fever-breeding" tenements, the ravages of disease and alcohol, the tragedy of homeless street urchins, the violence of "gangs," the sound of tubercular coughs, the oppressive atmosphere of sweatshops, and the "queer, conglomerate mass of heterogeneous elements" forming New York's diverse ethnic tapestry.[27]

Forty-three illustrations and fifteen halftone reproductions authenticated Riis's narrative. The stubbornly enigmatic stares of the destitute led many readers to feel morally implicated in the pathetic scenes. A typical example of Riis's photographic technique is "Lodgers in a Crowded Bayard Street Tenement: Five Cents for a Spot." It exposes a catacomb-like room clogged with immigrant "lodgers" and their modest possessions. Awakened in their wretched roost, the bleary-eyed tenants stare at the camera in the midst of pots, pans, boots, caps, clothes, trunks, duffel bags, and firewood. The precise details of such a mundane scene heightened its emphatic immediacy. But its real strength derived from the stark contrast with the norms of middle-class domesticity shared by most readers of *How the Other Half Lives*. The room's filth and claustrophobia spoke eloquently to the consciences of well-to-do suburbanites preoccupied with clean homes and wholesome families.

Riis, however, refused to let his pictures speak for themselves. "What then are the bald facts with which we have to deal with in New York?" he asked. "That we have a tremendous, ever swelling crowd of wage-earners which it is our business to house decently. . . . This is the fact from which we cannot get away, however we deplore it." Riis offered neither easy solutions nor the salve of consolation to bruised consciences. Unlike Bellamy, he envisioned no panacea for widespread poverty, nor did he see any real alternative to the competitive capitalist system. "We know now," he wrote, "that there is no way out; that the 'system' that was the awful offspring of public neglect and private greed has come to stay, a storm center forever of our civilization. Nothing is left but to make the best of a bad bargain." His own plan of action called for rigorous enforcement of building codes and the construction of new, sanitary, well-ventilated, and affordable housing, more city parks, and better playgrounds. Riis goaded the hesitant: "What are you going to do about it?" Some readers were moved to take immediate action. Theodore Roosevelt, then serving as president of the board of New York City's police commissioners, visited Riis's office and left a note: "I have read your book, and I have come to help."[28]

Confident that his well-intentioned purposes justified his intrusive methods, Riis rarely asked people for permission to enter their hovels and take their pictures. "Our party carried terror wherever it went," he recalled. "The spectacle of strange men invading a house in the mid-night hours armed with [flash] pistols which they shot off recklessly was hardly reassuring." Many of the slum dwellers understandably resented the uninvited efforts of Riis's "raiding party" to expose their lives to others, however noble their intention. Some people pelted Riis and his associates with rocks, others fled or demanded to be paid for their cooperation. Such reactions reveal again the fine line separating reformist realism from intrusive manipulation.[29]

Jacob A. Riis, "A Black and Tan Dive on Broome Street." *Jacob A. Riis Collection, Museum of the City of New York.)*

Riis illustrated the ethical ambiguities embedded in his brand of documentary photography in "A Black and Tan Dive," a picture of two white women and a black man in a darkened cellar saloon on Thompson Street where a fastidious Riis witnessed the "comingling of the utterly depraved of both sexes." The sudden flash of illumination invades their secret sanctuary and catches the patrons off-guard. Seated on a keg, the black man betrays both surprise and anger while the white woman on his left, her face smudged with grime, looks down with a visage suggesting sullen resentment or embarrassment. The hand of another black man (off camera) rests on her shoulder, providing a controversial nuance in an age when racial mixing was deemed an abomination. Most striking is the third figure, a woman with her back turned to the camera and a shawl draped over her head in a spontaneous effort to shield her eyes from the glare and protect her privacy from the invasive lens. The photograph gave Riis's middle-class audience voyeuristic access into an alluring world of sensual gratification and forbidden behavior.

Despite the intrusive nature of his methods, the crusading Riis did show

Americans how much of their contemporary social reality had been overlooked. From his pictures, he asserted, "there was no appeal." The poor of the inner city "compelled recognition." They were "dangerous less because of their own crimes than because of the criminal ignorance of those who are not of their kind." Riis's photographs revealed more than the degradations of slum life; they illuminated the dark side of the American dream. In the process they startled many into a keener sense of social responsibility. "I had but a vague idea of these horrors before you brought them home to me," James Russell Lowell wrote Riis. Similarly, the muckraking journalist and novelist Ernest Poole remembered how "hungrily" he read *How the Other Half Lives* because it revealed to him "a tremendous new field, scarcely touched by American writers."[30]

THE REVELATIONS CONTAINED in *How the Other Half Lives* and other sociological studies of the urban underclass at the turn of the century aroused what one reader described as a desire to discover "what life is and how it is lived by the people." Such revelatory journalism also helped provoke a national debate about poverty and social unrest that spilled over into imaginative literature as well. "All literature," reported *The Critic*, "has become permeated with sympathy for the under classes."[31]

Beginning in the late 1880s, many creative writers realized that their original assessments of the American dream reflected an apathetic naiveté. They now insisted that long ignored social facts deserved candid representation. Literature, declared Constance Fenimore Woolson, must no longer "refuse to deal with the ugly and commonplace and even the shockingly unpleasant. It is all in life and therefore not to be avoided." Similarly, Charles Chesnutt, the light-complexioned son of North Carolina free blacks, sprinkled his fiction with the salt of first-hand encounters with racial prejudice. After reading Albion Tourgée's novel, *A Fool's Errand* (1879), which described the condition of African Americans in the South during Reconstruction, Chesnutt asked, "Why could not a colored man who has lived among colored people all his life . . . write a far better book about the South than Judge Tourgée or Mrs. Stowe has written?" To him the answer seemed self-evident as he launched a career as a writer of local-color dialect stories set in the biracial South.[32]

Why did it take so long for "realistic" writers and artists to discover the reality of poverty and racism? The answers vary from individual to individual. In New York City, for instance, the emergence of thriving ethnic subcultures generated demand for literature about and by their own kind. The Jewish writer and editor Abraham Cahan, for instance, spent much of his career trying "to develop an interest in honest, realistic fiction." By the turn of the century, he could report that "now there is a large audience and deep appreciation on the east side for simple, direct transcripts from the life of the people." Most of the Jewish writers in New York City, he noted, were working people themselves who appreciated "the realities of life."[33]

Howells, on the other hand, discovered the "other half" in a more round-about way. He had initially minimized evidence of pervasive social strains because the very notion of rigid class divisions flew in the face of his democratic ideals. Unwittingly, he had ignored contradictory evidence. For him, indigence and prejudice were not so much unfelt as they were unseen. Hjalmar Boyeson observed that his friend Howells had become conservative without realizing it. Despite his truth-telling claims and egalitarian pretensions, Howells had focused his fiction on "things of small consequence" and had avoided any serious discussion of contemporary political and economic issues. Boyeson argued that Howells and other "sane and sunny" literary realists must now recognize that life in urban-industrial America did have tragic overtones that demanded new political options and literary stances.[34]

No one expressed more vividly the transforming shock of new social and economic facts than Howells himself. As late as 1886 he still believed that there was no "tragic element in our prosperity." That same year, however, his outlook began to change. His concern for social justice deepened and his commitment to reformist realism assumed heightened significance. More than any other single event, the trial of the Haymarket anarchists galvanized him into active criticism of the prevailing social order. In November 1887, Howells wrote an impassioned appeal to the governor of Illinois asking him to commute the sentences of the convicted Haymarket agitators. Although newspaper editors and his patrician friends excoriated him (Lowell said the anarchists were "well hanged"), Howells remained defiant, explaining that "my horizons have been infinitely widened."[35]

Howells thereafter came to view literature as a medium of social engagement. In 1886 he exchanged the editorship of the *Atlantic Monthly* for the same position with *Harper's Monthly* and began splitting his time between Boston and New York City. In Manhattan he found a city seething with class antagonisms. The same Howells who in 1886 had celebrated "the more smiling aspects" of American life admitted two years later that he had become "heartily ashamed of our competitive conditions." By 1889 he had discovered that "even here vast masses of men are sunk in misery that must grow everyday more hopeless, or embroiled in a struggle for mere life that must end in enslaving and imbruiting them." An enlightened Howells began to write about the dangers of unrestrained competitive capitalism. Little by little, step by step, he liberated himself from the genial optimism that had circumscribed his early fiction. He now began to fashion a more "critical realism" devoted to promoting social change.[36]

The most important influence on Howells's evolving radicalism was the Russian novelist and humanitarian Leo Tolstoy. "I can never again see life in the way I saw it before I knew him," Howells admitted. Tolstoy opened his eyes to the blinding fact of deeply embedded social inequalities and to the social responsibilities of the creative writer. "The old heathenish axiom of art for art's sake," Howells wrote in *Harper's Monthly*, "is as dead as great Pan himself, and the best art now tends to be art for humanity's sake."[37]

Howells, however, could never muster the moral courage or consistency of Tolstoy. "Words, words, words," he confided to Edward Everett Hale. "How to make them deeds. . . . With me they only breed more words." A few days earlier, writing to Henry James, he elaborated on his warring emotions. "I should hardly like to trust pen and ink with all the audacity of my social ideals; but after fifty years of optimistic content with 'civilization' and its ability to come out all right in the end, I now abhor it, and feel it is coming out all wrong in the end, unless it bases itself anew on a real equality." Yet Howells could not bring himself to relinquish the fruits of social inequality that he and his family enjoyed. He concluded his letter to James with a painful admission: "Meantime, I wear a fur-lined overcoat, and live in all the luxury my money can buy."[38]

Unable to practice Tolstoyan simplicity, Howells tried to use literature to widen the bounds of social sympathy. Fiction, he claimed, needed to give greater voice to the common folk, "the people who do the hard work of a nation, who really earn its living," who "seem by no means happy in proportion to the national riches and prosperity." To educate himself to the daily horrors of inner-city life, Howells began to explore on foot the slums of Manhattan and Boston. In his "revolt against unreality," he wanted to "see life whose reality asserts itself every day in the newspapers with indisputable force." Through his fiction he wanted to help readers see the sobering truths that he had witnessed. "Our well-to-do classes," he wrote, "are at present engaged in keeping their eyes fast shut to the facts of life of toil."[39]

In 1888 Howells wrote *Annie Kilburn* "to set a few people thinking"—and acting. Annie Kilburn, the thirty-one-year-old daughter of a distinguished judge, returns to live in her family's summer home in Hatboro, Massachusetts, after an eleven-year sojourn in Italy. Her father has died, and she is determined to serve his memory by "doing good." Soon after Annie's arrival, a group of well-meaning town leaders recruits her to help them organize an ambitious charity project. With the proceeds from a community "theatrical," they hope to build a recreational center and reading room for the town's working class.[40]

At the same time, however, the project's steering committee decides to ban the laboring folk from the event's dinner-dance celebration. Such blatant "social exclusion" infuriates the town's Congregationalist minister, Julius Peck, a widower who himself had once been a mill worker. Money, he asserts, cannot buy the affection of the poor. The working people needed economic justice, not condescending charity.

The simmering controversy boils over when the Reverend Peck predicts an outbreak of class war unless Christians help create a social commonwealth where "all shall share alike, and want and luxury and killing toil and heartless indolence shall all cease together." Peck's advocacy of Christian socialism enrages members of the congregation. Although Annie and a small band of sympathizers rush to his defense, the minister resigns, stating his intention to teach school in Fall River, a distant mill village. There he would share the daily hardships of the working poor and learn from their simple insights into the human condition. Before Peck can implement his noble plan, however, he is killed in a railroad accident, a symbolic victim of runaway industrial imperatives. After the funeral, Annie takes custody of the minister's orphaned daughter, and she and others help the mill workers gain their recreation center, revealing in the process that they have learned the difference between patronizing charity and honest assistance. The novel ends in stereotypical fashion: Annie marries the town doctor and lives "in a vicious circle" in which she "mostly forgets, and is mostly happy."[41]

Annie Kilburn provoked strong emotions but gave readers no blueprint for changing the "real" social situation. The novel concluded with Annie at a moral dead end. Exactly what should the privileged do to promote the general welfare? Howells, said his young friend Hamlin Garland, should have wrestled more directly with "the social riddle . . . that we may judge what line of reform he thinks must be pushed next." The reviewer in *The Standard* framed the issue more bluntly:

> Alas for "Annie Kilburn!" It promised to be so magnificent a story, so *true* a fiction. . . . Was Mr. Howells afraid of his own story? Howells does not see, Tolstoi sees only dimly, that the problem fronting humanity is not how to do good to the poor, but how to do justice to them and give them a chance to do good to themselves.

In a letter to Edward Everett Hale, Howells confessed that in *Annie Kilburn* "I solve nothing." Though the novel might "set a few people thinking," he had yet to start "*doing* anything myself."[42]

In November 1888 Howells and his wife moved permanently to New York in a desperate effort to find effective medical treatment for their invalid daughter, Winifred. Eight years earlier, at age seventeen, she had suffered a nervous breakdown that proved stubbornly resistant to every prescribed cure. By 1888 Winny's "nervous prostration" (akin to anorexia nervosa) had made her a fifty-seven-pound invalid. Her despairing parents, both of whom suffered from chronic nervousness and had experienced periodic mental collapses themselves,

turned in desperation to the foremost "authority" on the treatment of neurasthenia, Dr. S. Weir Mitchell of Philadelphia. Despite forced feedings and efforts
at "re-education," however, Winny Howells died of heart failure on March 2,
1889, the very month when her father's new novel, A Hazard of New Fortunes,
began to appear in Harper's Weekly. "The blow came with terrible suddenness,"
Howells reported to a friend, "when we were hoping so much and fearing
nothing less than what happened." His daughter's death and his wife's shattered
nerves made "one mass of anguished egotism of me, and shuts me up from all
things outside myself in which I have lately been interested."[43]

Serial publication of A Hazard of New Fortunes both distracted Howells
from his sorrow and provided an outlet for his crystallizing social conscience.
By far his "most vital" novel, it paints a grand tableau of New York City's
"frantic panorama" in all its teeming diversity. Howells leads the reader from
elegant town houses to sordid tenements and introduces along the way affluent
charity workers, "Negro" servants, Spanish restaurateurs, reactionary capitalists,
striking workers, middle-class professionals, gracious socialites, desperate immigrants, revolutionary agitators, a struggling artist, and an unreconstructed Virginia planter. In this large orchard of characters, not all of them ripen as
intended, but A Hazard of New Fortunes does reveal Howells's "quickened
interest in the life" about him. In the process it betrays the ambivalent impulses
and evasive rhetoric that plagued his social vision.

The novel's complex, multiple plot revolves around the staff of a new fort-
nightly literary magazine in New York called Every Other Week. Fulkerson, the
magazine's entrepreneurial founder, convinces millionaire Jacob Dryfoos to
bankroll the new publication. A Pennsylvania Dutch farmer who struck it rich
discovering oil on his land, Dryfoos has lost the simple rural virtues of his youth
in a relentless pursuit of easy profits as a Wall Street speculator. He decides to
finance Every Other Week because Fulkerson agrees to appoint Dryfoos's rebellious son, Conrad, an "idealistic humanitarian," as its nominal publisher. The
elder Dryfoos could not "bear to have his boy hanging around the house doing
nothing, like as if he were a girl."[44]

Fulkerson entices Basil March (who bears a striking resemblance to Howells)
to leave Boston and become literary editor of the new publication. In a prolonged
search for a New York apartment, Basil and his wife Isabel discover the varied
life of the city. The new elevated train provides them with a mobile observation
platform, and they delight in the drama of daily surveillance:

> It was better than the theater . . . to see those people through their windows: a
> family party of work-folk late at tea, some of the men in their shirt sleeves; a
> woman sewing by a lamp; a mother laying her child in its cradle; a man with his
> head fallen on his hands upon a table; a girl and her lover leaning over the win-
> dowsill together. What suggestion! What drama! What infinite interest!

With ravenous eyes and active imaginations, the Marches transform each snap-shot glimpse into a dramatic vignette, often talking "afterward of the superb spectacle, which in a city full of painters nightly works its unrecorded miracles."[45]

The middle-aged Marches are decent folk who "loathe all manner of social cruelty," but they are also capable of considerable self-deception. Howells's treatment of the Marches' fluid perceptions of their world indicated his maturing sense of the real as something mental as well as tangible. Initially Isabel and Basil decide that there is a visible social divide separating the "respectable" classes from the "shabbiness" of the urban poor. They want an apartment that will provide both a domestic sanctuary from the "dirty and repulsive" tenements and a venue from which Basil can scavenge the urban landscape for picturesque literary material. "They liked to play with the romantic, from the safe vantage-ground of their real practicality, and to divine the poetry of the commonplace."[46]

Daily contact with the malodorous rubbish of the city eventually erodes the Marches' spectatorial detachment. Assaulted by the stench and swarm of the streets, Isabel "yearns for the society of my peers" and tries to erect a barrier of rationalization to fend off a guilty conscience. "I don't believe there's any *real* suffering—not real *suffering*—among those people," she tells Basil; "that is, it would be suffering from our point of view, but they've been used to it all their lives, and they don't feel their discomfort so much." Her husband understands, and he too, refuses to "sentimentalize" the poor. When Conrad Dryfoos urges Basil to use his magazine to "make the comfortable people understand how the uncomfortable people live," March acknowledges only that "those phases of low life are immensely picturesque."[47]

In the end, however, Basil March sees his perspective evolve from the aes-thetic to the ethical. Life was more than a picturesque spectacle; it was also a human drama, even a tragedy! He "could not release himself from a sense of complicity" with the poor, "no matter what whimsical, or alien, or critical attitude he took." His freshened sense of the daily travails of the laboring classes "grew more intense as he gained some knowledge of the forces at work—forces of pity, of destruction, of perdition, of salvation." While recognizing the impact of environmental forces, March insists that the individual retains a mysterious potential for moral freedom. "We can't put it all on the conditions," Basil tells his wife; "we must put some of the blame on character."[48]

March's view of the "social problem" as essentially an ethical dilemma rather than a structural or ideological issue clashes with the militancy of Berthold Lindau, the embittered socialist who serves as a translator for *Every Other Week*. Lindau's militant political stance infuriates the elder Dryfoos, who orders March to fire him. March refuses and threatens to quit, only to learn that Lindau has already resigned. Lindau, it turns out, is a threat not only to the reactionary

Dryfoos but to the liberal March as well, for he advocates a genuinely radical response to the social question.

The novel's climactic event serves as a clumsy deus ex machina extricating Howells from the snare of a plot grown too complicated for him to resolve routinely. During a violent trolley-car strike (based on an actual event in New York in the winter of 1889), Conrad Dryfoos is accidentally shot and killed while trying to protect Lindau from a police clubbing. A few days later Lindau dies from his wounds. The two deaths conveniently remove the most volatile figures from the narrative and enable Howells to conclude with his usual series of reconciliations whereby enemies are either killed or married. To atone for his callous behavior, a grieving Jacob Dryfoos pays for Lindau's funeral and then sells the magazine to March and Fulkerson.

In the midst of these traumatic developments, Basil notes that Dryfoos, despite being devastated by his son's death, had not changed his misplaced priorities. This prompts his wife to ask: "Then what *is* it that changes us?" March struggles for an answer: "I suppose I should have to say that we didn't change at all. We develop." This shocks Isabel. "Basil! Basil!" she cries. "This is fatalism!" Unable or unwilling to confirm such a deterministic notion, he consoles her with an upbeat assessment common to Howellsian fiction: "I don't know what it all means, Isabel, but I believe it means good." Yet he never explains how to bring about progressive social reform. In the end his outlook remains spectatorial.[49]

A *Hazard of New Fortunes* generated brisk sales and elicited high praise from many readers sympathetic to its inclusive scope and explicit theme of social concern. Mark Twain complimented his friend for preaching a "great sermon without seeming to take sides or preach at all." In dissecting the psychology of middle-class liberalism, Howells displayed his trademark skills: credible dialogue, lifelike character types, and a wealth of irony, wit, and humor. No novel up to that time painted so vibrant a portrait of New York City, and few writers gave such explicit attention to the new social forces and ills shaping modern American life. William James told Howells that his new novel was "wonderful. Never a weak note, the number of characters, intensely individual, the observation of detail, the everlasting wit and humor, and beneath all the bass accompaniment of the human problem, the entire Americanness of it, all make it a very great book. . . . The book is so damned humane!"[50]

More than any other of Howells's writings, A *Hazard of New Fortunes* revealed his effort to use literary realism as a means of exerting imaginative control over a society splitting at the seams. Throughout the tangled narrative, Howells tried to represent the perspectives of minds other than his own; characters in A *Hazard of New Fortunes* repeatedly struggle to understand what others are thinking and feeling. Militant reformers, however, stressed that Howells's fictional treatment of the social question fell flat. Charlotte Perkins Gilman confided to a friend that Howells, for all of his workmanship and broad sympathies, failed to evoke in readers a passionate commitment to structural change. His novels "awake no other emotion in the class portrayed than a pleased surprise at their own reproduction. Like a child with a looking glass."[51]

To be sure, Howells was no agitator, and his devotion to the *art* of fiction precluded propagandizing. Aware of the moral ambiguities embedded in all efforts to reform modern society, no matter how noble or enticing, he offered no comprehensive solution to prevailing ills except his own resilient hope that society would somehow right itself through the aroused consciences of individuals. In this sense, as Amy Kaplan has argued, *A Hazard of New Fortunes* "both fulfills and exhausts the project of [Howellsian] realism: to embrace social diversity within the outlines of a broader community and to assimilate a plethora of facts and details into a unified narrative."[52]

Although acknowledging the essential humanity of "others," Howells never fully engaged the "unsmiling" aspects of real life. Like the tenement dwellers featured in Jacob Riis's photographs, the destitute characters in his fiction rarely speak for themselves; they are usually seen through the eyes of middle-class observers. The poor in *A Hazard of New Fortunes* remain voiceless, shadowy figures, the subject of middle-class anxiety and concern, but never characters in their own right capable of self-understanding and self-expression. At one point, for example, Howells parodies himself when Basil March encounters a swarm of immigrants and wonders "what these poor people were thinking, hoping, fearing, enjoying, suffering; just where and how they lived; who and what they individually were." But March "did not take much trouble" to learn the answers to such questions.[53]

In *A Hazard of New Fortunes* Howells raised a basic question: What is the social obligation of the imaginative writer? His answer, as exemplified in Basil March, was to provoke sympathy and endorse reform yet confine his art to exploring the kind of ambivalent middle-class consciousness with which he was familiar. Howells had formed his conception of literary realism well before the "social question" had commandeered public concerns. Given his own background among the class of marginal artisans in antebellum Ohio, he initially assumed that he knew the "common" people—and up to a point he did.

HOWELLS'S WELL-INTENTIONED efforts to convey how the laboring poor themselves perceived and managed intractable realities were fraught with ambiguity. Nowhere was this more apparent than in his novel *An Imperative Duty* (1892). Rhoda Aldgate is an attractive young Bostonian who, in the midst of fending off several suitors, discovers that she is "of Negro descent." Filled with self-loathing because of her "blackness," she wanders aimlessly through the streets, projecting her frustration onto the African Americans she encounters: "She never knew before how hideous they were, with their flat wide nostriled noses, their out-rolled thick lips, their mobile, bulging eyes set near together, their retreating chins and foreheads, and their smooth, shining skin: they seemed burlesques of humanity, worse than apes, because they were more alike." In the end Rhoda marries a white nerve specialist and escapes the "taint" of her past by relocating to Italy.[54]

A few reviewers justifiably pilloried *An Imperative Duty* for its clumsy effort to evoke sympathy for blacks. The *Critic* recognized that Howells "likes the race . . . in theory and at a distance." A more telling assessment came from Julia Cooper, an African-American high-school teacher, writer, and feminist living in Washington, D.C. The "unanimous verdict" among her community was that "Mr. Howells does not know what he is talking about." Cooper, the daughter of a North Carolina slave who herself felt the sting of social exclusion, argued that "he had no business to attempt a subject of which he knew so little, or for which he cared so little." She labeled it an "insult to humanity and a sin against God to publish any such sweeping generalizations of a race on such meager and superficial information." Howells's point of view, Cooper charged, "is precisely that of a white man who sees colored people at long range or only in certain capacities." Like so many affluent whites living in northern cities, Howells saw "colored persons only as bootblacks and hotel waiters, grinning from ear to ear and bowing and curtsying for the extra tips." As a result, he seemed unaware "that there exists a quiet, self-respecting, dignified class of [African Americans] . . . of cultivated tastes and habits, and with no more in common with the class of his acquaintance than the accident of complexion, — beyond a sympathy with their wrongs, or a resentment at being socially and morally classified with them." Cooper concluded that Howells had studied his subject "merely from the outside."[55]

Art for Truth's Sake: The Arena *and Purposive Realism*

During the 1890s a younger generation of activists, writers, and artists sought to go beyond the moral timidities and spectatorial perspective of Howellsian realism. Not content to remain detached sympathizers, they, like Jane Addams and the settlement workers, ventured from their middle-class world to live and work in the inner cities in order to close the gap between social ideals and social realities. As editor Benjamin Flower stressed, "nothing so deeply impresses thoughtful people with the need of radical and fundamental social reform as personal contact with the miserable in the sloughs of want." Throughout the

1890s, Flower and the other mostly native-born Protestant reformers who joined him in organizing the Union for Practical Progress used a new magazine called *The Arena* to mount an aggressive reform campaign on behalf of the unemployed and powerless.[56]

The Arena preached a clear message: reformers must move beyond "mere sentimentality" and accumulate "incontrovertible data upon which to base our arguments." The magazine's motto expressed its purpose: "Art for Truth's Sake." The "best progressive" and "reformative thought of the age," Flower announced, "is found massing under the banner of realism," but the urgency of the moment demanded a more "robust realism" than that identified with Howells and James; modern America needed a "purposive realism" that encouraged people to "frankly face conditions as we find them."[57]

Born in Albion, Illinois, in 1858, Flower initially aspired to become a minister like his father, but after attending Transylvania University's Bible College and laboring briefly as an itinerant preacher, he returned to Albion to found the *American Sentinel*, a weekly magazine dedicated to a variety of reform causes. Its success whetted his appetite for activist journalism. In 1881 he relocated to Boston, where he joined an informal group of progressive writers and artists dedicated to social reform and "aggressive realism." Hamlin Garland, a member of the coterie, recalled Flower as a "small round-faced smiling youth with black eyes and curling hair. He was a new sort of reformer, genial, laughing, tolerant. Nothing disturbed his good humor and no authority could awe him." Perhaps most impressive was Flower's passionate concern for the poor and his unstinting devotion to the cause of social justice. Garland labeled him "organized altruism."[58]

Flower repeatedly visited Boston's slum districts to spend time with the residents and experience their sordid living conditions. "Within cannon-shot of Beacon Hill," he wrote, "are hundreds of families slowly starving and stifling; families who are bravely battling for life's barest necessities, while year by year the conditions are becoming more hopeless, the struggle for bread fiercer, the outlook more dismal." Flower's crusading zeal led him at the end of 1889 to found *The Arena*, and within a few years its circulation was over 100,000.[59]

Flower feared that the social tensions plaguing the nation would escalate into class warfare unless people accelerated the pace of reform. He thus resolved to use the *Arena* to "agitate, to compel men to think; to point out wrongs inflicted on the weak and helpless." The time for empty rhetoric was over; only by unearthing and confronting the "squalid facts" of modern existence could society progress. "Nothing is so wholesome as candor," Flower contended, and he resolved to help people "face the exact truth in all its hideousness" so as to "remedy the wrongs or cure the evils." Facing harsh truths, Flower believed, was the single most important element of any reform crusade: "Arouse the people and evil will disappear."[60]

Flower and other activists associated with *The Arena* differed from earlier city reformers in their emphasis on the environment's role in shaping human fate. Where conservatives asserted that poverty was a self-inflicted condition, the result of individual moral failings, genetic deficiencies, or intemperance, Flower and his associates argued that loathsome surroundings bred miserable people. The "bum," Flower wrote, was "a creature degraded by circumstance and evil conditions. . . . Instead of blaming and condemning him, poor fellow, we should look at the circumstances that made him what he is, and endeavor to remedy them." Flower's proposed "remedies" included stringent sanitary regulations, more humane vagrancy laws, child labor statutes, kindergartens, playgrounds, parks, and children's clubs in order to give "new and joyous significance" to lives "long exiled from joy, gladness, and comfort."[61]

To help spur the social conscience of the affluent and the influential, Flower recruited all available weapons of social enlightenment, including literature and the fine arts. "The most urgent need of the present," he stressed, "is the united influence of the press, the pulpit, the novel, and the drama in *acquainting people with the terrible facts as they exist.*" In 1894 Clarence Darrow, the prominent Chicago attorney, testified to the alliance between cultural realism and progressive reform in an *Arena* article entitled "Realism in Literature and Art." An impassioned defender of the rights of the lower classes, Darrow contended that the best way for the arts to promote social justice was to tell the truth about contemporary evils. The world "has grown tired of preachers and sermons; to-day it asks for facts." The best writers and artists were "telling facts and painting scenes that cause humanity to stop and think, and ask why one shall be master and another a serf." Over time, he concluded, such a "realism" of exposure would necessarily improve the social welfare, for "knowledge is power," and "wisdom only can make harmony."[62]

Prairie Realism

Flower's favorite among the young literary realists featured in *The Arena* was Hamlin Garland. "I know of no American writer," he announced, "who presents real life as vividly and powerfully as Mr. Garland. Even Mr. Howells, who possesses far more finish . . . lacks, it seems to me, something of the power in depicting life's everyday tragedies, which Mr. Garland possesses." Born in 1860 in a log cabin on rented land near New Salem, Wisconsin, Garland was the son of a restless farmer who repeatedly moved his family westward in search of "the perfect farm." But wandering from Wisconsin to Minnesota to Iowa to the Dakota Territory brought them only unending exertion. Rising before the sun each day, Garland and his brother milked cows, scrubbed horses, gathered firewood, cleaned stables, plowed fields, mended fences, and husked corn.[63]

In 1876 Garland's spirits soared when he enrolled at the Cedar Valley Seminary in Osage, Iowa, a glorified high school for farm children. He relished the chance to put down his hay fork and indulge his passion for reading. "With what sense of liberty, of exultation," he remembered, "we took our way down the road on that gorgeous autumn morning!" After graduating from high school, Garland tried teaching in Illinois, but a rocky year in the classroom convinced him that he needed more seasoning. So in 1883 he joined the land rushers headed for the Dakota Territory and staked a claim on virgin prairie in McPherson County, not far from where his parents had recently resettled.[64]

A "pitilessly severe" winter in a shack on the treeless plains, however, chilled Garland's enthusiasm for pioneering. In the fall of 1884 he exchanged his hardscrabble homestead for a railroad ticket and enough cash to get to Boston. There he spent the next three years living alone in a roach-infested boarding-house while educating himself about literature and the "social question." He was, he recalled, "a bewildered plainsman, a scared rustic in the midst of a gigantic metropolis." To pay the bills, he gave public lectures on literary topics and taught a class at the school of Oratory. The Boston Public Library became his college, and for twelve hours each day he feasted on its books and the ideas of its patrons. "My brain young, sensitive to every touch, took hold of facts and theories like a phonographic cylinder. . . . I learned a little of everything and nothing very thoroughly." Darwin and Spencer absorbed much of his attention, and he became convinced that the concept of evolutionary progress applied to both human society and the arts. Henry George's books and essays explained the causes of "poverty and suffering in the world," while Howells's columns in *Harper's Monthly* introduced Garland to the debate between "romanticism and realism."[65]

But it was *Leaves of Grass* that spoke most directly to Garland's developing literary outlook. The prose-poems were "full of light and heat, electric, virile, rugged." Nothing daunted Whitman. He was "the master of the real," convinced that "all American art and literature should be founded upon the actual, the present." Buoyed by Whitman's relentless optimism and democratic sympathies, Garland began corresponding with the "good grey poet" and in 1888 visited him in Camden. Whitman told a friend that the young westerner "seems started all right: is dead set for real things: is disposed to turn himself to the production of real results."[66]

A family reunion in South Dakota proved to be an especially sobering experience for the twenty-seven-year-old Garland as he surveyed with fresh eyes the drudgeries, drabness, and isolation of prairie farming. "I looked at the barren landscape . . . where every house had its individual message of sordid struggle and half-hidden despair. All the gilding of farm life melted away. The hard and bitter realities came back upon me in a flood." The sight of his mother, now stooped and gray and "imprisoned in a small cabin," especially infuriated him.

An embittered Garland wondered why "have these stern facts never been put into our literature?"[67]

On the way to South Dakota from Boston, Garland had stopped in Chicago to visit Joseph Kirkland, the lawyer and novelist. Kirkland described his guest in a letter to his family:

> Hamlin Garland spent Saturday night and Sunday forenoon with me. Enthusi-ast—country-boy—farmer's son—largely self-educated—wants to reform the litera-ture of our country! Reconstruct it on realistic basis. Hates Lowell, Holmes and other fossil representatives of classicism. Loves Walt Whitman and Howells . . . and *me*.

That fall, Garland, still "in a mood of resentment," began writing stories based on his own agrarian experiences, and their gritty veracity impressed others who read them in draft. Kirkland urged him to keep writing such sturdy fiction: "You're the first actual farmer in American fiction; now tell the truth about it." After returning to Boston, Garland joined the Anti-Poverty Brigade, "threw myself heart and soul into the farmers' movement," and resolved to tell "the truth about western farm life, irrespective of the land-boomer or politicians."[68]

As a frequent contributor to *The Arena*, Garland relished Flower's intense commitment to social justice. "B. O. Flower and I," he wrote a friend, "are getting to be chums. He's one o' my kind. Don't smoke, chew, drink. . . . He's concentrated moral purpose." Emboldened by Flower's intensity, Garland resolved to employ fiction to "hasten the age of beauty and peace by delineating the ugliness and warfare of the present."[69]

In Boston Garland developed what he called two great literary concepts: "that truth was a higher quality than beauty, and that to spread the reign of justice should everywhere be the design and intent of the artist." On this foun-dation of aesthetic and ethical assumptions, he built his own version of literary realism, or "veritism," which he defined as "the truthful statement of an indi-vidual impression corrected by reference to fact." Like the impressionist painters, he strove to go beyond merely "delineating a scene." He wanted to provide a "personal impression of a scene, which is vastly different." His realism fused hard facts with strong feelings. A writer should not merely "write of things as they are," but of things "as he *sees* them." Realistic writing "is *not* photography." Nor was it exotic. The veritist "chooses for his subject not the impossible, not even the possible, but always the probable." By emphasizing the author's *impres-sion* of observed reality and infusing it with a strain of ethical idealism, Garland developed a pragmatic realism designed to promote progressive social change. The veritist was "an optimist, a dreamer. He sees life in terms of what it might be, as well as in terms of what it is; but he writes of what is, and suggests what is to be, by contrast."[70]

Garland displayed his "veritism" in *Main-Travelled Roads* (1891) and *Prairie Folks* (1893), two collections of coarse-grained stories about "poor and weary" midwestern farm folk. Garland portrayed farm life "not as the summer boarder or the young lady novelist see it—but as the working farmer endures it. . . . Butter is not always golden nor biscuits invariably light and flaky in my farm scenes, because they're not so in real life." Like most other literary realists, he offered readers a statement of authenticity: "I am a competent witness and I intend to tell the whole truth."[71]

Garland populated his stories with rugged rural figures who cast grim, hard shadows, people he had encountered while living on the prairie. Sim Burns, for example, "was a type of the average prairie farmer." After waking each morning, he "thrust his dirty, naked feet into his huge boots, and, without washing his face or combing his hair, went out to the barn to do his chores." Living in a three-room, unpainted cabin, he "chewed tobacco and toiled on from year to year without any clearly defined idea of the future. His life was mainly regulated from without." The main business of such sodbusters was "to work hard, live miserably, and beget children to take their places when they died." Such bitter sentences cast a harsh glare that bleaches their color but also generates a warmth of feeling rarely seen in fiction of the day. Even though Garland's farmers lead dull, isolated, and barren lives of toil, they urge upon the reader a compelling form of courage—stoic, tenacious, redemptive in its perseverance.[72]

The anti-romantic stories in *Main-Travelled Roads* disclose the ambivalent impact of evolutionary naturalism on Garland's evolving social philosophy. His characters struggle against an implacable physical environment and exploitive economic system, but are not always simple pawns of such impersonal forces. Some accept their lot; others try to escape but lack the resources and knowledge to succeed.

"Under the Lion's Paw," the most famous and explicitly political of Garland's stories, focuses on the Haskins family, Iowa tenant farmers struggling to refurbish a dilapidated homestead in order to buy it for $2500. No sooner does Haskins raise the purchase money than Jim Butler, the ignoble land owner, announces that the price has risen to $5500. "Why, that's double what you offered it for three years ago," Haskins notes in astonishment. "Of course," Butler replies, "and it's worth it. It was all run down then; now it's in good shape. You've laid out fifteen hundred dollars in improvements, according to your own story." An incredulous Haskins exclaims: "But *you* had nothin' t' do about that. It's my work an' my money." Butler's retort is chilling: "You bet it was; but it's my land." An outraged Haskins threatens to skewer Butler with a pitchfork but relents for the sake of his children. He resigns himself to a life of tenancy under the "lion's paw" of economic inequality.[73]

In Garland's gallery of western characters, however, a few souls blessed with unusual fortitude and good fortune manage to break free of their inherited circumstances or social constraints. In "A Branch Road," Will Hannan, like Garland, returns home to Rock River after a seven-year absence and looks with enlightened eyes upon his native farm community. "He felt as young people seldom do the irrevocableness of living, the determinate, unalterable character of living." Will, however, refuses to submit to the status quo. Upon learning that his teen-age sweetheart is married to a brutal boor and living in squalor, he urges Agnes Dingman to escape with him to Europe: "God don't expect a toad to stay in a stump and starve if he can get out." Agnes balks. "It can't be helped, now, Will," for "life pushes us into such things." But Will insists that they must not "let a mistake ruin us"; she can surmount her wretched condition through an act of will. The long-suffering Agnes eventually curses fate and her loveless marriage, picks up her baby, and walks out with Will while her abusive husband is away.[74]

Garland's early prairie stories have their faults. As an unsophisticated writer from the provinces intent upon impressing the established literary community, he sprinkled his prose with stilted phrases and elegant references that detract from the authenticity of his subjects. Clumsy melodramatic devices and fortuitous incidents mar the endings of several of his stories, and he often offered more passion than insight into the complex causes of the "farm problem."

Taken as a whole, however, Garland's "prairie realism" marked a genuine departure in American literary history. He had a pictorial knack for conveying the color, texture, and speech of prairie culture, and he was one of the few writers for whom a strident political stance helped rather than hindered their fiction. Although several reviewers found Garland's realism too acrid and brutal, too unrelenting in its evocation of "grinding, unremunerated toil," others claimed he was simply telling the truth about a region underrepresented in fiction. Howells characterized *Main-Travelled Roads* as a "robust and terribly serious" collection of stories that would "strike many gentilities as coarse and common." Garland, he added, brought to life

> those gaunt, grim, sordid, pathetic, ferocious figures, whom our satirists find so easy to caricature as Hayseeds, and whose blind groping for fairer conditions is so grotesque to the newspapers and so menacing to politicians. They feel that something is wrong, and they know that the wrong is not theirs. The type caught in Mr. Garland's book is not pretty; it is ugly and often ridiculous, but it is heartbreaking in its rude despair.

Howells urged anyone "at a loss to account for that uprising of farmers in the West" to "read *Main-Travelled Roads*, and he will understand."[75]

One farm wife wrote Garland a letter of earnest praise: "You are entirely right about the loneliness, stagnation, the hardship. We are sick of lies. Give the world the truth." Other readers offered similar testimonials. "You have delineated my life," a farmer wrote. "Every detail of your description is true. The sound of the prairie chickens, the hum of the threshing machine, the work of seeding, corn husking, everything is familiar to me and new in literature." Garland, however, claimed that he intended his stories not so much for farmers but for city dwellers who naively believed in the pastoral myth. One urban professional reported that after reading "Prairie Heroine" in *The Arena* he had been "miserable ever since. I knew such things existed, but I never *felt what it meant before.*"[76]

With the publication of *Main-Travelled Roads*, Garland found himself "welcomed into the circle of American realists with an instant and generous greeting which astonished, at the same time that it delighted me." In 1892 he toured the farm counties in Iowa, stumping for Populist candidates, and published three propaganda novels in support of the movement. Hastily conceived and luridly partisan, they were, he later confessed, "too sour of temper, too drab of clothing, too preachy."[77]

Thereafter, for reasons not entirely clear, Garland abandoned reformist realism and began writing romantic adventure stories about cowboys and Indians in the Rocky Mountains. In assessing his friend's abrupt change, Howells speculated that Garland had always been a romantic at heart. He "only kept to reality" at the start of his career because he "did not know unreality." The failure of Garland's political novels also hastened his escape to the high country of romance. In 1893 he told his publisher that he was "ready to send out purely literary books. . . . I shall not repeat either my economic writing or literary and art reform. Having had my say I shall proceed on to other things." With a wife and children to support, a middle-aged Garland now yearned for both financial success and the acclaim of the genteel literary community. So he readily agreed to follow the advice of Richard Watson Gilder, the poet-editor of *Century* magazine, who urged him to give "us more stories in which the beauty of life appears."[78]

By the end of the century, Garland "found myself almost popular." He had become a prosperous author and a celebrity in the literary circles of Boston, New York, and Chicago, where he moved in 1894. Universities awarded him honorary degrees, and he won election to the American Academy of Arts and Letters. Enamored of his newfound recognition and affluence, the prairie Populist now unapologetically labeled himself "an intellectual aristocrat" and began wearing velvet coats, smoking Cuban cigars, and sipping brandy at exclusive men's clubs. Concern for the farm problem and the single tax gave way to buying a commodious home and car. He even began speculating in real estate. Late in life, however, Garland expressed regret at his transformation from bitter realist to trite romanticist. "I have as a man of family, been warped . . . from my ideals. As novelist and historian I have conformed . . . as we must all do if we are to maintain an apartment in New York City, and a house in the country with an automobile in attendance."[79]

Garland took some comfort in the realization that he "was not alone in this reaction from the ethic to the aesthetic. . . . The reform impulse was steadily waning in power with us all." By the end of the century, the urgency of the "social question" and the "farm problem" had greatly subsided. After William McKinley's victory over the agrarian reformer William Jennings Bryan in the 1896 presidential election, rising commodity prices dealt a mortal blow to the Populist movement. Meanwhile, the problems associated with urban poverty seemed less acute in the aftermath of a torrent of reform activity and regulatory legislation. As Jacob Riis reported in 1902, "the grip of the tramp on our throat has been loosened." When the nation took an imperialistic swim in the Caribbean and the Pacific as a result of the Spanish-American War, public attention shifted from domestic concerns to foreign adventures. "The blare of the bugle," explained a Populist, "drowned the voice of the Reformer."[80]

Romantic Revival

The reform impulse associated with literary realism added new vitality and conviction to the fictional representation of contemporary life. Yet public taste remained quite diverse—and mercurial. At the end of the century, romantic literature witnessed a popular resurgence that led the *Literary World* to conclude that social realism was on the wane: "the world is tired of Kodak pictures of the dreary commonplaces of life." Many readers, it seems, again wanted to be distracted from the disturbing problems of contemporary life. The editors of *The Outlook* took undisguised pleasure in announcing the resilience of romantic idealism:

> Not many years ago a goodly number of critics were declaring that the romantic, symbolic, and spiritual school in poetry, fiction, and the drama had had its day; that the influence of scientific thought and of modern commercialism had enthroned realism as a literary method for all time.

Now, however, realism had "lost its hold on the imagination of the world," and the "newer school" of writers was recording "the escape of the spirit out of a world of realities into a world of pure idealism."[81]

In 1900 Maurice Thompson, the novelist and poet laureate of Crawfordsville, Indiana, gleefully declared that popular interest had veered away "from the fiction of character analysis and social problems to the historical novel and the romance of heroic adventure." The best-seller lists were clogged with sentimental historical romances such as Thompson's *Alice of Old Vincennes* (1900), Charles Major's *When Knighthood Was in Flower* (1898), and Mary Johnston's *To Have and to Hold* (1899), a trifling story set in colonial Virginia. Such trends emboldened self-designated "idealists" to lay siege to the citadels of realism. The editor of the *Dial* sneered in 1898 that the novels of Howells and James "no longer produce any kind of a thrill; the force by which they once produced it is spent." He concluded that the "romantic revival is at fulltide, and contemporary literature bids fair to offer us once more the solace that it brought us of old."[82]

The cosmopolitan novelist Francis Marion Crawford, whose exotic tales about eloping Italian nuns, chivalric Russian counts, and New York debutantes followed such a formula and enjoyed brisk sales, questioned the very premises of literary realism: "What has art to do with truth? Is not truth the imagination's deadly enemy? If the two meet, they must fight to the death. It is therefore better . . . to keep them apart." Fiction should make people laugh and occasionally provoke them to tears, but it should not make them think "too serious thoughts." The novel, he stressed, "must deal chiefly with love.". It "must be clean and sweet" and "its idealism must be transcendent" so as to foster in the reader "agreeable illusions." Crawford also dismissed the notion that literature should serve as an instrument of social reform. "An individual cannot change the conditions of the society in which he is obliged to live," he maintained, "and must either conform to them or be excluded from intercourse with his fellows."[83]

Of course, the advocates of this romantic revival exaggerated their new influence, just as apostles of realism had earlier claimed to have swept their opponents from the field. "The war of opposed opinions is carried on with special liveliness," observed a writer in the *Atlantic Monthly*, with "neither side admitting defeat or check." Several realists, however, saw the shifting current of public taste and rushed to swim with the tide. The Chicago novelist Henry Blake Fuller abandoned contemporary subjects in favor of historical romances set in Sicily. In 1898 John W. De Forest published *A Lover's Revolt*, a sugary tale about life in antebellum New England. Sarah Orne Jewett also took a sabbatical from local-color realism in order to write *The Tory Lover* (1901), a romance set during the American Revolution. A commercial success but a critical failure, it provoked Henry James to implore Jewett to "*come* back to the palpable present . . . that throbs responsive, and that wants, misses, needs you, God knows, and suffers woefully in your absence."[84]

In 1894 a young reporter named Stephen Crane asked Howells to assess the romantic revival. "I suppose," he replied, "we will have to wait" a little longer for the realistic movement to emerge victorious in the continuing "battle of styles." Howells did not realize that his interviewer was one of a talented group of innovative young writers who were then forging a colorful and controversial new synthesis of realism and romanticism. For all of their stylistic and personal differences, Crane, Frank Norris, Theodore Dreiser, and Jack London shared a commitment to "realism." But they also agreed that Howells, James, and others had not gone far enough in representing the powerful forces shaping modern life. Now it was their turn to try to close the gap between literary representation and social reality. [85]

10

Savage Realism

"Youth makes a savage realist," Hamlin Garland declared in 1903. He was referring to novelist Frank Norris, but the label applied equally well to others among a group of untamed young writers—Abraham Cahan, Ellen Glasgow, David Graham Phillips, Stephen Crane, Jack London, and Theodore Dreiser—who at the end of the nineteenth century began exposing sordid facts and primitive passions previously unrepresented in American literature. "Life," London asserted in *The Call of the Wild*, was "full of disgusting realism" that deserved forthright depiction. In giving voice to unheard truths, he and others developed what one critic called an "ultra realism" that stressed social conflicts, political and corporate corruption, life-and-death struggles, confused mental states, carnal pleasures, and the culture of poverty. Although terribly uneven in ability and performance, these "savage realists" brought a tonic shock to American literature. "There is no doubt," Norris wrote shortly before his premature death in 1902, "that the estate of American letters is experiencing a renaissance."[1]

This new breed of realists, according to H. L. Mencken, the vinegary Baltimore journalist and literary critic, offered "not the old, flabby, kittenish realism of Howells's imitators . . . but the sterner, more searching realism that got under the surface" of human motivation. In exploring such subterranean truths, the "savage realists" developed a curious ambivalence toward Howells himself. While appreciating his efforts to build what Norris called "the foundation of fine, hardy literature," they scorned his "bloodless" pruderies. Virginian Ellen Glasgow spoke for many turn-of-the-century literary rebels when she recalled that "Mr. Howells was still the acting dean of a severely regimented American realism" when she began her career. "And all my life I had been a rebel against regimentation in any form. I had not revolted from the Southern sentimental fallacy in order to submit myself to the tyranny of the Northern genteel tradition." Her choice was clear: "One must either encounter reality or accept the doctrine of evasive idealism." She thereupon resolved to undertake a "fearless exploration into the secret labyrinths of the mind and heart."[2]

Dreiser felt the same. After visiting Howells he acknowledged that his distinguished host "had been an influence for good in American letters." Yet he pointed out that the author of A *Hazard of New Fortunes* "had no direct experience" with the "great misery" or "passions" prevalent among the poor and suffering. Howells's novels expressed his own genial optimism and genuine sympathy, but Dreiser felt that they used the lower classes as objects of bourgeois concern rather than as subjects worthy of attention in their own right. Dreiser also questioned whether a writer of such startling mildness and reticence could claim to be a true realist. After all, Howells, unlike Dreiser, had never suffered the indignities heaped upon the poor; he had never felt the shame of debt or the sting of want. Nor had he ever "known" a woman other than his wife. Dreiser concluded that Howells was one of the many "professional optimists and yea-sayers" in America who believed that "we must discuss only our better selves, and arrive at a happy ending."[3]

To his credit, Howells encouraged most of the "savage realists" to surmount his own moral scruples and surpass his essentially spectatorial efforts to portray the rough edges of contemporary life. "I was bred in a false school," he admitted, "whose trammels I have never been quite able to burst." But the "new realists," he observed in 1899, did not "shun any aspect of life." With a rude vigor, they explored the "passions and motives of the *savage* world which underlies as well as environs civilization." Yet despite the new elasticity in his outlook, Howells still disapproved of depicting adultery "without shame or sin." He worried that the young literary rebels would encounter stiff resistance from a reading "public like ours which prefers smooth things and likes its mental nutriment in the form of pap."[4]

The "new realists," however, were willing to sacrifice popularity for candor. Most of them began their literary careers as newspaper reporters or magazine correspondents, and their journalistic experiences revealed to them how brutal and extraordinary daily life could be. Reporting the "facts of life" also planted in them a commitment to fiction based on face-to-face encounters and first-hand experiences. They did not want to view life from a distance; they clutched it by the throat. "Life is action," Dreiser stressed, "and actions point their own morals better than any reasoning about them can do." The only way to grasp the "real thing" in this "world of struggle and discontent," he added, was to go "out among the people and into the places where misery gathers." Norris expressed the same experiential approach but added a chauvinistic twist. He questioned the relevance of popular women authors because they lacked contact with "real life."[5]

A *Cultural Crisis*

The "savage realists" shared a widespread concern at the turn of the century that the United States suffered from the malaise of "overcivilization" plaguing all of Western society. The French spoke of a pervasive cultural sickness — *la maladie de fin de siècle* — infecting the educated bourgeoisie. Urban-industrial development brought spreading material comfort but also moral complacency, triumphant secularism, and an uneasy sense of artificiality. Throughout the industrialized nations, the terms "decadent" and "degenerate" became popular synonyms for modern culture. Cassandras arose in every country to bemoan the fate of traditional values and to plea for more vital and authentic experiences than those offered by sedentary urban life and liberal Protestantism.

Many observers feared that bourgeois American men were losing their virility. In David Graham Phillips's novel *George Helm* (1912), a character regrets "how few real men there are — real live men" in an urban society obsessed with the pursuit of money and luxurious ease. Like Phillips, anxious editors, moralists, and political leaders charged that modern city life, with its devitalizing material comforts and frenzied pace, sapped residents of their stamina and self-reliance. A writer in the *Atlantic Monthly* cautioned that robustness among the "highly educated, sophisticated class" of men was being "dulled and weakened by civilization." The result was a generation of "over-sophisticated and effete" men whose evident feebleness invited the restive working classes and unruly immigrant hordes to become more assertive. As the "rich become effeminate, weak, and immoral," a doctor claimed, the "lower classes, taking advantage of this moral lassitude, and led on by their savage inclinations, undertake strikes, mobs, boycotts and riots."[6]

Women continued to receive much of the blame for the erosion of virility. "Women's authority over matters social," concluded the jurist-turned-novelist Robert Grant, "is far greater than it has ever been." Grant and other anxious males painted a gloomy portrait of a metropolitan society in which assertive women were taking control of the major conduits of cultural life. They exaggerated, of course, but even modest increases in female authority *seemed* substantial during the half-century after the Civil War. Dr. S. Weir Mitchell, the physician and popular novelist, complained that the monthly middle-class magazines "are getting so lady-like that naturally they will soon menstruate." At the same time, he and others observed, overprotective mothers and teachers were producing a generation of passive, dependent young men. A prominent Baptist layman bemoaned the distinctively "feminine note in religion as well as in education," a tendency that deprived faith of its "virility" and substituted "a sentimental idea of what ought to be for a candid recognition of what is." Such spiritual sentimentalism threatened to "enfeeble" the national will. "Let us have a masculine religion," he pleaded.[7]

In 1908 G. Stanley Hall, the eminent psychologist and president of Clark University, gave academic prestige to the cult of besieged masculinity when he reported that only 23 percent of American public school teachers were male. Feminine control of classrooms, he argued, was destroying traditional notions of discipline. Women teachers disdained the rod in favor of "coaxing and coquetting" students, even though adolescent boys "need occasional thrashings." Such "sugary benignity" and "mawkish, hysterical sentimentality" created "pampered, weakly" boys.[8]

Those concerned about the nation's diminishing virility also bemoaned the devitalizing nature of office work and home life in an increasingly urban society. Luther Halsey Gulick, the director of physical training at the Springfield YMCA Training School and co-founder of the Camp Fire Girls, blamed "the matter of our daily occupation—the way we get our living" as the leading cause of eroding physical health and mental vigor. The "children of parents who have led nervously intense and exhausting lives of cities are likely to be delicate and nervous." Most middle-aged businessmen, he reported, "have bodies that disgrace them. Everywhere you see fat, clumsy, unsightly bodies; stooped, flabby, feeble bodies; each and every degree of dilapidation and inefficiency."[9]

The Cult of the Strenuous Life

Among the many champions of masculine renewal and physical hardihood at the end of the century, the most prominent and pugnacious was Theodore Roosevelt. In 1899, as governor of New York, he told the Hamilton Club in Chicago that he wished "to preach, not the doctrine of ignoble ease, but the doctrine of the strenuous life, the life of toil and effort, of labor and strife." He expressed a sovereign contempt for the "timid man, the lazy man, the man who distrusts his country, the over-civilized man, who has lost the great fighting, masterful virtues." The American man must somehow recover the courage to participate in "righteous war," to "dare and to endure and to labor," while the woman "must be the housewife, the helpmate of the homemaker, the wise and fearless mother of many children." In promoting a more vigorous regimen for college men, Roosevelt stressed that "rough sports" as well as close-knit fraternities were the best means of inculcating "the virile, masterful qualities of the heart and mind."[10]

Roosevelt helped transform the ideal of the "strenuous life" into a national mania at the turn of the century. Millions of flaccid urbanites—both men and women—self-consciously set about restoring a sense of hardihood and adventure amid the energy-sapping routine of bourgeois life and sedentary labor. Other virility advocates touted the regenerative benefits of renewed contact with the wilderness. A back-to-nature craze seized the nation as frazzled city dwellers went hiking or bicycling, took camping trips, joined hunting and fishing clubs, or spent time on western "dude" ranches. Cowboy life exerted a mythical attraction for eastern patricians suffering from "overcivilization." Harvard graduates Frederic Remington, Owen Wister, and Theodore Roosevelt, for instance, found that life on the trail provided a therapeutic alternative to their metropolitan backgrounds. The values of wilderness recreation—fostering healthy bodies, "masculine" self-confidence, and renewed mental energy—also proved especially attractive to members of an urban middle-class suffering from nervous exhaustion. Their own supine dependence on specialists, financiers, technicians, and servants enticed many to seek the balsam tonic afforded by summer camps and other forms of rustic endeavor.[11]

Frank Norris vigorously promoted the "return to the primitive" life as an invigorating alternative to the decadent aestheticism he feared was infecting young writers. "Of all the fads," he wrote in 1903, "the most legitimate, the most abiding, the most inherent—so it would appear—is the 'Nature' revival." Now that the frontier had been tamed and young Americans were increasingly surrounded by urban artificiality and cosmopolitanism, they needed to retreat to the woods and "take a renewed grip upon life."[12]

Those eager to restore masculine potency also touted sports as effective antidotes to modern decadence. Physical culture advocates expanded the activities of the YMCA, organized the city playground movement, incorporated physical education classes into school and college curricula, and encouraged individuals to adopt a daily exercise regimen. Louis Sullivan enthusiastically endorsed the new physical fitness craze and the rush of social testosterone. He regularly lifted weights and wrestled at a Chicago gymnasium, and he rejoiced at the wave of interest in physical culture and outdoor recreation, believing that it "shows that we are not indefinitely to remain a nation of city-dyspeptics and weary melancholics." Stronger muscles, he and others assumed, would mean stronger nerves and ultimately a stronger culture. "Physical exercise," boasted Bernarr McFadden, publisher of *Physical Culture Magazine*, "is destined to effect the regeneration of the Caucasian race."[13]

Through athletic participation, one sports advocate concluded, "the man develops all the 'manly' attributes—glorious strength and skill and endurance." Another claimed that contact sports were "reminiscent of the physical age, of the struggle for survival, of the hunt, of the chase, and of war." Such analogies appeared frequently in discussions of intercollegiate athletics, which experienced phenomenal growth at the end of the century. In his 1893 address to the Cambridge Phi Beta Kappa Society, economist Francis A. Walker described students as having been molded into anemic young men by the "feminine" doctrines of "transcendentalism and sentimentalism." Now, however, football and other "manly games" were providing character-building surrogates for martial endeavor.[14]

Football became an immensely popular instrument of revived manhood. Willa Cather noted that intercollegiate athletics in midwestern colleges were "the one resisting force that curbs the growing tendencies toward effeminacy so prevalent in the eastern colleges." Football, she added, "is the deadliest foe that chappieism has. It is a game of blood and muscle and fresh air." By 1905 some 95 percent of all high schools sponsored football teams, and college social life began to revolve around the Saturday afternoon gridiron contest. The president of Western Reserve University found in football a metaphor for the struggle of life itself. "It illustrates the science and art of realism," he asserted. "It embodies actuality." Still others used Darwinian and Nietzschean themes to explain the benefits of gridiron grappling. *The Independent* justified football as a means of introducing participants and spectators to the harsh realities of social strife: "A certain degree of hardness of heart is necessary for success in the increasingly intense struggle for existence. Ruthlessness is the chief attribute of the superman, and his progenitors have already appeared upon the earth." What was good for players was also good for spectators. A "season spent in the grandstands" would rid the "mildest maiden" of any "vestigial squeamishness."[15]

Boxing was another "blood sport" that excited a generation of artists and intellectuals who had been weaned on the "survival of the fittest" emphases of Social Darwinism. As the very embodiment of adolescent masculinity, prize fighting offered a "regulated" form of primal struggle. Each city had one or more private boxing "clubs" whose nightly rituals—gambling, drinking, fighting, and cheering—offered a disreputable oasis of raw virility and masculine bonding in the midst of the Victorian era's suffocating gentility. When someone questioned Oliver W. Holmes, Sr., about the bloody mayhem associated with boxing, the Harvard poet-professor said he preferred its regulated gore to "the white-blooded degeneration to which we all tend."[16]

Professor G. Stanley Hall developed a compulsive interest in the "raw side of human life," and he never "missed an opportunity to attend a prize fight if I could do so unknown and away from home." His surreptitious efforts to join the boxing subculture reveal the attraction of a purely masculine arena insulated from "feminine" influence or genteel decorum. Thomas Eakins likewise found in the boxing auditorium a manly haven from the strictures of "effeminine society." He himself loved to wrestle and was "nothing loth for a little tussle." Several times a week he walked from his Philadelphia studio to the arena where prize fights were held, and in 1898–99 he completed three major paintings of scenes he had witnessed: *Taking the Count, Salutat,* and *Between Rounds.*[17]

During the last quarter of the nineteenth century, a small but prominent group of American imperialists used aspects of the cult of strenuosity to buttress doctrines of racial superiority, military adventure, and territorial expansion. They found in the Darwinian idea of natural selection a handy argument for imperialism. If mortal competition worked so well in the natural realm, would it not also benefit human society? Among nations, as among individuals, some argued, the fittest should prevail. In *American Political Ideas* (1885), John Fiske, the Harvard historian and popular lecturer on Darwinism, stressed the superior character of "Anglo-Saxon" institutions and peoples. The "English" race, he predicted, was destined to dominate the globe, in part because of its "strenuous" heritage and spirit. Another commentator sharpened the point when he claimed that the "governing races of to-day are races of sportsmen." In both England and America, such racial prejudices were often used to justify foreign conquest and colonial domination. It was time, many argued, for America to "flex its muscles" abroad. "All around us," Captain Alfred Thayer Mahan, the prominent naval strategist, wrote in 1897, "is strife; 'the struggle of life,' 'the race of life,' are phrases so familiar that we do not feel their significance till we stop to think about them. Everywhere nation is arrayed against nation; our own no less than others."[18]

America's prolonged domestic turmoil during the 1890s led Mahan and others to embrace war was a therapeutic alternative for a society suffering from social unrest and the anemia of modernity. "Idleness and luxury have made men flabby," declared a woman writing in the *North American Review* in 1894. She wondered "if a great war might not help them to pull themselves together." If, as literary critic Maurice Thompson wrote in 1898, "the greatest danger that a long period of profound peace offers to a nation is that of [creating] effeminate tendencies in young men," then war must become a social necessity. Such notions helped propel the United States into its "splendid little war" against Spain in 1898.[19]

The "Masculine" School of Writers

All of the elements associated with the doctrine of the "strenuous life"—the promotion of athletic endeavor, the back-to-nature movement, aggressive masculinity, militarism, theories of cultural "degeneration," racial superiority, and imperial acquisition—found expression in the lives and works of the "savage" realists. Like many among their literary generation, they dismissed Howellsian realism for not being realistic enough. It failed to acknowledge the daily violence and seething passions under the surface of a timid bourgeois society and thereby contributed to the *fin de siècle* cultural malaise.

Disdaining "effeminate realists" such as Howells and James, the "savage realists" initially worshiped Rudyard Kipling and his self-consciously virile fiction. The Bombay-born English writer dazzled readers with stories about British army life and male adventure on the high seas and in India. Kipling's accounts of hard-swearing soldiers and men of material affairs appealed to "strong men everywhere," said a writer in *The Outlook*. It also appealed to a strong young woman writer named Willa Cather. She adored Kipling's stories because they were "full of real men, young men, active and able bodied," and their atmosphere was "one in which men may love and work and fight and die like men." In glorifying the rougher aspects of existence, Kipling usually either ignored women or treated them with contempt. "A man married is a man marred," he once quipped.[20]

Like Kipling, the American "savage realists" craved more "strenuous" experiences than those available within the restricted bounds of Victorian gentility. They shared the widespread belief that American culture desperately needed an injection of manly vigor and authentic experiences. In discussing the younger generation of irreverent realists, literary critic Grace Colbron reported that masculinity "is the hallmark of this school—masculine interests in the theme, virile masculinity in the author's angle of vision."[21]

The prevailing attitude of the "savage realists," since echoed by Ernest Hemingway, William Faulkner, Norman Mailer, and James Jones, was that one could not be a male writer without first proving one's *manliness*. In this sense, Crane, Norris, Dreiser, London, and others were "masculine" writers in the very sap of that conceit: they took almost unctuous pride in their strong passions and blithely assumed that all other men should share their robust outlook and quest for intensity. "In the name of God," Crane pleaded, "let us have virility." Norris likewise called for "great, strong, harsh, brutal men—men with purpose who let nothing, nothing stand in their way." Norris and the other chest-thumping literary insurgents idealized a life of action, of athletic endeavor, sexual adventure, and the tooth-and-claw frictions of daily existence. Their activist outlook resembled that of a character in one of Norris's short stories: "He was not literary. He had not much time for books. He lived in the midst of a strenuous, eager life, a little primal even yet; a life of passions that were often elemental in their simplicity and directness."[22]

Jack London touted himself as a "man's man," a self-described "rampant individualist" who went "raging through life . . . like one of Nietzsche's *blond beasts*, lustfully roving and conquering by sheer superiority and strength." The illegitimate son of an itinerant astrologer, London led a gypsy existence in and around Oakland, California, developing in the process a "raw and naked, wild and free" outlook: at sixteen he shipped out with oyster pirates and acquired his first mistress, then served as a seaman on a sealing schooner, worked in a cannery, tramped across the country, spent time in jail for vagrancy, became a drunken thug, embraced both Marxian socialism and Spencerian materialism, spent a listless semester enrolled at Berkeley, dug for gold in the Yukon, and covered the Russo-Japanese War for the Hearst newspaper syndicate. London claimed that his rough-and-tumble experiences had stripped him of all senti-mentalism. His devoted reading of Kipling and Nietzsche reinforced his own juvenile theories of Anglo-Saxon domination and prompted his frequent bel-lowings about the necessity for "blood-courage." He feared nothing more than becoming as "nervous and soft as a woman."[23]

The Sea-Wolf (1904) offers a graphic illustration of London's compulsive virility. Humphrey Van Weyden, a scholar rescued from a floundering ferry in San Francisco Bay by Captain Wolf Larsen, lacks a robust physique and manly outlook. Before boarding the *Ghost* he had lived a sedentary existence. "Violent life and athletic sports never appealed to me. I had always been a bookworm." As a consequence, his "muscles were small and soft like a woman's," and he was "innocent of the realities of life." Larsen decides to transform Van Weyden into a "real" man. He forces him to become his cabin boy and thereafter exposes him to the vivifying challenges of brutal conflict and primitive conditions.[24]

Although not as enamored of a "call of the wild" as London, Theodore Dreiser also aligned himself with what one critic called the "vital and virile school" of fiction. Acutely self-conscious of his own physical weakness and lack of athletic ability, he drew a sharp distinction between the "masculine" strength of a realistic outlook and the "feminine" weakness of sentimentalism. Life was a battle, he declared in *"The Genius,"* and "sickly sentimentalists" would be "thrown out or ignored by all forceful society." In *Jennie Gerhardt*, Lester Kane refused to do anything except "look the facts in the face and fight." Literary critics sympathetic to such a combative ethic praised Dreiser for his uncompro-mising force. "A man's novelist has arrived at last," proclaimed a reviewer of *The Titan* (1914). Dreiser "has rescued literature from the feminization to which it had so long succumbed."[25]

A *"Naturalistic" Philosophy*

Although turn-of-the-century literary critics usually referred to these "savage" young writers as "realists," literary historians have since lumped together the most famous of them—Crane, Norris, London, and Dreiser—as "naturalists." The label has just enough validity to be ultimately misleading. Playwright Eugene O'Neill once pleaded for "some genius" who was "gigantic enough to define clearly the separateness" of "realism" and "naturalism." He recognized that the two perspectives shared more affinities than differences. Both naturalists and realists were mimetic in approach: they agreed with Henry James that the fundamental purpose of literature was to create "a direct impression of life."[26]

But the "naturalists" found James terribly out of touch with life's raw passions and hopelessly "effeminate" in his manners. Norris argued that the modern novelist must leave the Jamesean drawing room and "go to real life for his story" because the "complications of real life are infinitely better, stronger and more original than anything you can make up." The distinctive mark of the novel was its concern with the actual world, the world of fact. Its purpose was to tell the truth about contemporary life.[27]

The other turn-of-the century "naturalists" refused to focus their writing on the comfortable and well-to-do, the virtuous and the successful. They instead opened the curtains on a cruel theater of social extremes. Where the older realists emphasized what Norris dismissed as "the smaller details of every-day life, things that are likely to happen between lunch and supper," the young "naturalists" featured characters from the lower strata of society who engaged in violent, barbaric, lustful, and often criminal behavior. Naturalism, he maintained, took "the best" from "the Realists and Romanticists" by intermingling the commonplace and sensational aspects of life.[28]

The crucial distinction between realism and naturalism, however, was philosophical. In 1868 Emile Zola provided the interpretive premise for literary naturalism when he called for a new fiction that portrayed characters "completely dominated by their nerves and blood, without free will." Although America's "savage realists" never formed a cohesive "movement" and never fastened upon a program as codified as Zola's, they did raise "naturalistic" questions about the self. Where Howellsian realists portrayed fictional characters capable of exercising independent action, rational judgment, and moral responsibility, Crane, Norris, Dreiser, London, and others felt that people often found their choices *determined* by forces beyond their rational control. "Man is not a free agent," London told a friend, "and free will is a fallacy exploded by science long ago." Likewise, Dreiser said of Jennie Gerhardt: "She was never a master of her fate. Others invariably controlled." Instead of acting as free agents, characters in naturalistic stories confront overpowering environmental and economic forces, inherit controlling hereditary tendencies, and succumb to unconscious drives and sexual urges. "Simply put," Lee Mitchell has recently observed, "realism assumes that individuals are responsible for their lives, while naturalism offers up characters who are no more than events in the world."[29]

Mitchell's definition is too "simply put," for it creates an absolute distinction to which no writer consistently adhered, but it does pinpoint the essential *area* of difference between the older "realists" and the younger "naturalists." Born after the Civil War and reaching maturity when Social Darwinism had become fashionable among academics and writers, Crane, Norris, London, and Dreiser found in biological terms such as "survival of the fittest" and "the struggle for existence" self-evident truths that applied to human conduct as well. For them, as London wrote, the "facts of life took on a fiercer aspect."[30]

Perhaps the most famous fictional representation of the survival of the fittest metaphor is in Dreiser's *The Financier*. One day young Frank Cowperwood passes a fish market and notices a squid and a lobster in a holding tank. For the next several days he watches as the lobster gradually devours the squid. The drama reveals to him the riddle of existence: "Things lived on each other—that was it. Lobsters lived on squids and other things. What lived on lobsters? Men, of course! Sure, that was it! And what lived on men? he asked himself. Was it other men?"[31]

Such a "survival of the fittest" ethic leads many characters in naturalistic fiction to behave more like apes than angels. Sociologist Edward Ross gave added legitimacy to such notions when he argued in *Social Control* (1901) that "there is an unreclaimed jungle in man. . . . There are unpleasant, slimy things lurking in the. . . undergrowth of the human soul." The "savage realists" agreed with Ross that bestial inclinations were unruly residents of the human psyche, and they readily—too readily—invoked Darwinian similes to explain how people could easily regress to a savage state and outlook. Crane, for example, asserted that most Americans acted like "animals in a jungle." London described the United States as "a jungle wherein wild beasts eat and are eaten." Most of Norris's characters hear "an echo from the jungle" as they struggle to suppress their animalistic tendencies. Dreiser wrote in *The Titan* that the surface of life "might appear commonplace . . . yet a jungle-like complexity was present, a dark, rank growth of horrific but avid life—life at the full, life knife at hand, life blazing with courage and dripping at the jaws of hunger."[32]

To one degree of another, then, these "savage realists" employed in their fiction the basic elements usually associated with the "naturalistic" impulse: unwholesome social environments and lower-class characters, animal imagery, event-intoxicated prose written in the superlative degree, an amoral and mechanistic universe seemingly beyond human control or understanding, a welter of violence, capricious instincts, uninhibited lust, and wholesale bloodshed. Characters in their fiction are neither inherently good nor free agents; several lack self-forging initiative and degenerate rather than develop over the course of the narrative. Yet such "naturalistic" qualities do not alone define the new type of fiction produced by Crane, Norris, and Dreiser. In each case they expressed contrary tendencies and unique talents that warrant extended discussion.

11

A World Full of Fists

"I go through the world unexplained," Stephen Crane once observed. He has remained an enigmatic character, a prodigal prodigy whose exhausting genius and lusty adventures eventually overtaxed his fragile constitution. The most talented and self-conscious stylist among the late nineteenth-century literary rebels, he became an international celebrity in 1895 at age twenty-four, only to die of tuberculosis five years later. His irreverent intensity and unfulfilled promise combined to make him a talismanic figure for his literary generation. There was something exorbitant about Crane that made him special. More alive and alert than other people, he cultivated an idiosyncratic outlook that resists pigeonholing. Literary historians have since classified him as a naturalist, symbolist, impressionist, or imagist, and each label has an awkward legitimacy. Yet Crane himself claimed to be a realist. He referred to Howells as his literary "father" and described his literary goal as "the pursuit of truth."[1]

Crane's pursuit of truth, however, was far more brazen than that of his chosen mentor. "Of course, I am admittedly a savage," he told a friend in 1896. "I am by inclination a wild shaggy barbarian." Possessed of a fierce candor and a defiant preference for the "unvarnished" aspects of life, he thumbed his nose at virtually every genteel convention and explored sordid realities and extreme situations rarely mentioned in polite literature. Crane set much of his fiction in the midst of wars, slums, and life-threatening situations. He believed that life was a perpetual "battle," an unruly "world full of fists," and most of his novels and stories highlight a fierce struggle for survival within a "flatly indifferent" universe. Yet Crane was not simply a mechanistic determinist. Life to him was like a game of cards. People had no control over the cards they were dealt, but they did have some flexibility in the way they played them. For fiction to succeed, he realized, it must provide at least the illusion of free will. Characters in his writings thus often remain *responsible* for essential aspects of their lives, and he used the sharp blade of irony to lay bare their fantasies and conceits.[2]

CRANE WAS BORN in Newark, New Jersey, in 1871, the same year as Dreiser. The youngest of fourteen children, he grew up within a household permeated with piety. On his mother's side of the family, he recalled, "everybody as soon as he could walk, became a Methodist clergyman—of the old ambling-nag, saddle-bag, exhorting kind." His father, Jonathan Townley Crane, was a prominent Methodist minister who wrote sanctimonious tracts condemning dancing and drinking. He died when Stephen was eight. Thereafter, death became a familiar occurrence in the Crane household. By 1892 only seven of the fourteen children were still living.[3]

Well-read, ambitious, and a zealous crusader against vice, Mary Peck Crane tried to inoculate her children with spiritual fervor. Young Stephen, however, spurned his mother's "mildewed" religious convictions and moral rigidity; he preferred to worship the forbidden. In the fall of 1890, he entered Lafayette College and discovered that the "fellows here raise more hell than any college in the country." Crane's fall semester grades reflected his skewed priorities: he failed five of his seven courses. The faculty minutes tersely noted that Crane "was dropped from the rolls." Crane thereupon transferred to Syracuse University, where he starred as a catcher and shortstop on the baseball team. Although weighing only 125 pounds, he was a ferocious competitor. Teammates noted that he played with "fiendish glee" and was "giftedly profane."[4]

The athletic Crane was keenly interested in literature, and during his stay in the Delta Upsilon house at Syracuse he began writing in earnest. He published a short story in the college magazine and started what would become his first novel, *Maggie, a Girl of the Streets*, a powerful rendering of Irish immigrant life in lower Manhattan. Crane also worked as a part-time reporter for the *New York Tribune*, covering inner-city Syracuse and the police court.

After only one semester at Syracuse, however, Crane flunked out. He then took a job as a reporter for his brother's news agency in Asbury Park, New Jersey. Working as a journalist convinced him that "the most artistic and the most enduring literature was that which reflected life accurately." He learned "to observe closely, and to set down what I have seen in the simplest and most concise way." During his years as an apprentice reporter, Crane developed what he called "a little creed of art which I thought was a good one. Later I discovered that my creed was identical with the one of Howells and Garland."[5]

Crane's mother died in December 1891, and afterwards he moved in with his brother Edmund in Lake View, New Jersey, not far from Manhattan. There

Crane developed his own version of the strenuous life with all of its virile overtones. He was determined not to be a "coxcomb" or a "sissy," and he sought out ways to test his mettle and pursue the mystery of manliness. He went hunting and camping with his brothers, served as coach and bantam quarterback for the town's new football team, and fell in love with a married woman. But he devoted most of his passion to exploring life in the immigrant neighborhoods and theatre districts of the Bowery and the Lower East Side, which he visited almost daily.

Crane eventually moved to New York City, sharing a room first with a group of medical students living in a lower Manhattan flat overlooking the East River, then boarding with three impoverished painters in the old Art Students' League building at 143 East 23rd Street. As the months passed, Crane grew intimate with the inner city in all its moods and aspects, brutalities and charms. His urban wanderings led him to develop a notion common to the other "savage realists": life was more genuine, more intense, more "real" among the poor than the affluent. In Bowery bars and parks, he sat "soaking in knowledge of the reactions of the kind of men he liked to write about." Convinced that no creative writer had yet said anything "sincere" about Bowery life, Crane resolved to give it honest representation in fiction. The "nearer a writer gets to life," he stressed, "the greater he becomes an artist, and most of my prose writings have been toward the goal partially described by the misunderstood and abused word *realism*."[6]

Crane viewed literary realism as a means to express the "truth of life" around him, "without regard to cheers or damnation" from the critics. "A man is born into this world with his own pair of eyes," he observed, "and he is not at all responsible for his vision—he is merely responsible for his quality of personal honesty. To keep close to honesty is my supreme ambition." Sincerity of feeling and authenticity of expression became for Crane a substitute for transcendent belief. Yet he did not assume that he could simply "mirror" reality in words.[7]

Crane knew that literal realism was impossible. "A man is sure to fail at it, but there is something in the failure." His objective was not to produce an *imitation* of life but to offer a vivid *impression* of life as he saw and experienced it. He knew that realistic fiction at its best represented a translation of life rather than a transcription. To "get down the real thing" in print and enhance the actuality of his impressionistic narratives, he once stood in a bread line with a group of unemployed men during a blizzard and used the experience as the basis for his story, "Men in the Storm." In preparation for "An Experiment in Misery," he slept in a Bowery flophouse.[8]

Crane also used his Bowery experiences to integrate color and texture into *Maggie, a Girl of the Streets*. After finishing the manuscript in 1892, he offered it to several magazine editors, but they balked at its unrelenting grimness. So in March 1893 he arranged to have 1100 copies of *Maggie* published at his own expense. The newsstands, however, showed no interest in the book. Eventually Crane gave away a hundred or so copies, often to settle debts. In an inscription to one young woman, he wrote: "This story will not edify or improve you and may not even interest you but I owe your papa $1.30 for tobacco." Most copies of the first edition of *Maggie* gathered dust in a corner of the author's room.[9]

Crane was both destitute and depressed. "I am going to chuck the whole thing," he told E. J. Edwards, an editor for the *New York Press*. "I have worked for two years, living with tramps in the tenements on the East Side so that I could get to know these people as they are, and what is the use? In all that time I have received only $25 for my work." Crane then left the newspaper office and began walking down Broadway. He had taken only a few steps when a newspaper reporter named Curtis Brown grabbed his shoulder and said, "Crane, what do you think? William Dean Howells has read your book, and he says it's great." Crane stammered a confused reply, leading Brown to reiterate: "I say that Howells has read your book, and he compares you with Tolstoy, and he is going to say so in print." Crane "grinned like a woman in hysterics, and then went off to take up his vocation again." In his rave review of *Maggie*, Howells warned that Crane's realism might be too realistic. There "is so much realism of a certain kind in it that unfits it for the general reading, but once in a while it will do to tell the truth as completely as *Maggie* does."[10]

The novella deals with what had become a hackneyed theme in sentimental literature: the descent of an innocent slum girl into prostitution and suicide. Yet Crane gave the familiar story a new twist. Instead of pointing a moral, he left key issues unresolved. Maggie Johnson grows up in a "gruesome" Rum Alley tenement suffused with violence rather than love. Her "gnarled and leathery" mother, Mary, is a drunken, blaspheming brute, her father a sullen derelict, and her brother a brawling truck driver known for his "chronic sneer." Curses, blows, and the sound of splintering furniture punctuate the household's chaotic routine.

Weary of her warring family and enervated by her job making shirt collars in a prison-like sweatshop, Maggie succumbs to romantic fantasy when a swaggering sixteen-year-old bartender and street fighter named Pete shows interest in her: "Here was a formidable man who . . . had a contempt for brass-clothed power; one whose knuckles could defiantly ring against the granite of law." Pete liberates the vulnerable Maggie from her squalid routine. He takes her to beer halls, the Central Park Zoo, wax museums, and cheap melodramas that nourish her illusions. Eventually they decide to live together. With Pete, she imagines a "rose-tinted" future.[11]

After only a few weeks of cohabitation, however, Pete summarily abandons Maggie in favor of an older, more experienced woman. "But where kin I go?" she pleads. Pete tells her to "go teh hell!" A dazed Maggie returns home, only to be greeted by the "scoffing laughter" of her mother and brother. Mary Johnson, sodden with gin and sentimental self-pity, banishes Maggie from the apartment as the other tenement dwellers look on with leering curiosity. Forced to fend for herself, the broken-hearted Maggie becomes a prostitute. One rainy evening, after failing to solicit any business, she aimlessly wanders from "glittering avenues" into the "blackness of the final block," and, apparently, drowns herself in the East River.[12]

Was Maggie a victim of a vicious environment? poverty? self-deception? social hypocrisy? Scholars still debate the various possibilities. Crane himself stated that he "had no other purpose in writing *Maggie* than to show people to people as they seemed to me." He disavowed any reform agenda or even didactic intent; he saw himself as a literary artist, not a social crusader. "Preaching," he once argued, "is fatal to art in literature." If there is any moral or lesson in a story, "I do not point it out. I let the reader find it for himself." Yet in a copy of *Maggie* that he presented to Garland, Crane penned the following "preachy" inscription: "It is inevitable that you be greatly shocked by this book but continue, please, with all possible courage to the end. For it tries to show that environment is a tremendous thing in the world and frequently shapes lives regardless. If one proves that theory, one makes room in Heaven for all sorts of souls (notably an occasional street girl) who are not confidently expected to be there by many excellent people."[13]

Literary historians have cited the first part of this statement as proof positive that Crane was a "naturalist" preoccupied with the controlling force of environment. Yet Crane the "realist" believed that individuals often share responsibility for their fate. The forces causing Maggie's downfall are largely circumstantial rather than deterministic. Individual responsibility was possible, even in the Bowery, but it was not often exercised, he concluded, for slum life involved a struggle for survival in which most people were armed with the inadequate weapons of ignorance, innocence, and self-delusion. The poor rarely snatched the chance to become authors of their own fate; too many of them lacked backbone and vitality. Crane once told an admirer that his stories had "tried to make plain" that "the root of Bowery life is a sort of cowardice. Perhaps I mean a lack of ambition or to willingly be knocked flat and accept the licking."[14]

In exploring the moods and motives of Bowery denizens, Crane was as much a psychological realist as he was a mechanistic naturalist. To him, what went on within a person's mind was as significant as the nature of outside circumstances. Individuals never simply "mirrored" external reality. People perceived the same "real thing" differently. In *Maggie*, characters create their own mental worlds, drawing their own different portraits of the same "reality." They view life through the "blurred glass" of fantasy or the "mist of muddled sentiment." Maggie "blossomed in a mud puddle," but her own false notions of the good life betrayed her. The world does not live up to her dreams, and it refuses to satisfy her desires. In this regard, Crane said that he wanted to understand why people could not see things "with a true eye."[15]

By rendering scenes rather than narrating events, by focusing on how individuals misapprehend or ignore stubborn facts, Crane nudged literary realism onto a new plane of immediacy and indeterminacy. He displayed an uncanny ability to make ordinary scenes seem *extraordinary*, as in his plangent description of a tenement courtyard:

> from a careening building, a dozen gruesome doorways gave up loads of babies to the street and gutter. A wind of early autumn raised yellow dust from cobbles and swirled it against a hundred windows. Long streamers of garments fluttered from fire-escapes. In all unhandy places there were buckets, brooms, rags, and bottles. In the streets infants played or fought with other infants or sat stupidly in the way of vehicles. Formidable women, with uncombed hair and disordered dress, gossiped while leaning on railings, or screamed in frantic quarrels. Withered persons, in curious postures of submission to something, sat smoking pipes in obscure corners. A thousand odors of cooking came forth from the street. The building quivered and creaked from the weight of humanity stamping about in its bowels.

Such passages indicate the sensory appeal and impressionistic quality of Crane's prose. Using prose startling in his freshness, he enabled readers to imagine themselves as spectators at the scene of his fictions. Said the reviewer for the *New York Times:* "Mr. Crane cannot have seen all that he describes, and yet the reader feels that he must have seen it all. This, perhaps, is the highest praise one can give the book."[16]

In painting word "pictures of his time," however, the youthful Crane indulged in stylistic excesses that older realists had steadfastly avoided. Much of the dialogue and the action in *Maggie* seems artificial and exaggerated, sown with startling metaphors and screaming diction that call attention to the writer rather than the story. Yet his hyperactive prose had purposes other than exhibitionism. By unveiling the elemental desires and desperate circumstances of the slums, Crane sought to make the "unreal real." He thus keyed most of the scenes in *Maggie* at the highest pitches of emotion and expression: people shout, bellow, roar, wail, drink, and brawl. Crane's own experiences in Manhattan had convinced him that "the sense of a city is *war*," and he repeatedly refers to the Bowery as a battlefield. So much furniture is destroyed during Mary Johnson's liquor-induced rampages that readers have rightly questioned how these desperately poor people could have afforded such weekly losses. Yet underneath the book's frenetic clamor and breathless prose resides a kernel of credibility that awed and appalled contemporary readers.[17]

Maggie stretched the limits of realism to the breaking point. Although Howells and others had featured the inner city and the urban poor as accessories to their middle-class dramas, Crane was one of the first writers to focus on the lower working class and do so with first-hand knowledge and little sentimentalism. Throughout *Maggie*, as well as in its sequel, *George's Mother*, he mocks the pious assumptions of the cult of domesticity. Instead of the Johnson home serving as a sanctuary of familial love and Christian devotion, it is a habitat in which primal instincts reign. Maggie's massive-shouldered mother is a homewrecker rather than a nurturing Madonna, and her children enter the apartment with more trepidation than relief. At one point Jimmie crawls home "with the caution of an invader of a panther's den." The family dinner table resembles a Neanderthal cave after a successful hunt: "The babe sat with his feet dangling high from a precarious infant's chair and gorged his small stomach. Jimmie forced, with feverish rapidity, the grease-enveloped pieces between his wounded lips. Maggie, with side glances of fear of interruption, ate like a small pursued tigress."[18]

Although the first edition of *Maggie* went virtually unnoticed, the book was resurrected after the smashing success of Crane's Civil War novel, *The Red Badge of Courage* (1895). D. Appleton and Company released a slightly revised version of *Maggie* in 1896, touting it as a "real and strenuous tale of New York life," and it now garnered enthusiastic praise. The *Nashville Banner* labeled *Maggie* a "magnificent piece of realism," while *Godey's Magazine* called it "the strongest piece of slum writing we have." After acknowledging the book's "aggressive and pitiless realism," a Boston literary critic recommended it to readers "who are not afraid to face the realities of existence, and who are willing to look upon humanity at its worst." Crane's "refusal to sentimentalise" led a London reviewer to list him as a member of the "sternly realistic school." Howells deemed *Maggie* "great."[19]

In *Maggie, George's Mother, The Red Badge of Courage,* and many of his short stories, Crane revealed a frantic urgency to endow life with meaning in the face of forces that coerce the self. He embraced neither the absolute determination of Zola's naturalism nor Howell's sunny faith in free will, but instead combined elements of both perspectives. He saw the human condition as inescapably problematic: mixed motives, partial insights, and romantic illusions conspire to distort perceptions of reality, while brute instincts and environmental conditions exert constant pressure against the rational self. The physical universe of Crane's fiction is indifferent to individual fate; it offers no simple or decisive answers to profound moral questions; there is no providential God intervening on behalf of his characters.

Yet even though people live in a web of discomfort and mean circumstances, they are not merely helpless victims. Those who face stubborn facts and shed themselves of delusive ideals, who recognize life as a field of combat and take the offensive through defiant gestures and bold deeds, endow the absurdities and cruelties of life with meaning and lay claim to a serene dignity. "I do not confront [life] blithely," he once wrote a friend. "I confront it with desperate resolution. There is not even much hope in my attitude. I do not even expect to do good. But I expect to make a sincere, desperate, lonely battle to remain true to my conception of my life and the way it should be lived."[20]

Frank Norris

In 1896 a young San Francisco writer named Frank Norris wrote a review of *Maggie* in which he questioned its fidelity to reality. The author, he charged, seemed too distant from his subjects. Two years older than Crane, jealous of his success, and possessed of the same guzzling confidence and raw vitality, Norris concluded that the author of *Maggie* was too self-conscious a stylist. With Crane, he stressed, "it is the broader, vaguer, *human* interest that is the main thing, not the smaller details of a particular phase of life."[21]

For his own part, Norris claimed to be indifferent to matters of style. Like so many other literary naturalists, he was constantly denying the "literariness" of his own work. "It's the Life that we want," he asserted, "the vigorous, real thing, not the curious weaving of words and the polish of literary finish." His disavowal of stylistic concerns was intended to reassure readers of his sincerity and accuracy. This is not fiction, he wanted to convey; it is real. These are simply the facts of the matter without any literary costuming. Despite his disclaimers, however, Norris was a self-aware stylist. He employed highly colored and abundant prose, declaratively brash in its tone and evidently straining after dramatic effects. Most of his novels buckle under the stress of their overwrought manner. As the poet John Berryman once quipped, Norris's verbose style resembled that of "a great wet dog."[22]

Norris was in his own words a "romantic realist." Like Crane, he could not limit himself to the representation of the commonplace occurrences of middle-class life. He instead depicted the fierce extremes of existence, relying overmuch on coarse simplifications, gratuitous violence, and emotional hyperbole. Only late in his foreshortened life (he died at thirty-two) did he begin to show signs of literary maturity. As the *New York Herald* observed upon learning of Norris's death, "American literature [has] lost its greatest potentiality in fiction."[23]

Frank Norris represents a classic American excess story. People who knew him well said that they never quite figured him out, but everyone agrees that his family background holds the key to his personality. Born in Chicago in 1870, the son of a wealthy jeweler-merchant and a former actress, Norris moved with his family to San Francisco in 1884. His father was aloof and practical-minded, the stereotypical bourgeois preoccupied with his "masculine" business affairs. His mother, on the other hand, epitomized the smothering cult of refined domesticity. In a calculated effort to mold Frank into a young man of culture, she outfitted him in velvet suits, read to him from Dickens and Scott, arranged for dancing lessons, and supervised his confirmation in the Episcopal Church.[24]

In 1890 Norris entered the University of California at Berkeley, and he relished his freedom from maternal supervision. Like Vandover, the young artist in his first novel, *Vandover and the Brute*, he had come to "chafe under his innate respect and deference for women, to resent and despise it." At Berkeley, Norris lived in the Phi Gamma Delta house, played in all-night poker games, drank beer at Hagerty's saloon, courted attractive women, and patronized prostitutes. Attending university football games fed his passion for violence. On the playing field he watched "red-blooded he-men" engage in noble strife, displaying in the process that "dogged, stubborn spirit that . . . absolutely refuses to be defeated."[25]

Social Darwinism was then the rage at Berkeley, and its tooth-and-claw premises nourished Norris's fascination with "primordial instincts." Geologist Joseph Le Conte, his favorite professor, gloried in the "upward movement" of human development. At the same time, however, he warned students that people possessed two temperaments, one bestial and one civilized. A person's animal nature provided a vital "life force," but if allowed to run amok it could also unleash criminal or deviant tendencies.

Norris took from Le Conte's teachings two basic premises. First, he assumed that there "is a moral order inherent in nature and its laws" and that God represented the "life force" both in nature and in people. The operations of the natural world and the social system, therefore, must necessarily contribute to the upward spiral of evolutionary development. Whatever *is*, in other words, must be right in the ultimate scheme of things. Second, Norris became keenly aware of humankind's powerful sexual drive and atavistic tendencies. Indeed, Le Conte's emphasis on the tension between the self's "higher" and "lower" impulses helped Norris understand his own struggle with "brutish" behavior and irrational desires.[26]

As a boisterous college student Norris embraced the sophomoric notions of Anglo-Saxon racial superiority common at the time. He delighted in the discovery that "somewhere in the heart of every Anglo-Saxon lies the predatory instinct of his Viking ancestors." Writing in defense of fraternity hazing rituals, he argued that a "good fight" would wake in a young man "that fine, reckless arrogance, that splended, brutal, bullying spirit that is the Anglo-Saxon's birthright. . . . The life of men in the world is one big 'rush' after all, where only the fittest survive and the weakest go to the wall."[27]

Although enamored of fraternity life and football games, Norris claimed that he had entered Berkeley "with the view of preparing myself for the profession of a writer of fiction." Classmates remembered that he always seemed to have one of Zola's novels tucked in his coat pocket, and he often signed his letters, "The Boy Zola." Norris shared Zola's fascination with the coarse and violent aspects of life among the lower classes. Writers, he stressed, were "more apt to find undisguised human nature along the poorer unconventional thorough-fares." Life seemed more elemental, more "real," among laboring folk. Norris's own ramblings in San Francisco's working-class districts revealed to him a culture fermenting with struggle, tragedy, sensuality, and grotesque characters. In 1893, for instance, he took keen interest in newspaper reports of the brutal murder of Sarah Collins, a kindergarten cleaning woman killed by her sottish husband, an incident Norris would latter use as the plot for *McTeague*.[28]

Norris failed to earn a degree from Berkeley, but he gained admission to Harvard as a "special student" in the fall of 1894. In Cambridge, he wrote his first serious fiction under the direction of a young professor named Lewis E. Gates, an enthusiastic champion of "the great world of fact." Writers, said Gates, should not shy away from the "commonplace and from the crude" aspects of contemporary life. They should instead use a "microscopic eye and the inge-nious instinct for detail" to interpret "life's puzzling complexities of motive, character, and passionate action." Gates called for an "imaginative realism" that would explore the nether "regions of the Actual." Under his tutelage, Norris wrote several sketches that would later be incorporated in his novels.[29]

In 1898 Norris relocated to New York City to join the staff of *McClure's Magazine*. Within a month of his arrival, he arranged to visit Howells and found him to be "one of the most delightful men imaginable." Norris appre-ciated Howells's pathbreaking literary efforts and his personal encouragement, but in "A Plea for Romantic Fiction," he chastised Howells and James for having limited their realism to the "drama of a broken teacup, the tragedy of a walk down the block, the excitement of an afternoon call, the adventure of an invi-tation to dinner." He pleaded instead for stories that featured strange characters and bizarre plots, stories that exposed "evil" and "sordid" truths, stories that dealt with primal "motives and emotions" rather than "merely externals." Norris wanted writers to explore "life uncloaked and bare of convention—the raw, naked thing, that perplexes and fascinates—life that involves death of the sudden and swift variety, the jar and shock of unleashed passions."[30]

Most of all, Norris pleaded for writers who were "real" men: "Let us have men who are masculine, men who have other things to think of besides fooling away their time in ballrooms." Such macho bluster seems compensatory, the mark of Norris's own physical weakness and inner insecurity, a shield against critics and a barrier between the hardy self he had fashioned and the "sissified" self his mother had cultivated. Norris's hysterical promotion of manly vigor also reflected his distaste for the *fin-de-siècle* aesthetes who were generating excited attention in the press. He contemptuously dismissed Oscar Wilde and his Amer-ican imitators as "long-haired," effete "posers" who wore "salmon pink" clothes. These effeminate promoters of "art for art's sake" were "sexless creatures who cultivate their little art of writing as the fancier cultivates his orchid."[31]

Norris saturated his own fiction with "manly" themes—brute instincts and single-minded drives—and in the process violated many realistic conventions. *Vandover and the Brute* (written first but posthumously published in 1914), *Moran of the Lady Letty* (1898), *McTeague* (1899), *Blix* (1899), and *A Man's Woman* (1900) feature primitive brutality, uncontrolled sexuality, gross materialism, and garish symbolism. As one reviewer pointed out, Norris strove "to give reality to the unusual" and in the process send unsettling tremors through bourgeois society. Yet he often provided "realism without reality." His novels suffer from a fatal moral simplicity and display a young man's pleasure in the exaggerated and fantastic, the criminal and mad.[32]

Characters in Norris's adventure stories are clumsy symbols of primal forces. They feature "brutal jaws," "huge fists," and "enormous strength." Several of his female characters are "she-men," masochistic creatures of unrelenting passion and burly physique who find their greatest satisfaction in being physically and sexually assaulted. They prefer men with "lots of evil" and become "a man's woman" by exchanging their autonomy for love and sexual pleasure. Norris's male characters, on the other hand, tend to be either decadent sophisticates or brutish simpletons who succumb to sensual and sadistic desires and rarely indulge in mental reflection. In *A Man's Woman*, for example, Ward Bennett embodies the evolutionary heritage of the human species. His lower jaw is "huge, almost to deformity, like that of a bull-dog." At one point he contemplates killing and eating his dog: "He would live; he, the strongest, the fittest, would survive."[33]

Norris's fiction offered a theatre of feral extremes. While creating characters so grotesque that they defy credibility, he also provided long stretches of "realistic" description of places and things. With an anthropologist's eye for the telling physical detail and the cultural ambience of particular neighborhoods and social groups, he graphically brought to life California's varied landscapes. Norris's novels contain great chunks of commonplace and even trivial material, prose inventories of people's possessions, clothes, and furnishings, all designed to impress upon the reader the shaping impact of the physical environment.[34]

Yet Norris insisted that the serious writer must go beyond a concern for descriptive "accuracy" in order to convey a "dramatic sense of truth." He refused to equate "truth" with "what life actually is" because "reality" was simply "what it looks like to an interesting, impressionble" writer or artist. Time and again, Norris's commitment to tell a "rip-roaring" story and his preoccupation with deterministic forces led him to spice his fiction with outlandish allegories, melodramatic incidents, and capriciously cruel encounters. Herein lies the basic problem with his mixing of "romance" with "realism"; it is utterly inconsistent. There is something childish, almost inane, about Norris's tortured toughness, his love of climactic struggles, and his mania for violence. Characters in two of his novels have one or more digits or hands removed, and by the end, most of the others are mutilated, dead, destitute, or grieving. Such excesses coarsen rather than enrich the puzzle of life. In sum, his cleverness was often captivating but rarely revealing.[35]

Norris's most consistently "naturalistic" and cohesive novel is *McTeague: A Story of San Francisco* (1899). It tells the story of an alcoholic miner's son who has become a self-taught, unlicensed dentist on Polk Street in San Francisco. The "absolutely stupid" McTeague is a Teutonic behemoth with a "huge shock of blond hair" and "immense limbs." His hands "were hard as wooden mallets, strong as vices. . . . Often he dispensed with forceps and extracted a refractory tooth with his thumb and finger." McTeague's mental powers pale in comparison to his physical strength, and his banal existence follows a regular pattern: six days of work followed by a Sunday of eating, smoking, and "gorged" sleep. His one ambition in life is to acquire a huge gold tooth to serve as a display for his business.[36]

McTeague's vapid routine abruptly changes when his best friend, Marcus Schouler, brings his sweetheart, Trina Sieppe, to the "Dental Parlor" to have a broken tooth repaired. Her frail beauty captivates McTeague. While hovering over his anesthetized patient, he suddenly "leaned over and kissed her grossly, full on the mouth. The thing was done before he knew it."[37]

Norris leaves no doubt that forces beyond McTeague's conscious control determine his actions: "Below the fine fabric of all that was good in him ran the foul stream of hereditary evil, like a sewer. The vices and sins of his father and of his father's father . . . tainted him." Yet at the same time that Norris preached such genetic determinism, he acknowledged that random chance also affected human fate. After surrendering to McTeague's advances, Trina struggles to make sense of what has happened. "Did she choose him . . . of her own free will?" No, she decides, love is "ruled by chance alone." Norris adds that "neither of them was to blame. From the first they had not sought each other." Chance and "mysterious instincts" had brought them together.[38]

Trina's wedding to her "big bear" created one of the most notorious marriages in American literary history. Readers were shocked to encounter two such loathsome characters whose behavior is instinctual and spontaneous rather than reasoned and deliberate. Primal passions brought them together, and morbid avarice eventually tears them apart. A relentless struggle for money, in fact, becomes the dominant motif of the story. After winning $5000 in a lottery, Trina becomes a surly miser who loves "her money with an intensity that she could hardly express."[39]

McTeague, too, rapidly degenerates. Initially his marriage had sparked an ambition to better himself. He bought a Prince Albert coat and a silk hat, began reading the newspaper, and even joined the "Polk Street Improvement Club." His efforts at self-improvement, however, collapse when a vengeful Marcus, angry because his friend had stolen his "girl," tells the authorities that McTeague had never graduated from a dental college. Forced to close his business, McTeague reverts to his slovenly ways. When the pitiless Trina refuses to support him with her cherished gold, he develops the disgusting habit of biting the tips of her fingers, "crunching and grinding them with his immense teeth, always ingenious enough to remember which [fingers] were the sorest." Yet such physical abuse arouses in Trina "a morbid, unwholesome love of submission. . . to the will of an irresistible, virile power."[40]

This gruesome study of mutual disintegration concludes on a note of pure melodrama. A drunken McTeague forces his way into Trina's room, beats her to death, takes the remaining gold, and flees to the mining country. Pursued by a posse that includes Marcus Schouler, a desperate McTeague tries to cross Death Valley on a mule. Eventually, Marcus corners his prey. The two men, both deranged by heat and thirst, engage in furious combat:

> McTeague did not know how he killed his enemy, but all at once Marcus grew still beneath his blows. Then there was a sudden last return of energy.
> McTeague's right wrist was caught, something clicked upon it, and the struggling body fell limp and motionless with a long breath.
>
> As McTeague rose to his feet, he felt a pull at his right wrist; something held it fast. Looking down, he saw that Marcus in their last struggle had found strength to handcuff their wrists together. Marcus was dead now; McTeague was locked to the body. All about him, vast, interminable, stretched the measureless leagues of Death Valley.

When friends and reviewers chastised Norris for such an overdrawn ending, he countered that the primary purpose of literature was to "produce an interesting story—nothing more."[41]

In this regard Norris succeeded. The *Literary World* announced that *McTeague* would place Norris "in the first rank among our writers, beside Mary Wilkins [Freeman], and Howells, and Stephen Crane." Readers who shuddered at the novel's brutalities kept turning the pages. The literary editor of the *Louisville Times* found the book "strong, virile, throbbing with life," while his counterpart on the *Washington Times* said that "nothing stronger in the way of realistic fiction is to be found in American literature."[42]

Howells was equally effusive. Of course, he acknowledged, "polite readers" would not like the book, but those willing to have their horizons broadened would benefit from *McTeague's* "strong" presentation of life's underside. He marveled at the novel's dense particulars and its fidelity to the actual events Norris had witnessed on Polk Street. *McTeague* "abounds in touches of character at once fine and free, in little miracles of observation, in vivid insight, in simple and subtle expression." In concluding his review, however, Howells highlighted the difference between Norris's "Zolaesque" naturalism and his own version of realism: "His true picture of life is not true because it leaves beauty out. Life is squalid and cruel and vile and hateful, but it is noble and tender and pure and lovely, too."[43]

Howells's review so flattered Norris that the young author wrote him a letter in which he agreed with virtually "every one of your criticisms." Thereafter, he resolved to abandon the atavistic motifs of his early novels and attempt a modern epic that would speak to contemporary issues and incorporate some of the finer things" in life. Norris told a friend that he was taking "myself and my work much more seriously."[44]

In early 1899 Norris drew up the blueprint for a trilogy of "novels with a purpose" centered on the production of wheat—how it was raised in California, brokered and sold in Chicago commodity markets, and shipped cross-country and around the world by railroad and steamship lines. He titled the first novel in the series *The Octopus* and decided to base it on an actual incident: a bloody shootout in the San Joaquin valley between wheat ranchers and a sheriff's posse acting on behalf of the Southern Pacific Railroad. The ranchers had been farming land leased from the railroad. Intending eventually to buy the property at a cost of $2.50 to $5 an acre, they improved the land by building extensive irrigation systems and soon were producing bumper crops. When the railroad informed the ranchers that the price for the land had risen to $40 an acre, they organized armed resistance. Eight men were killed in a confrontation at Mussel Slough, and soon thereafter the remaining ranchers were jailed before being forced off their lands.

Norris spent four months in San Francisco and the San Joaquin valley doing field research before returning to New York in 1899 with a sheaf of notes, newspaper clippings, interviews, and visual impressions. *The Octopus*, completed in December 1900, represented his most ambitious novel. A vast, sprawling book, it swarms with diverse characters and uses a vivid palette and poetic imagery to illustrate the colorful panorama of western life and the cycle of wheat production.

Although filled with secondary themes and subplots, *The Octopus* essentially describes how the railroads spread their steel tentacles not only across the landscape but also around the economic life of an entire region. Like other hated "trusts," the Southern Pacific Railroad monopolized California's agricultural economy by regulating the ownership of land and the stream of production and distribution. In *The Octopus* a group of wheat ranchers band together to thwart the railroad's efforts to seize their lands. They are not simple yeomen but land barons who preside over fiefdoms of wheat, agrarian entrepreneurs who display "no love for the land. They were not attached to the soil." The ranchers engage in speculative ventures and financial chicanery not unlike the activities of the railroad owners. Yet Norris confuses matters when he casts these bonanza farmers as "The People" victimized by the railroad. Ultimately, a bloody confrontation between the two factions results in the death of several ranchers. Others are arrested, freight rates raised, and ranches seized. "Into the prosperous valley, into the quiet community of farmers, that galloping monster, that terror of steel and steam had burst . . . leaving blood and destruction in its path."[45]

In the process of telling this dramatic story, Norris introduces many of the "real" issues affecting the California economy. Foreign markets, watered stock, high freight rates, railroad regulation, ethnic prejudice toward Chinese, Mexican, and Portuguese laborers, worker unrest, blacklisting—all play a role in the clash of interests between ranchers and railroads. In this respect, *The Octopus* was one of the first American novels to examine monopolistic capitalism in detail. As usual, however, Norris could not content himself with a straightforward handling of actual economic and social issues. Instead he tried to graft a subplot of mystical romanticism onto the novel's basic story, and he thereby muddled the book's message.

One of the main characters in *The Octopus* personifies Norris's own warring impulses. Presley, the neurasthenic eastern poet slumming with the wheat ranchers, dreams of producing an epic integrating "that huge romantic West that he saw in his imagination" with the grim, unlovely, unyielding reality of life-and-death struggle between the farmers and the railroad owners: "On one hand it was his ambition to portray life as he saw it—directly, frankly, and through no medium of personality or temperament. But, on the other hand, as well, he wished to see everything through a rose-colored mist—a mist that dulled all harsh outlines, all crude and violent colors."[46]

Norris himself tried to view life through both lenses simultaneously, and the result is a blurred image. In *The Octopus* he portrayed a mythic conflict between a vital natural order and the abstract power of money. The problem for individuals is to align themselves with the most beneficent "forces of life." Near the novel's end, Shelgrim, the shadowy head of the railroad, explains to Presley why the ranchers lost their fight: "You are dealing with forces, young man, when you speak of Wheat and the Railroad, not with men." The laws of supply and demand and the life processes of nature and its products, he stresses, determine personal destinies. Nature was not to blame for such an inexorable system of force. "There was no malevolence in Nature," Presley decides, ". . . only a vast trend toward appointed goals." With Zola and Spencer, he concludes that the law of strife and struggle and the immense capacity of Nature (as symbolized by the wheat) to reconstitute itself ensures the greatest good for the greatest number. "The individual suffers, but the race goes on."[47]

This sanguine ending has baffled readers because its benign fatalism contradicts the major events and overriding tone of the novel. Such buoyant absolution of a railroad combine that had cheated landowners, bribed judges, juries, and legislators, evicted tenants, murdered farmers, and disrupted families falls flat. *The Octopus* is fundamentally the story of stubborn people refusing to submit to the railroad. The arrogant young rancher Annixter, for example, realizes that he and the other farmers are warring against terrible odds, but he nevertheless decides to resist rather than submit. In doing so he liberates himself from what had been a crabbed selfishness and discovers the redemptive power of love. "I'm tired of fighting for *things*—land, property, money," he declares. "I want to fight for some *person*—somebody else beside myself."[48]

Despite his penchant for overdone symbolism and his inability to blend romanticism with realism, Norris revealed in *The Octopus* a maturation of insight. Howells pronounced it "a great book, simple, somber, large, and of a final authority as the record of a tragical passage of American, of human events." He was amazed that such a young writer could produce a novel "so vast in scope while so fine and beautiful in detail." Benjamin Flower was equally impressed. He told the socially conscious readers of *The Arena* that *The Octopus* was a "strong, compelling, and virile. . . work of genius" that would "quicken the conscience and awaken the moral sensibilities of the reader."[49]

Norris did not live to finish his wheat trilogy. After seeing *The Pit* through serial publication, he died in San Francisco of a perforated appendix and peritonitis on October 25, 1902. He had yet to start *The Wolf*, the final volume in the series. Although Norris died young, he, like Crane, had peered into dark corners of the self and society that few others were willing to confront, and he had exposed realities that were disgusting, stupid, vicious, and vulgar. Yet his silly masculine bravado and deterministic theories negated his refreshing candor. Norris's definition of naturalism as the exploration of the extraordinary led him into self-deceiving flights of mythic fantasy. Shortly before his death, he offered a characteristically misleading self-assessment: "I told them the truth. They liked it or they didn't like it. . . . A grim and bitter truth! But no one can dispute its power!"[50]

Looking Life in the Face—Theodore Dreiser

While working as a reader for Doubleday, Page and Company during the summer of 1900, Norris received an unsolicited manuscript from an aspiring novelist named Theodore Dreiser. The story so intrigued him that he read it in one sitting. After finishing *Sister Carrie* (1900), he decided that it was the best novel he had ever read. The next day Norris convinced his editorial associates to offer Dreiser a contract. "Mark my words," Norris told a friend, "the name of Frank Norris isn't going to stand in American literature anything like as high as Dreiser's."[51]

Norris recognized that the untutored Dreiser was fully awake to the energies, conflicts, and illusions of his own time. In *Sister Carrie* and his later novels, Dreiser exposed readers to the hard realities of urban poverty, loveless sex, and

personal defeat. He was, as a New York critic declared, "our arch-realist, unrestrained and unafraid." Dreiser depicted individuals driven by material desires,
wrestling with their consciences, misled by their illusions, and buffeted by the
winds of chance and environment. Neither as impressionistic as Crane nor as
hyperbolic as Norris, he graphically exposed the coarse elements of common
experience and tinctured them with bursts of pathos.[52]

Dreiser remains an enigma. How could a virtually uneducated newspaper
reporter become a novelist of such titanic power and the linchpin of twentieth-
century American literary history? He was not a polished craftsman. His novels
are filled with stretches of turgid prose pocked with clumsy diction, pompous
phrases, and broken-backed sentences. Yet we still read his works because his
unburnished fiction radiates unmistakable power and explores matters which no
other novelist had taken seriously. As a young woman said of *Jennie Gerhardt*
(1911), it "compels one to recognize the truth about life."[53]

Dreiser was the nation's first major writer not born of Protestant Anglo-
Saxon stock, the first to be raised a Catholic, and one of the few to have
experienced prolonged poverty. The propelling force of his life seems to have
been the very capaciousness and tenacity of his ego. He warred all his life with
conflicting emotions and modes of behavior. Like Whitman he contained mul-
titudes. The public Dreiser was a massive, overpowering personality, yet in
private he was shy, vulnerable, and confused, a brooding, introspective man
who felt a massive dread of failure, an intense distrust of others, and shame at
his own shamelessness. Although Dreiser accorded his fictional characters an
extraordinary degree of sympathy, he was himself a man of clammy self-regard
who treated other people with contempt and abuse. Friends and foes alike agreed
that he was often a scoundrel, a liar, a bully, and an adulterous lecher. Yet the
fact remains, wrote H. L. Mencken, his devoted friend and most incisive critic,
"that he was a great artist, and that no other American of his generation left so
wide and handsome a mark upon the national letters."[54]

Late in life Dreiser confided to one of his mistresses that he was still "trying
to find himself." In fact, he used most of his novels and his many autobiographical writings to explore the mysteries of his own complicated personality. In
each case his ruminations led him back to his singular childhood and family
life. Dreiser, it seems, was always an outsider longing for the respectability and
material "success" that he and his family failed to achieve. His father, John
Paul Dreiser, was a German Catholic immigrant who arrived in Ohio in 1844
at age twenty-three. Soon thereafter, he married Sarah Schänab, the illiterate
sixteen-year-old daughter of a Mennonite farm family. Their union, however,
seemed cursed from the start. Sarah's father disowned her for marrying a Catholic, and her first three children, all boys, died in infancy. John Dreiser developed a thriving woolen business, only to see his uninsured mill burn to the
ground.[55]

After the mill fire, the Dreiser family (eight children) moved to Terre Haute, Indiana, where Theodore was born in August 1871. While a feckless John Dreiser worked sporadically at odd jobs and used the strap on his rebellious children, Sarah struggled to keep the family together in the face of misfortune and deprivation. The Dreiser clan developed a gypsy existence, moving from one small Indiana town to another (Vincennes, Sullivan, Evansville, Warsaw, and briefly to Chicago) in search of affordable housing and a regular source of income.

Despite Sarah's arduous efforts, family tensions grew more acute, and the children began to stray from the fold. Theodore's two older sisters, Sylvia and Emma, were, he wrote, "a pair of idlewilds, driven helplessly and persistently by their own internal fires." Sylvia bore an illegitimate baby that Sarah Dreiser raised as one of her own. At the same time, Emma developed a reputation as a trollop and was banished from the family home. She landed in Chicago and soon became involved with L. A. Hopkins, a middle-aged cashier of a downtown saloon who, it turned out, had a wife and children. After Mrs. Hopkins discovered his adultery, he absconded with $3500 from his employer and fled with Emma to New York. Hopkins eventually lost his job, turned to other women, and lapsed into indigent lethargy. (This episode would provide the scaffolding for *Sister Carrie*.)[56]

By the age of sixteen Theodore was a tall, thin, wall-eyed, and buck-toothed boy yearning for the glamorous life in a large city. "To go to the city," he later wrote, "is the changeless desire of the mind. . . . It is the magnet which no one understands." In 1887 the sensuous energies of metropolitan life lured Dreiser to Chicago, where he boarded with his sisters. He quickly fell in love with the city. "It was so strong, so rough, so shabby, and yet so vital and determined." Here he could be "somebody" and also indulge his sensory and sensual pleasures. Life in gaslit Chicago was "avid and gay, shimmering and tingling."[57]

A rapt observer of human obsessions and appetites, Dreiser envied the finery and status enjoyed by others. His "thoughts were of luxury—fine clothes, an elegant home, and beautiful women—and his keen sense of relative deprivation intensified his longings and aroused his envy. "My eyes," he remembered, "were constantly fixed on people in positions far above my own." He found himself "growling at the rich for enjoying pleasures which I could not enjoy, while seeking eagerly to be wealthy myself so that I could do the same." Yet he coupled his ravenous appetite for worldly goods and loose women with an equally intense compassion for people living in straitened circumstances. As a literary critic noted, Dreiser was "a romantic, a realist and a mystic all in one."[58]

Dreiser's first real break occurred in the summer of 1892 when the *Daily Globe*, Chicago's smallest newspaper, hired him as a reporter. As a feature writer for the Sunday edition, he roamed the city's slums, conversing with alcoholics, beggars, and prostitutes who gave him "insight into the brutalities of life." Dreiser's success at the *Globe* led to other newspaper jobs in St. Louis, Toledo, Cleveland, Pittsburgh, and, finally, New York City, the goal of "all really ambitious people." Dreiser described himself at the time as a "realist in thought" eager to explore the "rough, raw facts of life." He longed for the day when readers "could look life in the face."[59]

Dreiser especially relished his experiences in Pittsburgh, where he arrived in April 1894. "What a city for a realist to work and dream in!" He had never seen such lavish wealth or perplexing poverty. In the Pittsburgh public library, Dreiser gorged himself on the novels of Balzac. Here was a writer who, like himself, had lost faith in a providential God and had succumbed to the city's alluring materialism and sensualism. Balzac, he wrote, provided "intensely beautiful" portraits of "raw, greedy, sensual human nature and its open pictures of self-indulgence and vice."[60]

At the Pittsburgh library Dreiser also discovered the writings of Herbert Spencer and other social theorists who treated conventional religion as a form of superstition. Spencer's skeptical materialism ambushed Dreiser's optimism and destroyed the remnants of his Catholicism. Such "scientific" theories "nearly killed me," he recalled. They "took every shred of belief away from me; showed me that I was a chemical atom in a whirl of unknown forces." Where before Dreiser believed there was an underlying purpose to existence, he now concluded that life was a "grim" struggle without plan. Chance, circumstance, and an array of "chemic" impulses shaped human fate. "If progress is necessary," he stressed, "it can be had only by competition, and competition cannot be had without strife. And strife cannot be had without misery."[61]

Dreiser himself, however, displayed a ferocious will to succeed in a God-abandoned world. "I may talk pessimism," he once admitted, "but I never cease to fight forward." Haunted by a fear of poverty and a sense of inferiority, narrowly self-centered and ambitious, this supposed determinist strove mightily to direct his own destiny.[62]

Dreiser left Pittsburgh in November 1894 and arrived in Manhattan. His brother Paul provided a bed and helped him become the editor of *Ev'ry Month*, a new magazine of the arts. Editing *Ev'ry Month* gave Dreiser a forum to express his developing ideas on a variety of topics, and he often took his social philosophy straight from Spencer. "It is only the unfit who fail—who suffer and die," he observed. Some were born strong and some weak, the strong devoured the weak, the weak "nibbled" at one another, and their predatory palates ensured widespread human suffering.[63]

Yet for all of his sincere efforts to understand the mystery of life, Dreiser was an utterly confused analyst and a poor philosopher. He read just enough of Spencer to develop an arrogant agnosticism and an aura of profundity, but his inconsistent statements betray the essential ambivalence at the heart of American literary naturalism. Dreiser would become famous for invoking scientific theories to explain his deterministic literary creed and for crafting reductive fictional characters driven by impulses and forces they neither created nor mastered. Yet at other times he sounded like a muckraking activist indicting inhumane or dishonest social practices.

After resigning from *Ev'ry Month* in 1897, Dreiser wrote dozens of articles for other middle-class magazines such as *Munsey's, Ainslee's, Cosmopolitan, Leslie's Weekly,* and *Success.* He prospered in his new role as a "hack writer" of "saccharine . . . sentiment and mush," and a steadily rising income enabled him to build "my private library of American Realism." Works by Crane, Norris, Garland, Howells, and Henry Blake Fuller began to fill his shelves and shape his outlook toward fiction. Gradually, Dreiser developed two literary personalities. As a freelance journalist he wrote whatever editors would buy. And as an aspiring novelist he began making plans for a new style of naturalistic realism remarkable for its candor.[64]

Sister Carrie

During the summer of 1899 a friend convinced Dreiser to write a work of fiction that would expose the "hard, brutal" facts of life. He resolved to provide readers with "a true picture of life, honestly and reverentially set down, whether it offends the conventions or not." In early 1900 he completed the manuscript of *Sister Carrie* and submitted it to Harper and Brothers. Although the editor called it a "superior piece of reportorial realism," he predicted that its "immorality" would shock and alienate "feminine readers who control the destinies of so many novels." Dreiser next gave the manuscript to Norris, who successfully pressed it upon his colleagues at Doubleday, Page and Company. When Frank Doubleday returned from a European vacation and read the proof sheets, however, he ordered publication stopped because of the novel's "indecent" heroine. At Norris's urging, Dreiser threatened to file suit for breach of contract, and Doubleday grudgingly agreed to print one thousand copies.[65]

Norris sent copies of *Sister Carrie* to reviewers, and he "strained every nerve" to sing the book's praises to critics. His friend Isaac Marcosson, the young literary editor for the *Louisville Times,* wrote one of the first reviews of *Sister Carrie.* "Out in the highways and hedges of life," he observed, "you find a phase of realism that has not found its way into many books. . . . It reeks of life's sordid endeavor; of the lowly home and the hopelessly restricted existence. Its loves, joys, its sorrows, are narrow. There is little sunshine. It is plain realism." At last, he exclaimed, a novel addressed the "other side of the social scale." Such uncensored fiction was what "must be read and which we must have." Frank

Doubleday, however, thought otherwise. He refused to promote or advertise *Sister Carrie*, and only 456 copies were sold.[66]

A reviewer for the *Toledo Blade* highlighted the problem when he said that *Sister Carrie* was "too realistic, too somber to be altogether pleasing." It was not "a good or wholesome" book for women to read, warned the *Chicago Advance*, "for Sister Carrie is not nice or clever or bright or kind-hearted or respectable." Its truth-telling, explained a writer in the *St. Louis Mirror*, "out-Howells Mr. Howells." Indeed, soon after *Sister Carrie* appeared, Dreiser met Howells at the Harper offices, and the aging "Dean" of American letters brusquely told the new novelist that he "didn't like *Sister Carrie*." Howells's curt disapproval, Doubleday's betrayal, the novel's mediocre sales, the cater-wauling of the moral sentinels among the "tea and toast crowd," and his own failed efforts to return to freelance journalism helped send Dreiser into a psychological tailspin.[67]

By 1902 Dreiser was suffering from neurasthenia and literary paralysis, afflicted with insomnia, hallucinations, and morbid self-doubts. He wandered the streets, slept in flophouses, and considered drowning himself in the East River. Only through the timely intervention of his brother did Dreiser regain his grip on life. Paul Dreiser enrolled Theodore at the Olympia, a sanitarium near White Plains, New York. For the first time in his life the spindly Dreiser "learned what true exercise was." He began to gain weight, strength, and self-confidence. By early 1904 he was ready to return to magazine editing. "After a long battle," he wrote a friend, "I am once more the possessor of health. . . . I have fought a battle for the right to live and for the present, musing with stilled nerves and a serene gaze, I seem the victor."[68]

Dreiser's new work as a magazine editor brought him a substantial income, and in 1907 he bought a third interest in the B. W. Dodge Company, a newly formed publishing house in New York. He then "took the liberty" of reissuing *Sister Carrie* himself. This time the novel attracted many rave notices. Several reviewers now likened Dreiser to Balzac, Tolstoy, and Zola. The *New Orleans Picayune* deemed *Sister Carrie* "the strongest piece of realism which we have yet met with in American fiction." Yet defenders of genteel morality still resisted Dreiser's efforts to broaden the horizon of American fiction. In discussing the "realism" of Crane, Norris, and Dreiser, for instance, the editor of *Putnam's Monthly* compared their fiction to the "contents of a garbage can." He reminded the "new realists" that "sunshine and fresh air are as real as disease and dirt," and he encouraged readers to protest the publication of such "morbid" stories and to "plead for a return to the novel of an earlier day."[69]

Readers today find it hard to understand how *Sister Carrie* could have provoked such controversy. The novel never describes any actual sexual activity. Nor does it preach any controversial political opinions or radical social criticism.

What made *Sister Carrie* so shocking was that it reversed the usual literary treatment of the fallen woman. In the conventional novels of the day, women who indulged their sexual passions outside of marriage found themselves sternly punished. Even Stephen Crane felt the need to have Maggie Johnson commit suicide out of a sense of moral disgrace. Carrie's illicit behavior, however, eventually brings her fame and fortune as a Broadway actress. Dreiser does not even portray her conduct as sinful.

Sister Carrie opens with an archetypal scene: eighteen-year-old Carrie Meeber, an ambitious young woman from a small town in Wisconsin, carries all her worldly possessions and naive dreams on board a train and heads to Chicago, intending to live temporarily with her sister. Like tens of thousands of other young migrants from rural America, including Dreiser himself, she sees in the "great, pleasing metropolis" both economic opportunity and social excitement. Chicago was for Carrie—and Dreiser—a symbol of seductive force. It is a "giant magnet," a throbbing "wonder" whose "blare of sound and roar of life" create a spectacle of material display and desire that coerces her attention. Pretty, ignorant, and graceless, devoid of firm moral convictions and driven by unappeasable cravings, Carrie is taut with wonder and anticipation as she explores the immigrant-swollen city. She is a "waif amid forces," a "lone figure in a tossing, thoughtless sea" of surging humanity and seductive commodities.[70]

Carrie did not come to Chicago to live in squalor. Nor was she interested in becoming part of a morally upright and frugal home insulated from the corruptions of the marketplace. Instead she wanted to indulge her sensual fancies and societal ambitions. She imagined herself among those "counting money, dressing magnificently, and riding in carriages." In her untutored and "deluded" eyes, "things . . . are realities."[71]

More than any other writer of his age, Dreiser perceived how consumer goods and popular entertainments often became more "real" for people than the mundane actualities of their daily routine. Money to Dreiser was as dynamic a force as sexual passion—and more ubiquitous. Many of his characters live in a dream world of fetishized commodities and theatrical display. To Carrie, for instance, clothes, furnishings, and goods are far more important than human relationships. While applying for a job at The Fair, one of Chicago's massive new department stores, Carrie "passed along the busy aisles, much affected by the remarkable display of trinkets, dress goods, shoes, stationery, jewelry. Each separate counter was a show place of dazzling interest and attraction. She could not help feeling the claim of each trinket and valuable upon her personality." This "drag of desire" defines Carrie's pursuit of happiness. "Ah, money, money, money! What a thing it was to have. How plenty of it would clear away all those troubles." Dreiser's relentless development of this material imperative helps differentiate him from older realists. "Where Howells identifies character with autonomy," Walter Benn Michaels has written, "Dreiser . . . identifies it with desire, an involvement with the world so central to one's self that the distinction between what one is and what one wants tends to disappear."[72]

Born with the "gift of observation," Carrie bases her self-esteem as well as her assessment of others on what they wear or ride in or own. That so many city dwellers have "fine clothes" and live in stately homes heightens her sense of want and helps explain her decision to become the well-kept mistress first of Charles Drouet, a dapper traveling salesman, and then his more polished friend, George Hurstwood, the manager of an elegant saloon.[73]

When Carrie first meets Drouet, her eyes and thoughts fasten upon his clothes, jewelry, and fat wallet. He quickly discerns Carrie's material priorities and, during their second meeting, presses two "soft, green, handsome ten-dollar bills" into her hand. To her the money represents "something that was power in itself." Later, having become Drouet's mistress, Carrie savors his affluence during dinner: "As he cut the meat, his rings almost spoke. His new suit creaked. He was a splendid fellow." But not as splendid as Hurstwood, a man of greater wealth who lures Carrie away from Drouet. When conversing with Hurstwood, she heard not so much his words as "the voices of the things which he represented. How suave was the counsel of his appearance. How feelingly did his superior state speak for itself."[74]

The bourgeois Hurstwood sees in Carrie an enchanting alternative to his shrewish wife, selfish children, and dreary domestic routine. He yearns to "possess" her as he would any other attractive new commodity. Yet he waffles with ambivalence until chance intervenes to direct his actions. While closing the saloon one night, he discovers the safe still open and begins to fondle the "soft, green stack" of "ten thousand dollars in ready money." Then the safe accidentally clicks shut with the money still in his hands. Since he cannot put it back, he decides to reach "for Paradise" by fleeing to Canada with the money—and Carrie.

Carrie accompanies Hurstwood to Montreal and then to New York City. In Manhattan Hurstwood buys part-ownership of a third-rate saloon. The fugitive lovers rent a flat on 78th Street, hire a servant, and settle into a life of domestic routine. As time passes, however, Hurstwood's income diminishes while Carrie's desires increase. She "needed more and better clothes" to compare favorably with her neighbors. Walking along Broadway, she sees women "spending money like water," while she is forced to scrimp and save simply to maintain a "commonplace" existence. After Hurstwood's saloon shuts down, he begins an inexorable slide into squalid self-hatred. Initially, Carries find him more pitiful than contemptible. She forgives his sins but cannot tolerate his dependence, for her own dreams are at stake. After finding work as a chorus girl in a Broadway "opera," Carrie abandons her misbegotten lover.[75]

Moving from one flophouse to another, eating in soup kitchens, and begging for change on the streets, Hurstwood becomes a pathetic remnant of his former self. His degeneration is so intense, so desperate and overwhelming, that he displaces Carrie as the focal point of the novel. When Hurstwood commits suicide in a cheap lodging house, his plaintive last words speak volumes: "What's the use?"[76]

In the meantime, Carrie has become an acclaimed actress and attained all of her romantic goals: wealth, fine clothes, jewelry, a carriage, celebrity status. At the end of the novel, however, she is sitting alone in a rocking chair in her posh suite, pondering the emptiness of her theatrical success. For all of the "tinsel and shine" she has gained, happiness still eludes her. Carrie has yet to feel the serenity of true fulfillment. Dreiser implies that she never will, for she has fully absorbed the illusions of a consumer culture that can provide "neither surfeit not content." She seems fated to "dream such happiness as you never feel."[77]

In a fit of sentimentalism, Dreiser concludes the novel with a syrupy peroration: "Oh, Carrie, Carrie! Oh, blind strivings of the human heart! Onward, onward, it saith, and where beauty leads, there it follows." Such archaic clichés, exclamatory punctuation, and overwrought emotion illustrate a frequent problem in the fiction of the "savage realists": how to bring stories dealing with complex emotions and fluid social forces to a robust resolution for readers accustomed to uplifting conclusions. Despite his fatalistic tendencies, Dreiser could not leave the reader staring into an abyss of hopelessness. The manuscript of *Sister Carrie* originally closed with Hurstwood's final words before committing suicide: "What's the use?" Dreiser's decision to add the section describing Carrie in the rocking-chair may have been a sop to the expectations of women readers, or, as several critics have argued, it may have been a parody of sentimental conventions. A third possibility seems more plausible: Dreiser never reconciled his own realistic and sentimental tendencies.[78]

Setting aside the purple ending, what is it about *Sister Carrie* that makes it so significant and so moving? To begin with, it is very much an urban novel with a strongly realized sense of place. As a connoisseur of the cityscape, Dreiser had a remarkable ability to recreate the physical surfaces and social customs of urban life. By working detail upon detail into the grain of his narratives, many of his vivid descriptions linger in the mind: the Chicago shoe factory where Carrie briefly worked, Drouet's and Hurstwood's residences, Fitzgerald and Moy's saloon, Broadway's theatre district, and the Brooklyn streetcar strike. The cumulative effect of such authenticating sites and incidents, said one reader, "helps greatly to heighten that sharp impression of reality which it makes upon us."[79]

But it is Dreiser's closely related and observed characters who endow *Sister Carrie* with its peculiar force. By the turn of the century, he had decided that society had reached an intermediate stage of development in which people were "no longer guided wholly by instinct" and "not yet guided by reason." Like "a wisp in the wind," the "untutored" man acted "now by his will and by his instincts, erring with one, only to retrieve by the other—a creature of incalculable variability." In *Sister Carrie* Dreiser portrays characters fumbling and drifting through their days, struggling with inner waywardness and turbulent yearnings they can neither articulate nor suppress. Yet they never seem to realize the nature of the forces arrayed against them, nor do they successfully surmount them. The message of *Sister Carrie*, it seems, has little to do with illicit sex and everything to do with the traditional assumption of both classical republicanism and Christian theology: money offers a false—and often corrupting—path to happiness.[80]

The Drama of Desire

In *Sister Carrie* Dreiser echoed Crane's "realistic" recognition that relentless aspiration for unattainable ideals often becomes a self-destructive dream. He himself betrayed an unsatisfiable hunger for "success," and the stream of defeated hopes, lost ambitions, and unquenchable desires courses through his later novels as well. Dreiser's own frustrated craving for wealth, his lust for sexual pleasure, his ambivalence toward conventional moral codes, and his admiration for "big, raw, crude, hungry men who are eager for gain—for self-glorification" provided the themes for his other early novels: *Jennie Gerhardt* (1911), *The Financier* (1912), *The Titan* (1914), and *The "Genius"* (1915).

In *Jennie Gerhardt*, for example, Lester Kane, scion of a wealthy Cincinnati family, is torn between his love for Jennie, the pretty, innocent scrubwoman, and his fear of losing his social status and his financial security. Initially he takes extraordinary measures to support Jennie as his mistress, only to abandon her when faced with the "'armed force of convention" and the threat of disinheritance. In explaining his decision, Kane tells Jennie that "all of us are more or less pawns. We're moved about like chessmen by circumstances over which we have no control. . . . The best we can do is hold our personality intact. It doesn't appear that integrity has much to do with it."[81]

Integrity also has little to do with the actions of Frank Cowperwood, the buccaneering capitalist and supreme egotist of *The Financier* and *The Titan*, who describes himself as a "selfish and ambitious materialist." Modeled after the real-life business tycoon Charles Yerkes, Cowperwood follows the same deluded notion of the good life that goaded Carrie Meeber. "We think we are individual, separate, above houses and material objects generally," Dreiser observes, "but there is a subtle connection which makes them reflect us quite as much as we reflect them." Cowperwood acquires mansions, a priceless art collection, a yacht, a wife, and a string of sybaritic mistresses, but contentment eludes him. Likewise, Eugene Witla, the "realistic" painter in *The "Genius,"* finds satisfaction neither in his widely acclaimed art nor in his frenzied pursuit of women. In the end, his illusions dashed, "he did not know what to believe. All apparently was permitted, nothing fixed."[82]

In each of these autopsies of misplaced desire, Dreiser displayed a quality lacking in Norris: an ability to write from inside others. Although his characters are largely creatures of circumstance driven by desires they did not create and cannot control, they are more than puppets. Dreiser's fiction derives much of its potency from the empathy he conveyed for his helplessly flawed characters and his hunger to understand their corrupting vision of the world. He does not cheer or chide Hurstwood, Carrie, Jennie, or Cowperwood; he is at one with their tangled aspirations and tormented spirits. If his naturalistic emphasis on determining forces seems at times too one-dimensional, his presentation of people forever in pursuit of false but alluring goals compels interest.

Despite Dreiser's seemingly dispassionate outlook, his novels ache and their pain is genuine. Like Crane, he realized that a ruthless adherence to a survival-of-the-fittest outlook made compassion irrelevant, and, to him, sympathy was a necessary human emotion in a heartless world. "Let no one underestimate the need of pity," Dreiser wrote in *The Financier*. "We live in a stony universe whose hard, brilliant forces rage fiercely." At times, he also refused to accept the fatalism inherent in a deterministic world view. Both will and accident, fate and choice, shape his work. If his stories had a moral, he once told an interviewer, it was that "all humanity must stand together and war against and overcome the forces of nature."[83]

Although initially enamored with Herbert Spencer and others who touted the certainties of science and the extrapersonal forces shaping the natural world and human society, Dreiser discovered that both life and literature resisted the relentless predictabilities demanded by the experimental researcher. Scientific models of behavior, like religious dogmas, bludgeoned the novelistic truth of character into prescribed shapes that contradicted the actual flux of human experience. In other words, a deterministic emphasis on "fate" and "facts" was not a sufficient path to fictional truth. Unlike Cowperwood, who "believed only what he saw," Dreiser decided that "the real thing" was often more potent as a mental force than a physical entity, and its fluidity defied any universal model of human conduct. "Life," he wrote in *The Financier*, "cannot be put into any mold, and the attempt might as well be abandoned at once."[84]

Dreiser sensed more deeply than any of his literary peers the distinctive psychological consequences of metropolitan life, the way the city's corrupting sensuality and pulsing social energies excite and ultimately engulf vulnerable individuals. "What a tangle life was," one of his characters concludes. Dreiser's emphasis on human relationships as the crossroads where morality and immorality vie for access to the will revealed his agnosticism about life's ultimate purpose and his uncertainty about how to grasp and understand social "reality."

Unlike the positivistic social scientists, he did not embrace the giddy notion that the patient accumulation of tangible "facts" would always lead to "truth." Facts in his view did not lead anywhere in particular; they were merely useful tools to portray individuals acting within the marketplace of social relations. His own objective was to weave together the truth of fact and the truth of motive into a seamless fabric.[85]

In sum, Dreiser never let his fascination with "naturalistic" theory erode his loving interest in the actual human drama. At his best and most authentic he was a penetrating observer of the tragicomedy of self-delusion. Dreiser spoke from experience as both a dreamer trapped in his own web of illusion and as a vigilant observer of human obsessions and appetites. By impressing upon readers the dark side of the American dream, by demonstrating the pervasiveness of misplaced want and overweening ambition, and by highlighting the disparity between reality and appearances, he raised new questions about the blind striving of the will and the inscrutable riddle of life. His fiction, Mencken said, gave the reader "a powerful effect of reality, stark and unashamed. It is drab and gloomy, but so is the struggle for existence."[86]

Dreiser's pathbreaking efforts made possible a new candor in American fiction. Sinclair Lewis testified that Dreiser "cleared the trail from Victorian and Howellsian timidity and gentility in American fiction to honesty and boldness and passion of life." Dreiser and the other "savage realists," Lewis added, introduced a new element within American fiction: "people with clenched fists, people saying a great many impolite things, people highly discomforting to the cultured and the nice by raucously demanding that they have some share in the purple and the linen." A year later literary critic Randolph Bourne could report that "American taste since 1900 has grown up to and indeed far beyond the frankness of *Sister Carrie*."[87]

For all of their stylistic faults and excesses, Dreiser and the other literary naturalists unearthed sobering new facts about contemporary American life. Through their obstinate disregard of polite sentiments and moral conventions, they revealed to middle-class readers the innate cruelty of life and its ironic indifference to individual fate. By exposing the comfortable to the reality of violent households and repulsive persons, they reminded readers that the supposed moral benefits of poverty may come at the expense of humanity itself.

Perhaps most important, the naturalists questioned the very notion of the autonomous self capable of moral judgment and independent action. What most sharply differentiated Dreiser, Norris, and Crane from Howells, James, Jewett, and Wharton was their recognition of the overwhelming power of economic forces and nonrational impulses. To one degree or another, the naturalists imposed upon the world of observed fact an austere assumption about the deterministic nature of existence. This led them to go beyond a realism of simple facts, literal objects, and evident moral choices. They saw in such concrete things symbols of overarching forces which help explain their penchant for abstraction, allegory, and melodrama. Their philosophical outlook, in other words, led them to their distinctive sensationalism and crude symbolism. "Terrible things must happen" in a naturalistic story, as Frank Norris explained. The characters "must be twisted from the ordinary, wrenched out from the quiet, uneventful round of every-day life, and flung into the throes of a vast and terrible drama that works itself out in unleashed passions, in blood, and in sudden death." Such extraordinary realism offered Americans an acrid reminder that much of life involved, as Crane insisted, "a world of fists."[88]

12

Ash Can Realism

I n *The "Genius"* (1915) Theodore Dreiser told the story of a lustful young painter named Eugene Witla who, like himself, migrated from a small midwestern village to Chicago and then to New York City. Witla worked first as an illustrator for newspapers and magazines before becoming a celebrated "realistic" painter known for his "new, crude, raw and almost rough method" of portraying the prosaic charm and daily spectacles of urban life. Whatever he touched "seemed to have romance and beauty, and yet it was real and most grim and shabby." An art dealer praised Witla's "astonishingly virile" work: "Thank God, for a realist."[1]

Dreiser modeled Witla in part after himself and in part after Everett Shinn, a precociously brilliant graphic artist who had worked with Dreiser on the staff of *Ainslee's Magazine*. Shinn was one of a handful of iconoclastic young painters in New York who belonged to what critics called the "Ash Can School" because of their fascination with squalid scenes along inner-city streets. More sympathetic observers labeled them the "New York Realists." In addition to Shinn, these kindred spirits included Robert Henri, William Glackens, George Luks, John Sloan, and George Bellows. Like Dreiser, Norris, and Crane, they shared an unrelenting ambition to express the coarse facts and surging energies of urban America. Both groups displayed an earnest commitment to unvarnished "truth" rather than decorative "beauty," a rebellious disdain for the genteel tradition, and a masculine preference for "life" over "art."

Dreiser, who knew most of the "New York Realists," rejoiced at their rising significance. Henri, he announced in *Broadway Magazine* in 1907, "has a school of followers who go with him to the tenements and alleyways to paint— Glackens, Sloan, Shinn, to name a few of them." Their ambition was to "paint every day New York life," and they believed that "a Hester Street pushcart is a better subject than a Dutch windmill." Dreiser then made a prescient prediction:

by painting urban realities long hidden from view, the "Ash Can" artists would exercise a radiant force in American painting.[2]

Robert Henri

The "Ash Can School" first coalesced during the 1890s in the studios of the Pennsylvania Academy of the Fine Arts, where most of them were students. Their chief instructor was Thomas Anshutz, one of Thomas Eakins's most gifted pupils who had taken his place as teacher of the anatomy and life classes. Anshutz shared his mentor's disgust for romantic affectation and academic formalism. In his own work he tried to "make as accurate a painting of what I see in front of me as I can." Yet while recognizing that art was "based on knowledge and knowledge on facts," he stressed that serious painters forge "truths" out of facts rather than simply copy them.[3]

Robert Henri remembered Anshutz as "a great influence, for he was a man of the finest quality, a great friend, gave excellent advice, and never stood against a student's development." Henri brought to the Pennsylvania Academy a unique background. Born in Cincinnati in 1865 and raised on a Nebraska farm, he was the son of John Henry Cozad, a tempestuous professional gambler and real-estate speculator ever alert for the main chance. In 1882, however, his deal-making caught up with him. During a bitter argument with a knife-wielding cattle rancher, Cozad shot and mortally wounded his assailant. Fearing vigilante justice, he fled the state with his wife and two sons. They eventually settled in Atlantic City, New Jersey, changed their names, and claimed that the boys were adopted. Robert Henry Cozad thus became Robert Henri (which he pronounced Henrye).[4]

In 1886 Henri enrolled at the Pennsylvania Academy. Eakins had been forced to resign eight months earlier, but the spirit of the "great master" still permeated the building. Soon after his arrival, Henri went to see Eakins's *The Gross Clinic* at Jefferson Medical College and pronounced it "the most wonderful painting I had ever seen." From the older students at the Academy, Henri learned that Eakins was "a man of great character" whose "vision was not touched by fashion. He cared nothing for prettiness or cleverness in art." Eakins studied humanity in an honest and frank way that evoked the "real stupendous romance in real life." He saw "no need to falsify to make romantic or to sentimentalize to make beautiful." Henri decided that Eakins was "the greatest portrait painter America has produced," a courageous pioneer whose efforts to paint the "real thing" filled art students such as himself with "courage and hope" and provided a model of greatness to emulate.[5]

After two years of study at the Pennsylvania Academy, Henri succumbed to

"Paris fever" and went to study in France. In Paris he read Balzac and Zola, stood enthralled before the canvases of Courbet and Manet, and aligned himself with the Impressionists. The sight of Monet's *Haystacks* sent him into a frenzy of excitement: "What realism!—and too it is not brutal realism—it has the sentiment of nature which is never brutal."[6]

Henri returned to Philadelphia in 1891 and enrolled again at the Academy. His three years abroad had confirmed his youthful interest in a realism that was both "outward and *inward*." A painting, he believed, should record a "real" scene as filtered through the temperament of the artist, and he saw in city life not only compelling material but masculine force as well. Yet Henri's impressionistic realism did not excite several mossbacks among the art community who panned his cityscapes on display at the annual Academy exhibition in 1892. Henri, however, was unfazed. "I don't care how much kicking they do," he told a friend. "The backs of tenement houses are living documents" deserving of artistic representation.[7]

During the early 1890s, Henri served as president of the Charcoal Club, an informal group of forty-three aspiring artists who had studied at the Pennsylvania Academy but had come to resent its classical curriculum and high tuition. The Charcoal Club provided its members with a well-lit studio, a nude model, and six easels at half the price of the Academy's evening school. For six months, before it fell victim to the economic depression, the Charcoal Club served as the city's liveliest center of artistic innovation and discussion. After the club's collapse, Henri kept its insurgent spirit alive by hosting a weekly open house at his Walnut Street studio. There he developed intense friendships with a group of young artists then working as illustrators for Philadelphia newspapers: John Sloan, William Glackens, George Luks, and Everett Shinn.

Although only a few years older than his ensemble of young admirers, Henri was their acknowledged master, a worldly cultural rebel who led them in earnest discussions of art, literature, politics, and social issues as well as rowdy burlesques and all-night poker games. In everything he did there was a touch of charisma. Henri's colleagues cherished his willingness to help cultivate their own latent talents. Sloan testified that Henri "could make anyone want to be an artist. I don't think I would have become a painter if I had not come under his direction."[8]

Henri was an unabashed Americanist. "As I see it," he once declared, "there is only one reason for the development of art in America, and that is that the people of America learn the means of expressing themselves in their own time and in their own land." In exhorting his apprentices to use their real-life experiences as subjects for their art, Henri alerted them to Whitman's courageous example as a man who "found great things in the littlest things of the world." Sloan recalled that Henri shared with the group Whitman's "love for all men, his beautiful attitude toward the physical, the absence of prudishness," all of which "represented a force of freedom."[9]

Six years younger than Henri, Sloan worked as an illustrator first with the *Philadelphia Inquirer* and then the *Philadelphia Press*. He cherished the fellow-ship of other newspaper artists such as Shinn, Glackens, and Luks, all of whom studied under Anshutz at the Academy. For these skilled draftsmen, newspaper work served the same purpose as it had for Crane and Dreiser: it led them to see life up close, as an immense, sprawling, kaleidoscopic affair, often sordid and ugly, but always interesting. Sloan said that his journalistic work alerted him to the "beauty in commonplace things and people." Shinn added that urban journalism provided a training ground for visual "memory and quick perception."[10]

Henri convinced these young newspaper illustrators to pursue careers as serious painters of the "common life" they knew so well. In appealing for an indigenous American art, Henri often sang the praises of Eakins and Homer, and he stippled his commentaries with aphoristic imperatives. "Don't bother to paint art," he ordered, "—get your motives from life." Paint "what you feel. Paint what you see. Paint what is real to you." They followed his advice and eventually developed a bold new strain of urban realism using a dim, low-keyed palette and agitated brush strokes. Sloan explained that he and his friends "chose our colors from observation of facts and qualities of the *things* we painted."[11]

In promoting the unidealized treatment of familiar subject matter, Henri convinced his friends to reject what Shinn called the "effete, delicate, and supinely refined" tradition of academic formalism. Under Henri's tutelage, Sloan testified, "we came upon realism as a revolt against sentimentality and artificial subject matter and the cult of 'art for art's sake.'" The "Ash Can" painters fastened upon urban realism not because they were anxious about the threat of unruly social change but because the energies of contemporary Amer-ican life excited them. They were more concerned with expressing the charm of city life than with kindling social indignation or exploring the causes of poverty and injustice. Their vision of society was organic and inclusive, and their tone was more upbeat than strident, more descriptive than prescriptive, more enthusiastic than fearful.[12]

True, Sloan flirted with socialism after 1910, but his *artistic* perspective was only mildly political. He questioned the depth of his social commitments when he confessed that at heart "we were snobs about labor." Sloan, Glackens, and the other "Ash Can" painters reflected a virile form of urban realism that por-trayed the "smiling aspects" of poverty and minimized roiling class tensions and evidence of social injustice. Their animating impulse was an overwhelming empathy rather than an impulse to reform. "The members of the 'Ash Can School,'" as Glackens's son observed, "were not painting for social comment, or for political propaganda, or to reform the world. Their concern was only with reality, with life."[13]

New York Realists

Eventually, all of Henri's Philadelphia "gang" relocated to New York City— Luks and Glackens in 1896, Shinn in 1901, and Sloan in 1904—but Henri's arrival in 1900 was the key development. In that year the New York art critic Sadakichi Hartmann called for artists courageous enough to reflect the "sights and scenes of our own times." Henri eagerly took up the challenge. In New York he discovered a city alive with activity. The subway system was then under construction, as were dozens of new hotels, apartment buildings, and skyscrapers. New electric lights brought glitter and sparkle to the formerly gas-lit streets, theatres, and roof-garden restaurants. The first automobiles began to compete with horse-drawn cabs and hansoms, and the first moving picture theatres attracted throngs of crowds.[14]

In Manhattan Henri and "the crowd" from Philadelphia took frequent walks, dined out at Mouquin's or the Café Francis, visited each other's studios, attended football games, boxing matches, and movies together, and met for drinks at McSorley's, Sweeney's, or the Haymarket. While the other four made their living as illustrators for middle-class magazines, Henri plotted strategy for a collective assault against the National Academy of Design, the citadel of genteel art taste. "I really do believe that the big fight is on," he wrote in 1902.[15]

Henri was distinctively "modern" in his recognition that there "are many ways of seeing things" and in his insistence that a "realistic" painter should strive for more than the detached fidelity of a camera image. Many academic artists, he asserted, filled their canvases with the pride of fact, but their "pretty" paintings remained as emotionally dead as the surface of a mirror. What mattered most to Henri was the immediacy and intensity of the artist's *reaction* to life. Facts must be transmuted into feelings as well as forms. "A work of art," Henri repeatedly stressed, "is not a copy of things. It is inspired by nature but must not be a copying of the surface. You have to make your statement of what is essential to you—an innate reality, not a surface reality."[16]

Henri thus abandoned the meticulous detailing and glossy finish so prized by the "Academy" painters in favor of a more spontaneous approach intended to lift things above the plane of visual facts. The art that Henri produced during his early years in New York might best be described as dark impressionism. Unlike Eakins, he made no preparatory studies, did not outline his figures in charcoal, and only rarely retouched a canvas. Instead he worked rapidly with a loaded brush and a dark palette, emphasizing mass rather than line. The result

Robert Henri, *Snow in New York*, 1902. *(National Gallery of Art, Washington, D.C., Chester Dale Collection.)*

was striking. His early cityscapes have the "unfinished" effect of sketches; they impart only the general outlines and elemental forms of a scene. "The real motive, the real thing attained," he repeated, "is the revelation of what you can perceive beyond the fact."[17]

Snow in New York (1902), for example, depicts with slashing brush strokes a snow-clogged street walled in by dingy brownstone townhouses and skyscrapers. A solitary street lamp looms over the pedestrians and horse-drawn wagons struggling through the slush. To render the "effect" of the dreary scene, Henri consciously avoided precise detailing and bright colors. Only a splash of red in the woman's shawl and a daub of yellow mud on the wagon wheel relieve the brown and gray tones dominating the canvas. In conveying the chilling atmosphere of a harsh winter day, he challenged both conventional techniques and public expectations. Realism, Henri demonstrated, could effectively combine surface truth with subjective mood.

The drab bleakness of *Snow in New York* was enough to excite the bile of traditionalists, but its slapdash style, heavy impasto, and purposely restricted palette made it even more objectionable. Still, Henri remained defiant. In a letter to his parents after his first solo exhibition at Macbeth's gallery, he explained that the critics falsely believed "that art is the business simply of reproducing *things*—they have not learned yet that the *idea* is what is intended to be presented and the thing is but the material *used* for its expression." He remained committed to a "freer field of expression" and a realism "more suggestive than descriptive."[18]

For all of his considerable artistic talent, especially as a portraitist, Henri proved to be more significant as an agitator and teacher than as a painter. This was especially true after 1903, when he began teaching at the New York School of Art. William Merritt Chase had founded the school seven years earlier, and he remained its head instructor and guiding genius. Initially, Chase and Henri shared both a dislike for "dainty" painters and a sense of mutual respect, but within a few years they could not abide one another.

Their strained relationship resulted from both clashing personalities and differing artistic perspectives. By the end of the century, Chase had become, as one critic wrote, the epitome of "academic realism." He firmly believed that art should represent contemporary life—but only its most beautiful aspects. An extraordinarily facile painter himself, he served as president of the Society of American Artists, belonged to the National Academy of Design, served on numerous exhibition juries, and garnered dozens of awards and medals. Chase was a dapper man who wore tailored morning coats and striking cravats, sported a bristling Vandyke beard and a black-ribboned monocle, and loved to parade down Fifth Avenue with his white wolfhound. He was also a stimulating, if lordly, teacher who preached and practiced an epidermal realism of dazzling surfaces and delicate details. In his eyes, verisimilitude was the highest virtue of painting: students should allow "no intermission between the hand and head."

Henri, on the other hand, was a much younger and more down-to-earth personality. Warm and unpretentious, he ate lunch with his students and lingered with them after class to smoke a cigar. Henri promoted both the creation of a distinctive American "style" of painting and the democratization of the art world. "We must get over the idea that the study of the fine arts is to be relegated to the few," he stressed in reference to the authoritarian tradition of the National Academy of Design. One of Henri's pupils remembered that he taught them to disdain the romantic "novelists of the day and to side with Theodore Dreiser when he came along." Henri had "no patience with escapists." Where Chase preached art for art's sake and stressed conventional technique, Henri promoted art for life's sake. Chase, however, complained that the subjects selected by Henri and his students were "vulgar," and he questioned his colleague's "unorthodox" teaching methods and social informality. An unyielding Henri countered that the school should not strive to fit students "into the groove of rule and regulation" but should be a place "where personality, originality of vision, idea, is encouraged, and inventive genius" stimulated.[19]

That a disporportionate number of students flocked to Henri's classes only heightened the tension between Chase and himself. The younger man's magnetic personality and reputation for innovation attracted a platoon of talented young artists: George Bellows, Rockwell Kent, Glenn Coleman, Edward Hopper, Walter Pach, and Guy Pène du Bois. Henri's impact upon the school, Du Bois recalled, was like that of a "rock dashed, ripping and tearing, through bolts of patiently prepared lace." Henri's class was "the seat of sedition among the young," for the instructor "waved the red flag of realism in the faces of an idealistic New York" art community. Under his tutelage, the students "discovered that snow was not necessarily white or need not remain so, that city streets could be crowded and disorderly, that people frequently broke the taboos of good taste."[20]

Unlike Chase, who assigned his students still-life portraits of fish and water pitchers, Henri encouraged pupils to join him and his "gang" of Philadelphia art friends in exploring areas of city life unrepresented in American art: the Bowery, the Haymarket on Sixth Avenue, the Lower East Side. Henri, recalled one of his female students, assumed "that the poor were nearer the realities of life."[21]

Henri also promoted a self-consciously "masculine" style and preached the virtues of a strenuous life. He was determined to prove that "real men" could become great painters. "Be a man first, an artist later," he counseled. Pène du Bois remembered that Henri wanted students who had "guts," who were "fighters." Eager to have his students paint in "the straightforward unfinicky manner of the male," Henri forbade them from using small brushes because they symbolized a "feminine" approach to art. Henri also encouraged vigorous exercise. He himself loved to play golf with Sloan and attend boxing matches with Bellows, and he encouraged his pupils to play basketball or box during their spare time and to do chin-ups on the studio door lintels when the model was taking a break. His students organized their own baseball team and challenged the National Academy of Design and the Art Students League to do the same.

Henri boasted that his team's "brave realists" never lost a game to the representatives of "insipid academism." Several encounters between the two teams
erupted into brawls, and on one occasion the police had to be called to restore
order.[22]

Like many literary realists and naturalists, Henri and his colleagues felt a
keen sense of besieged masculinity. Shinn, for instance, told a reporter that one
by one the "masculine ramparts are crumbling before the onslaughts of the
modern woman." In a similar vein, Pène du Bois boasted that Henri's "vital,
manly" coterie of artists was determined to displace the "long-haired, dreamy-
eyed, velvet-coated aesthete." Bellows revealed the same concerns when he said
he strove to inject into his art "manliness, frankness, [and] love of the game."[23]

Before long Henri and his circle combined to form the first really vital
movement dedicated to the development of a distinctively American art. In
1907 Samuel Swift, the art critic for *Harper's Weekly*, took excited notice of the
"school of Robert Henri" forming in New York. He admitted that Glackens,
Shinn, Luks, Sloan, Coleman, and Bellows were rugged individualists with
quite distinct artistic personalities, but they held evident qualities in common.
"These painters convince us of their democratic outlook. They seek what is
significant, what is real, no matter whither the quest may lead them. They give
no hint of 'slumming' among either rich or poor—they are at home and at ease
no matter what or whom they paint." What most impressed Swift was the
"affirmative and stimulating" energy of their art and its "manly" tone. "There
is virility in what they have done, but virility without loss of tenderness; a manly
strength that worships beauty, an art that is conceivably a true echo of the
significant American life around them."[24]

During the first decade of the new century, each of the "New York Realists"
surpassed Henri's own efforts to represent on canvas the infinite variety and
inherent beauty of their immediate urban surroundings. They looked upon life
with open eyes and inclusive attitudes, as well as spectatorial affection. "I simply
accept it," Coleman stressed. "I love it. I think a great many things about it. I
do not condemn it. I do not comment on it." Shinn explained that their quest
to find subjects more vital than the "gardenias and orchids" painted by the
Academicians led them to the Bowery, "where people were real by default of
riches."[25]

Like Dreiser, the "New York Realists" rebelled against Victorian prudery
and insisted that the urban landscape was a fit subject for artistic representation.
Realistic art, Glackens stressed, entailed presenting the "facts of life as a permanent record—a record at once analytical, humorous, often satirical, and
always unfailingly sincere—an art not saved for the embroidery of life."[26]

The "Ash Can" painters, however, showed no interest in the fatalistic determinism of "naturalistic" philosophy. In the early stages of their careers, they

were much closer to Howells in believing that individuals could shape their own destinies through an act of sustained willpower. "Endlessly, tirelessly," wrote Luks, "the struggle for betterment goes on" among the city poor. Most slum dwellers, he said, were always striving to better their condition. "The people do overcome their poverty and pass into other spheres." This sanguine view of social mobility helps explain why the Ash Can painters produced such sanitized versions of urban subjects.[27]

"The Eight"

After only a few years in New York, Henri had developed a reputation as one of America's most "progressive" artists and most exciting teachers. His paintings were selected for three Paris salons, and his six one-man exhibitions garnered much interest and considerable praise. In 1905 he was elected a member of the National Academy. Yet although he gained a grudging acceptance among the art elite, many of his more rebellious friends and students did not. Their struggle to win access to exhibition space and art dealers finally provoked a showdown that would break the Academy's stranglehold over American art.

Early in 1907 Henri was one of thirty jurors for the Academy's Spring Exhibition. His own entries fared well in the initial screening, but the jury rejected each of the works submitted by Glackens, Sloan, Luks, and Shinn. Henri gagged at such encrusted traditionalism. Incensed at the intransigence of the jury members and their snide treatment of his friends, he withdrew two of his own paintings from the show. "The Academicians," John Sloan told a reporter, ". . . have a keen resentment for anything that is inspired by a new idea, and have an equally keen appreciation for everything that follows the old hide-bound conventions."[28]

On April 4, 1907, Sloan recorded in his diary that he "went to a meeting at Henri's to talk over a possible exhibition of the 'crowd's' work next year." They decided that the only way to free themselves from the Academy's "monocled idealism," as Shinn reported, was to form their own independent movement. To that end Henri suggested that they organize an exhibition featuring works by himself and seven other rebel painters—Glackens, Luks, Shinn, Sloan, Maurice Prendergast, Ernest Lawson, and Arthur B. Davies. The latter three insurgents were quite different in style from the others, but they shared an irreverent distaste for the Academy. Sloan said that the Henri "crowd" refused any longer to have their works "sifted through the beards of the academicians." For all of their confident bombast, however, they knew that the risks of going their own way were high. "We are going to get an awful roasting from some of the papers," Glackens warned his wife.[29]

The landmark exhibition of "The Eight" opened on February 3, 1908, at the Macbeth Gallery on Fifth Avenue. For two weeks, thousands of curious art lovers crowded in to see the sixty-three canvases making up the controversial show. "The thing is a splendid success," Sloan noted in his diary; "they tell me that 300 people were coming in every hour." As Glackens predicted, however, several art critics savaged the show. "Vulgarity smites one in the face at this exhibition," snorted a writer in *Town Topics*. "Is it fine art to exhibit our sores?" Others complained of "nausea" after viewing the "crude" works. "These so-called artists," announced one journalist, "are a black gang who paint half-stripped prize fighters, barkeepers, dissolute folk of the night, immigrants sipping their coffee in dingy restaurants, children of the streets, cabmen, dock rats and . . . the decrepit houses in New York's crumbling East Side slums."[30]

Yet a surprising number of critics offered the insurgents enthusiastic praise. James Huneker wrote in the *New York Sun* that the exhibitors were all "realists inasmuch as they paint what they see, let it be ugly, sordid, or commonplace." In their quest for "sincerity," they refused to see the world through "the studio spectacles of their grandfathers." Instead they painted "with masculine power" an array of arresting new subjects: "coal holes, wharf rats, street bums, the teeming East Side and the two rivers." True, these "valiant pugilists of paint" were "often raw, crude, harsh, but they deal in actualities." Huneker compared the slashing urban realism of the Henri "gang" to the fiction of Dreiser and Norris, though he claimed that the artistry of the painters surpassed that of the writers.[31]

The success of "The Eight" convinced the National Academy to offer them their own gallery at the Academy's next exhibition. When the show opened, the *New York Times* announced in its headline: "New Spirit Seen at the Academy." Glackens told a friend that "the Academy has turned over a new leaf and has accepted everybody this year." Thereafter, Henri and his group shared the New York cultural spotlight with the academic idealists such as Kenyon Cox. Students rushed to take lessons from them, and museums clamored to exhibit their works. These bold men, noted Mary Fenton Roberts in *The Craftsman*, professed "one very manifest and definite purpose, to paint *truth* and to paint it with strength and fearlessness and individuality."[32]

John Sloan

Of all the painters associated with the "Henri circle," John Sloan had perhaps the most inclusive vision. He was a promiscuous observer of the life around him. Like Whitman, he saw in the grimy city a landscape of robust beauty and sensual allure, a peopled world full of pathos and humor that forever pricked his unexampled curiosity. Sloan once described himself as an "incorrigible window watcher" with a "little peeper instinct." He and his suffragette wife Dolly could view "all the world" from their top floor studio apartment on West 23rd Street in Chelsea, bordering the seedy Tenderloin district. In surveying such daily scenes, Sloan strove to enter the minds and feel the emotions of his subjects. His art reveals the clarity and radiance of humanistic realism at its best. "The real artist," he contended, "finds beauty in common things."[33]

With a sort of intense, hushed voyeurism, Sloan delighted in catching people unawares: hanging out the wash, cooking a late supper, reading a Sunday paper, riding the Elevated or ferry, drying their hair, conversing at a corner bar, turning out the light at bedtime, sunning at the beach, soliciting prostitutes, urinating in the gutter, parading down Fifth Avenue sidewalks. After viewing such incidents, he would return to his studio and draw from memory the person or scene he had witnessed. His fugitive pleasure in observing life's myriad incidents and furtive comedy, his unerring sense for the humanly dramatic moment, and his respect for the ephemeral and evanescent gave to his art its distinctive authenticity and photographic spontaneity. To catch people in unmonumental acts of living—that, indeed, was the great motive for Sloan's art. "Keep your mind on such homely things, such deep-seated truths of reality," he advised art students, and there will be no temptation to indulge "in the superficial."[34]

Hairdresser's Window (1907), reveals Sloan's peering perspective. It shows a cluster of people on a city sidewalk gazing upward into an open window as a homely hairdresser performs her craft on a young woman. Among the spectators is a trio of laughing young girls whispering together about the woman customer. In a diary entry, Sloan recounted the inspiration for the canvas: "Walked up to Henri's studio. On the way saw a humorous sight of interest. A window, low, second story, bleached blond hair dresser bleaching the hair of a client. A small interested crowd about." The painting provoked some observers to question the need for such "ugly" realism. "Do we hate this woman so realistically presented to us?" asked the transplanted Irish artist-poet John Butler Yeats. "Is that the artist's intention? She is hideous. . . . She has yellowed hair and thick lips and a short turned up nose and a many folded chin and short fat arms." Nevertheless, for all her grossness, "we realize that she is shrewd and strong and knows every girl's secret." Yeats noted that Sloan, while "looking everywhere for visions of tenderness and beauty, refuses to shut his eyes to facts."[35]

The perennial enchantment of Sloan's paintings and drawings grows out of their simple elements. Here a glimpse, there an incident, and suddenly the real stands out, arresting in its unappreciated familiarity. Sloan had a cunning knack for apprehending the sudden gesture, the fleeting pose, and the socially revealing scene. His oils and lithographs are alive with sharp bits of observation, genuine feeling, and sly touches of refreshing humor. By warmly illuminating ordinary moments in the lives of others, he implied that sympathetic scrutiny is funda-

John Sloan, *Hairdresser's Window*, 1907. *(Wadsworth Atheneum, Hartford, Connecticut. The Ella Gallup Sumner and Mary Catlin Sumner Collection Fund.)*

mentally a moral act. "Be kind to human beings," he counseled his students. "Don't make caricatures. It is easy enough to be cruel. Find the worthwhile things first."[36]

Sloan's admittedly "sympathetic" attitude entailed "interpreting" rather than simply "seeing" reality. What most impressed him during his daily walks or spying episodes was the robust energy and stoic resiliency of the working poor. In his diary in 1906, he reaffirmed his middle-class conviction that the affluent had no monopoly on happiness:

> Walked through the interesting streets on the East Side. . . . Doorways of tenement houses, grimy and greasy door frames looking as though huge hogs covered with filth had worn the paint away and replaced it with matted dirt in going in and out. Healthy-faced children, solid-legged, rich full color to their hair. Happiness rather than misery in the whole life. Fifth Avenue faces are unhappy in comparison.

Such comments reveal how Sloan and others among the "Ash Can" group verged on the romantic fallacy that people who are poor and oppressed are good and noble *because* of their degraded living conditions and that their sufferings and struggles somehow make them more virtuous than others.[37]

Sloan's greatest challenge was to inject scenes with emotion without indulging in sentimentalism. "There is a profound difference," he declared, "between sentimentality and sensitivity." In referring to one of his prints showing a mother and child sleeping on the roof of a tenement on a torpid summer night, he supposed that "there are evidences of 'social consciousness' in this plate, but I hope no sentimentality."[38]

Sloan provided a documentary record of urban life that gives to his work an historical value aside from its artistic merit. His numerous depictions of working-class women, for example, represent a major contribution to American social history. In April 1909 he recorded in his diary a scene viewed from his apartment window: "I have been watching a curious two-room household, two women and, I think, two men, their day begins after midnight, they cook at 3 A.M." This bit of surveillance inspired *Three A.M.* (1909), a painting of two women roommates in their kitchen during the middle of the night. One is cooking a meal while "gossiping" with the other, who sits at a table drinking tea. That one of the women is shown smoking a cigarette and attired only in a nightgown was enough to provoke the ire of traditionalists, and it was rejected by the National Academy.[39]

Sloan often doused his representations of life's daily ironies with a wry sense of humor. *The Women's Page* (1905), one of his favorite etchings, shows a rotund woman sitting in a rocking chair in a cluttered flat. Clad only in a sordid nightgown, her hair unkempt and feet bare, she is shown reading the women's fashion page of the newspaper while her son plays on the bed with a cat. Sloan remarked that his purpose was simply to express the "ironical attitude that my mind assumed in regard to what my eye saw." The woman's utter absorption in the alien world of the fashionable elite offers a powerful contrast to the disheveled state of her own life. Although the burlesque is powerful, Sloan had a peculiar ability to strip his satire of malice. His sympathy was even stronger than his wit, and it ensured that the blowsy woman retains an essential dignity.[40]

Many historians have attributed Sloan's interest in such working-class subjects to his socialist convictions, but he vigorously denied any connection. Although a member of the Socialist party after 1910 and a contributing editor of the socialist magazine *The Masses* from 1912 to 1914, he drew a rigid boundary between his explicitly political cartoons and his professional art. "When propaganda enters into my drawings," he explained, "it's politics not art—art being merely an expression of what I think of what I see." In practice Sloan did not always abide by such clear distinctions, but for the most part he strove to separate his politics from his art. "I am more interested in the human beings themselves than in the schemes for social betterment." Sloan felt that artists with an explicit political agenda unintentionally looked "down on people." In this regard, he enjoyed Stephen Crane's fiction because the author of *Maggie* was "human rather than socially conscious." He "really loved the down and outs on the Bowery out of his love for all life."[41]

Yet where Crane stressed the ubiquitous strife and violence in the inner city, Sloan highlighted familial and social cohesion. His domestic scenes reveal the squalor and deprivation of tenement life but betray none of the brutality that punctuated the Johnson household in *Maggie*. Sloan portrayed sturdy, dignified people, not deluded or resigned victims of impersonal forces and unfortunate circumstances. He wanted to individualize the city crowd, to reveal that "each face [is] beautiful and individual," to "show here and there a detached bit of life which has the power of suggesting the whole turbid current." Observers appreciated his tender detachment. "John Sloan is not telling any story or preaching any moral or painting just for art's sake," noted Mary Fanton Roberts. "He has no criticism of life; the most commonplace thing interests him."[42]

Although Sloan's figurative abilities were modest, he excelled at creating an evocative mood. Perhaps his most powerful painting in this regard is *The Wake of the Ferry* (1907). The setting's simple, yet masterful design features a solitary, anonymous woman leaning against the rail at the back of a ferry. The weather is gray and stormy. Its intimidating bleakness creates a melancholy effect as the shadowy passenger peers out across the churning backwash at a tugboat and ship headed out to sea. The pitching ferry forms a sharp diagonal that accents the woman's lonely, vulnerable countenance and shows Sloan's ability to convert incident into insight and engage the empathy of the viewer. "You cannot escape entering into her thoughts, into the sorrows that have come to her," said *The Craftsman*.[43]

Although crowded with an abundance of narrative detail, most of Sloan's New York scenes are more than simple transcriptions of fact. They enable the viewer to feel the emotions of the depicted moment and to marvel at Sloan's sincerity and depth of feeling. In his bountiful appetite for life, in his recognition that even the humblest people have a role which they enact, elaborate, and live by every day, Sloan achieved the type of democractic realism Whitman had celebrated and Howells never quite realized.[44]

George Bellows

Although George Bellows did not participate in the famous exhibition of "The Eight," he was intimately associated with the band of rebellious painters and eventually became the most versatile and important representative of the "Ash Can" school. Born in 1882 in Columbus, Ohio, George Wesley Bellows was the only child of conservative midwestern parents. His father was a prominent building contractor and partisan Republican, and his mother was a devout Methodist. The elder Bellows encouraged George to become a banker while his mother hoped that he would enter the ministry, but their son developed other plans.[45]

At Ohio State University, Bellows starred on the baseball and basketball teams, sang in the glee club, and immersed himself in fraternity life. He seemed relentlessly determined to earn and display his "manliness," an ambition that helps explain why he fastened upon Crane's *Red Badge of Courage* as his favorite book. Although his classmates assumed that the "swashbuckling" Bellows would become a professional baseball player, he resolved to make his way as an artist.

In September 1904, after only three years in college, Bellows moved to Manhattan and enrolled in the New York School of Art. To gain spending money he played semiprofessional baseball, earning five dollars a game plus expenses. As a new student at the School of Art, Bellows quickly gravitated to Henri's earthly cordiality and informality. "My life begins at this point," he wrote in his journal. Within a few weeks, student and teacher had developed a mutual affection. Bellows referred to Henri as "my father" who taught him to paint his surroundings and in the process express life's inherent "dignity, humor, humanity, kindness, order." Bellows also fell in love with New York. He displayed an outlander's sense of naive wonder at the metropolitan scene, and he became an avid city watcher, savoring the flux and turmoil of the streets, the rushing current of life and energy in Manhattan. Neither a profound nor penetrating thinker, Bellows was a rapt observer who always kept "his eyes open for some hitherto untold piece of reality to put on his canvas." Like Sloan, he felt that the artist must be a "reverential, enthusiastic, emotional spectator, and then the great dramas of human nature will surge through his mind."[46]

As Henri's most precocious student, Bellows quickly gained notice among the larger New York art community. He had too much talent to ignore. In 1909 the twenty-six-year-old Bellows was elected an associate of the National Academy of Design, the youngest artist ever so honored. By then he also had his own studio and students.

The power of Bellows's personality gave a peculiar force to his art. Tall, shambling, prematurely bald, and big-hearted, he loved bay rum and Corona cigars, and he mixed easily with the male patrons of Sharkey's Athletic Club and the YMCA. "There is so much that is effeminate in the American school of painting," observed Henry McBride, art critic for the *New York Sun*, "that the mere manliness of Mr. Bellows's style is enough to distinguish him" from the aesthete. A Boston critic said that Bellows "is a real man, with 'pep' enough for half-a-dozen."[47]

Bellows took a keen pleasure in discussing his own work. He told interviewers that he sought an impassioned realism shot through with human aspiration and vital force. He wanted to get "hold of something real, of many real things," to paint the "cruelty of life as well as its joy." He thus gravitated to the human energies evident in East Side tenements, North River docks, the YMCA, prize fights, circuses, and Central Park. "I paint New York because I live in it," he observed in phrases echoing Henri, "and because the most essential thing for me to paint is the life about me, the things I feel to-day and that are a part of the life to-day."[48]

Boxing was one of Bellows's recurring interests because it offered the most concentrated emotion of any human activity except actual combat. It was also readily accessible. Across the street from his Broadway studio a former pugilist named Tom Sharkey operated a private club which staged boxing matches. Bellows portrayed prize fights in six paintings, sixteen prints, and two dozen drawings. Unlike Eakins, however, he relished the violence of boxing more than the opportunities it afforded for painting the male anatomy.

One of the best of his boxing scenes is *Both Members of This Club* (1909). The drawing is sloppy, the brushwork loose, and the bravado excessive, but the frenzied scene of two bloodied boxers, one black (Joe Gans) and one white (Kid Russell), caught in a violent collision exudes a brutal force and creates a mesmerizing spectacle. Highlighted against a black background and lit by a hellish glare, the boxers are dramatic performers, entertaining an audience of club "members." The leering bloodlust of the caricatured spectators, vividly reminiscent of Goya, adds an almost surreal element to the acrid smell of smoke and sweat and the naked savagery of the fighters in the ring. Bellows stressed that he wanted to convey how the emotions and energies of the nightmarish scene affected him *as a spectator*, and to do so he designed the painting with a "you-are-there realism" so as to offer the viewer a ringside seat.[49]

After seeing Bellows's early boxing paintings, art critic James Huneker placed the artist "in a class by himself." His "muscular painting" hits the viewer "between the eyes." This is exactly what Bellows wanted to do. He strove to

George Bellows, *Both Members of This Club*, 1909. *(National Gallery of Art, Washington, D.C., Chester Dale Collection.)*

perfect what he called an "emotional realism." Like Henri, he wanted to produce something vibrant, something physically alive; he could not rest content with a mere "copy" of what he had witnessed. He succeeded as an artist to the extent that he made "other people feel from the picture what I felt from the reality."[50]

Bellows coupled his masculine taste for "the big facts of life" with a zest for democratic humanism. Like Howells he cherished the ideal of harmonious diversity. "Democracy," he once observed, "is an idea to me, it is the Big Idea." In *The Cliff Dwellers* (1913) Bellows encapsulated what he believed was the lively spirit of a working-class neighborhood and in the process testified to his own romantic faith in an inclusive American dream. Based on a drawing he had done for *The Masses* entitled *"Why Don't They Go to the Country for a Vacation?,"* the painting, stripped of the original drawing's social commentary, depicts a tenement-lined street so crowded with people that the trolley car in the middle distance struggles to pass through the knot of humanity.[51]

At first glance *The Cliff Dwellers* appears to be a scene of mass confusion and suffocating congestion, but on closer inspection it reveals individuals and small clusters of people engaged in a variety of activities on a steamy summer day. Customers surround a peddler's push-cart, kids tussle or play leap-frog in their underwear, tired women nurse babies, scold children, carry home a bucket of beer, converse with each other, and hang out laundry. Other "cliff dwellers" take in the busy scene from balconies, windows, stoops, and fire escapes. The almost caricatural simplification of human forms captures the turmoil and energy of the diverse crowd. Blacks and whites, men and women, young and old, combine to form a mosaic of struggling, laughing humanity whose energies spill over the canvas. Although obviously staged and a bit melodramatic, here was the middle-class version of democratic realism at work: so many hopes, passions, lives tingling with vitality and spirit in the midst of squalid conditions.

Viewers of *The Cliff Dwellers* differed over whether Bellows was condemning slum conditions or celebrating the resilient humanity of the tenement residents. One art critic claimed that *Cliff Dwellers* would cause "the smug to feel con-

George Bellows, *Cliff Dwellers*, 1913. *(Los Angeles County Funds, Los Angeles County Museum of Art.)*

scious of the brutal facts of life" while another concluded that whether "it is a sermon or an achievement is hard to say—perhaps it is both." Such ambivalent reactions cheered Bellows. Like Sloan, he developed a keen concern for those living in destitute poverty and suffering under terrible working conditions, and he, too, aligned himself with the socialists and contributed highly partisan drawings to *The Masses*. At the same time, however, he refused to inject bald propaganda into his art. "As a painter," he once told a journalist, "I am not a preacher; I am not trying to uplift or teach. I am merely trying to do the best work of which I am capable."[52]

In 1911 a female art critic labeled Bellows "the most successful young painter in New York" and asked him to isolate the cause of his rapid success. "Others paved the way," Bellows acknowledged, "and I came at the [right] psychological moment." Like the others among the "Henri gang," Bellows abhorred the merely "pretty" in art and repeatedly sinned against prevailing "academic" standards. He, too, saw life as a diverse drama, often sordid and ugly, occasionally absurd and tragic, but always interesting. His strength was theirs as well: he drew what he saw, easily and naturally, and he saw much. His immense portfolio of paintings and lithographs included intimate portraits of his wife and two daughters, seascapes, and "manly" renderings of the whole spectrum of turn-of-the-century urban society—street urchins, businessmen, Central Park strollers, tenement dwellers, East River dock workers, wealthy polo players, battered prize fighters, and religious revivalists. He drew from such sources, if not beauty in the conventional sense, a vibrant intensity that creates its own allure.[53]

The Triumph of New York Realism

Today it is hard to understand the furor generated by the "New York Realists." When viewed almost a century later, their works evoke a tranquil charm. Yet their initial impact was profoundly unsettling. Like a gust of fresh air, Bellows, Sloan, Henri, and the others helped rid American art of much that was stale and moldy. By 1910 Huneker could announce that the "pendulum has swung from the insipidity and conventionality of the Academy to the opposite extreme." In the process of overturning prevailing convention, the "Ash Can" painters made the city a fit subject for art. "Of later years," the *New York Times* reported in 1911, "we have been waking up to the fact that, from an artistic point of view, this metropolis of ours is anything but ugly." The majority of New York artists were now "enthusiastic upholders of the theory that modern New York is a beautiful city, skyscrapers and all."[54]

The "Ash Can" painters also reinvigorated Whitman's hope that art could enhance social sympathy by exposing different classes, races, and ethnic backgrounds to one another. Their peopled art struck observers as being peculiarly alive, packed with human beings and human drama, energetic and occasionally brutal yet at other times surprisingly tender and poetic. In referring to the efforts of the "Ash Can" painters to give artistic representation to the unappreciated and excluded elements of American society, a writer in *Town Topics* said that the "inspiration that is filling all the world's keen minds with the hope of greater justice is the inspiration of the Realistic School."[55]

While disrupting the polite complacency of the Academy and winning over the art community to the inherent beauty and significance of the cityscape, the "Ash Can" artists revealed the growing complexity of the realistic agenda. Real life, as Sloan often stressed, must provide the inspiration for art, and, as such, art represented a reflection of life, "but the temper and rhythm [of a work of art] are products of the creative mind." Merely photographic art was inert. In this regard, Sloan's great enemy was the unfeeling camera, with its mindless imitation of visual surfaces. By simply copying a scene or a model, the "photographic" artist allowed the creative "impulses" to come from the subject alone. The result was anemic art. "Don't be afraid to be human," he told students in urging them to go beyond mere "eyesight work." They should willingly incorporate their personal reaction to scenes they observed. What "you know helps the mind to dominate the eye in seeing." The great artist "never allows facts to interfere with the truth" of a scene *as witnessed.* The true work of art "tries to *say* the thing rather than to *be* the thing."[56]

In other words, Sloan and his "Ash Can" colleagues agreed that great art began with concrete facts, but culminated with an expressive idea. Bellows, for instance, said that his objective was "to understand the spirit and not the surface" of a scene. "Art is not the result of pure reason," Sloan likewise asserted. "There must be some feeling, emotion, excitement about life." Reality in this sense resided not so much in the thing depicted as in the thing felt. With William James, the "Ash Can" painters believed that the mind engendered a subjective truth upon objective facts. "The sense of reality," Sloan explained, "is a mental product which may be carried in the memory and furnishes all necessary creative impulse." No sooner does a visual stimulus enter the eye than the mind begins processing it into a thought and giving it a new form. He recognized that artistic memory is mischievous; it changes and selects things, fastens on a subject's formal possibilities, penetrates into the inner life of visible facts. By refusing to "finish" their paintings according to "academic" standards, Sloan and his friends revealed the canvas for what it was—a painted surface—rather than the "real thing" itself. They presented, as Mary Fanton Roberts wrote in *The Craftsman*, the "illusion of realism" rather "than a photographic interest in detail."[57]

By acknowledging that the act of painting represents a negotiation between what is seen and what is felt, between object and impulse, image and mind, Sloan, Bellows, Henri, and their band of like-minded rebels did not try to evade the distinction between the imaginative and the imaginary or between the imaginary and the seen. Instead they produced a new type of "realistic" art that operated on two levels: a representation of things seen and a meditation on the act of imagination.

By producing "human documents" in paint, the "Ash Can" painters thrust themselves onto center stage of American cultural life. But their time in the limelight was short-lived. No sooner had the "New York Realists" gained stature and recognition than a new challenge emerged on the cultural horizon in the form of a European-inspired "modernism." An art critic sounded the clarion as early as 1908 when he observed that anyone recently returned from Paris or Munich would call the "Ash Can" painters "very interesting but far from revolutionary." The true "revolutionaries," he announced, were in Europe, and it would not take long for their radical new notions of art and "reality" to make their way across the Atlantic. [58]

Part V

An Epoch of Confusion

"Reality," Ralph Waldo Emerson once observed, "has a sliding floor," and during the early twentieth century it slid wide open. No sooner had realists and naturalists gained access to center stage of American culture than they found themselves being shoved down an epistemological trapdoor by a new "modernist" sensibility that denied the significance of tangible facts and the relevance of mimetic representation. The modernists, Hamlin Garland warily observed in 1914, "aim to push us from our stools." At once a temper, a mood, and a movement, modernism emerged first in Europe at the end of the nineteenth century and became a pervasive international force during the decade before World War I, rupturing accepted beliefs about science, philosophy, psychology, and the arts. "Since the fall of the Bastille," wrote art critic Henry McBride, "nothing has been so astonishing." In assaulting the stolid certainties of bourgeois morality and representational art, modernism fragmented human understanding into a variety of conflicting realities. As the French painter Paul Gauguin acknowledged in 1911, the upheavals of modernism had produced "an epoch of confusion."[1]

13

The Modernist Revolt

odernism arose out of a widespread recognition that Western civilization was entering an era of bewildering change. New technologies, new modes of transportation and communication, and new scientific discoveries combined to transform perceptions of reality and generate dramatic new forms of artistic expression. The modernist world was one in which, as Karl Marx said, "All that is solid melts into air." What once were perceived as formidable absolutes about the physical universe dissolved under the pressure of perplexing new scientific theories. Researchers in a variety of fields refuted several of the most fundamental concepts—Euclidean geometry, Aristotelian logic, and Newtonian physics. In the process, the very solidity of reality evaporated in a steam of doubt about the truth of appearances. As the French mathematician Henri Poincaré declared, "all that is not thought is pure nothingness."[1]

Poincaré and other theoretical physicists and mathematicians altered the image of the world in ways that seemed to conspire against reason itself. Sir Isaac Newton had posited a universe governed by immutable laws rather than divine commands, laws which the scientific method could ultimately uncover and manipulate. This deterministic, orderly cosmos, ticking away with the predictability of a great clock, and looking the same to every observer, began to implode in 1883 when the Viennese physicist Ernst Mach declared that "the world consists only of our sensations." There was, he asserted, no common "reality" independent of human acts of perception.

The revelation of theoretical physics that space, time, and mass had no objective existence ruptured conventional notions of a stable and uniform reality. Albert Einstein and others revealed that what people see is relative to their place and speed. "Many things once thought real have vanished," noted an Illinois physicist. What "we can call real is only that which we create." People began to doubt that things *were* as they appeared, that there was an accessible reality "out there" capable of being accurately registered. "What is this kind of reality?" asked a journalist. "Is it the kind we attribute to a thing or the kind we attribute to a mind?"[2]

Quantum mechanics and relativity theory revealed that mind and nature— subject and object—were interrelated rather than separate realms. Nothing was what it seemed in this relativist or quantum world, and notions of truth grew increasingly ambiguous and self-defining as the new century advanced. A sup- posedly "immutable and unique reality," concluded the Spanish philosopher José Ortega y Gasset, ". . . does not exist: there are as many realities as points of view."[3]

The shared beliefs about the nature of things that were so central to nine- teenth-century realism were breaking down. People could no longer take for granted that their impressions of the phenomenal world held true for others. "One must never forget," declared Gertrude Stein, "that the reality of the twen- tieth century is not the reality of the nineteenth century, not at all." She and other cultural experimentalists "lost faith in what the eyes could see." Where nineteenth-century realists took for granted an accessible public world of tangible surfaces and bounded events that could be readily observed and accurately rep- resented, self-willed modernists viewed the "real" as something to be created rather than copied, expressed rather than reproduced.[4]

This subjectivist challenge to scientific positivism and representational art found a growing number of adherents at the turn of the century. "Is it not time," asked an art critic in 1901, "that we renounce the heresy that it is the function of art to record a fact?" Many American modernists answered yes. In transcend- ing a preoccupation with visible facts and recognizable subjects, they fashioned an aesthetic justification of life as a spiritual substitute for conventional religion and morality. "Nowadays, when so many people no longer believe in super- natural things," the poet Wallace Stevens reported to a friend in 1909, "they find a substitute in the stranger and more freakish phenomena of the mind— hallucinations, mysteries and the like."[5]

Streams of Consciousness—William and Henry James

The subjective emphases of the new physics found a ready parallel in the philo- sophical speculations of William James. Heretofore, he observed in 1904, people had taken for granted that "the sciences expressed truths that were exact copies of a definite code of non-human realities," but the "enormously rapid multi- plication of theories" since that time had demonstrated that "even the truest formula may be a human device." Reality, he had come to realize, was not simply an evident *thing* or a universal *truth*; it was something much more elusive.[6]

James decided that the process of facing facts was much more complicated than previously assumed. People did not simply see a reality "out there" in its autonomous state. They saw what they were conditioned to think of as being real. Reality was thus "more intricately built than [Newtonian] physical science allows." James now preferred "to call reality if not irrational then at least non-rational in its constitution." Consciousness continually interacted with sight to produce perception. Did not hallucinations, James asked, produce a real effect on people? Was not the same true of dreams? spectral visions? mystical states? passing glimpses? blurred impressions? Did not such images and phantoms *become* real when they make people sweat, cry, shiver, or rejoice? In other words, perceived reality encompassed the actual and the possible or imaginary, the concrete and the abstract, the objective and the subjective. "Anything is real," James posited, "which we find ourselves obliged to take account of in any way."[7]

Of course, critics then and since have balked at the elasticity of this provisional notion of truth rooted in personal experience. James showed little respect for formal philosophical distinctions or systematic reflection, and he ignored many logical conventions in his effort to reinstate "the vague to its proper place in our mental life." Thoughts, he repeatedly maintained, were not discrete building blocks operating according to a rational calculus. "Consciousness does not appear to itself chopped up into bits," James wrote in *Principles of Psychology* (1890). "Such words as 'chain' or 'train' do not describe it fitly. . . . It is nothing jointed; it flows. . . . In talking of it hereafter, let us call it the stream of thought, of consciousness, or of subjective life."[8]

A similar fascination with the subjective "stream of consciousness" flowed through the late fiction of Henry James. Scholars still debate the reasons for his heightened interest "just in the mind," as he described it. Some cite two personal calamities: the death of his beloved sister Alice in 1892 and, two years later, the suicide of Constance Fenimore Woolson after she failed in her efforts to transform a friendship with James into a romance. Others highlight James's humiliating failure as a dramatist or stress his intimate exposure to his brother William's writings and speculations.

Still another factor was James's recognition that the burgeoning culture of consumption required a new sensitivity to the psychological values associated with goods. By the turn of the century, increasingly sophisticated marketing and advertising techniques were contributing to the fragmentation of consciousness at the center of cultural modernism. In earlier times, people were encouraged to counterbalance their propensity for self-absorption and self-indulgence by engaging in civic or spiritual activities. Now, a runaway market economy catering to individual fantasies was eroding communal bonds and religious belief. More and more Americans, it seemed, were displaying a veritable passion for commodities. As Madame Merle asserts in James's *The Portrait of a Lady* (1881), "we're each of us made up of some cluster of appurtenances." Tangible things had powerful symbolic effects. The modern self had largely become a function of what people owned or wore. It "overflows into everything that belongs to us—and then it flows back again. I know a large part of myself is in the clothes I choose to wear. I've a great respect for *things!*" To the newly affluent, rather than the well born, James declares in *The Spoils of Poynton* (1897), "things were of course the sum of the world." Things, gushes Mrs. Gereth, "were our religion, they were our life, they were *us!*"⁹

Whatever the reasons for his more acute preoccupation with the fluidity of consciousness, Henry James at the end of the century adopted a new narrative style that pared away conventional action so as to delve more deeply into the psychology of characters and their relationship to things. His earlier assumptions about reality and representation now struck him as being terribly naive. "It's all tears and laughter as I look back upon that admirable time,"he remarked, "in which nothing was so romantic as our intense vision of the real." He now soberly admitted that "the measure of reality is very difficult to fix." Capturing the "real world" in artistic works was a much more complicated process than he and Howells had first imagined. Like his brother William, he decided that reality is not a static condition but a fluid process, not a being but a becoming. The subjectivity of consciousness could not be reduced to objective features of reality. Consciousness could not only reflect the world "out there"; it could also project its own subjective notions of significance onto the screen of experience.¹⁰

In short stories such as "The Real Thing," "The Figure in the Carpet," and "The Turn of the Screw," and in novels such as *The Sacred Fount* (1900), *The Wings of the Dove* (1902), *The Ambassadors* (1903), and *The Golden Bowl* (1904), James led readers on a conscious journey into the interior of his characters' minds. Reality for them is largely a function of their *ideas* of reality. James demonstrated that figments of the imagination could shape human behavior just as much as actual events and circumstances. In fact, what people thought or imagined often exerted more influence over their actions than practical concerns or daily routines.

In *The Ambassadors*, for example, James relentlessly analyzes the consciousness of Lambert Strether, a genteel Massachusetts widower-editor who travels to Paris at the behest of his fiancée, a wealthy widow named Mrs. Newsome. His mission is to retrieve her son Chad, who has refused to return home to manage the family's business interests because he is engaged in a love affair with a married woman. After meeting Chad and his beguiling French mistress, however, Strether succumbs to the seductive allure of liberated life on the Left Bank. His fixed notions of right and wrong dissolve among the aesthetic and sensuous delights of this French "land of fancy." The enticing shop windows excite in Strether dreams of material splendor and self-indulgence. "It hung before him this morning, this vast bright Babylon, like some huge iridescent object, a jewel brilliant and hard, in which parts were not to be discriminated nor differences comfortably marked. It twinkled and trembled and melted together, and what seemed all surface one moment seemed all depth the next."

These collisions of experience with sensibility provide the novel's animating dynamic. Strether, the strolling visitor, tries to read the luminous text of Parisian life with the spectatorial detachment of the realist, only to find that the city's glittering surfaces entice him to want "things he shouldn't know what to do with." Seeing, he realizes, is not simply a "mirroring" process; it can also become a medium of irrational desire as well as a tool of self-transformation. Reality in *The Ambassadors* is thus much more than a series of social transactions and physical settings; it is largely what Strether thinks about what he sees and experiences. He becomes an observer observing himself making observations rather than a passive mirror of external events. What does he see? Does he observe the world as it really is or does he fashion what his self desires to see? Ah, he sighs, if he could only live in his dreams.[11]

For Strether, the visible world repeatedly penetrates his inner life as his fugitive sensations and impressions force him to question his inherited sense of values. It now "made more for reality" for him to abandon his conventional past and immerse himself in the sensuality of Paris, where the very "air had a taste of something mixed with art." Eventually, however, Strether realizes that the mind harbors guilt as well as desire. In the end he encourages Chad to stay in France and pursue his amorous affairs, while he, the failed ambassador, returns to Massachusetts alone, but with a freshened sense of life at its fullest and most voluptuous.[12]

In centering *The Ambassadors* on Strether's exhausting mental gymnastics, James revealed his own fascination with the shifting "stream of consciousness" and his desire to use fiction as a mediating medium between a sensitive mind and a mercurial world. Individuals, he decided, were "deeper, more substantial, more self-directing" than the "conditioned" types featured in the novels of Zola and other literary naturalists. The challenge of a modern realism was to "penetrate" the ways in which "real things" shaped mental outlooks.[13]

Bergsonism and Cinematic Reality

Henry and William James discovered in the French philosopher Henri Bergson a powerful ally in their study of the unseen realm of psychic activity. In a series of immensely influential and exquisitely vague books written during the first decade of the new century, Bergson assaulted the positivistic assumption that people gain knowledge solely through the rational processing of sensory data. Positivistic science explored only the outer shell of reality. The "really" real, he asserted, was not solid matter or mental images but a living process of "creative evolution" whereby an *élan vital*—an immaterial life force—engineered the progressive evolution of human society. Reality in such an evolutionary context "flows," Bergson argued in terms echoing William James; "we flow with it; and we call true any affirmation which, in guiding us through moving reality, gives us a grip upon it and places us under more favorable conditions for acting." Like the quantum physicists, Bergson believed that the very act of trying to analyze feelings or ideas altered their nature because "the feeling itself is a being which lives and develops and is therefore constantly changing."[14]

Intuition, Bergson declared, enabled people to *know* reality by living *within* it in a sort of time incapable of being measured by a clock. "There are no things" in such a realm, Bergson declared, "but only actions." Bergsonian truth was thus created rather than discovered, experienced rather than captured, penetrated rather than mirrored. Reality resided in mobility, in things in the making, in changing states of mind and meaning. The genuine philosopher, he concluded, "will see the material world melt back into a single flux, a continuity of flowing, a becoming." Bergson readily acknowledged that the world of matter often operated according to natural "laws," but he insisted that metaphysical speculation based on "intuitive" feeling offered a stronger tool with which to apprehend "the real."[15]

Bergson's nebulous assertion that "it is only through ideality that we can resume contact with reality" enthralled receptive thinkers in Europe and the United States at the turn of the new century. The French novelist Marcel Proust attended Bergson's lectures at the Sorbonne, became his close friend and his nephew by marriage, and applied his insights in writing *Remembrance of Things Past* (1913–27), a magisterial seven-novel work exploring the relationship between time lost and time recalled. Proust had become convinced that "a literature which is content with 'describing things,' with offering a wretched summary of their lines and surfaces, is, in spite of its pretension to realism, the furthest from reality, the one which impoverishes and saddens us most." In narrating his own autobiographical story, Proust struck an inspired balance between description of the "realistic" world of appearances and exploration of the subterranean world of his own involuntary memories, visions, and emotions, states of consciousness so fugitive and complex that they could only be suggested rather than depicted. Proust's nameless narrator discovers that "true reality" resides in his own psyche rather than in the social world.[16]

Bergson's "vitalist" emphasis on an "inner reality" operating within the "stream of time" generated intense interest in the United States. During an American lecture tour, the French philosopher spoke to huge crowds and garnered front-page newspaper coverage. Columbia University awarded him an honorary degree, and bookstores could not keep his writings in stock. After reading Bergson's *Creative Evolution* (1907), William James wrote the author an effusive letter: "O my Bergson, you are a magician, and your book is a marvel." James felt that he and Bergson were "fighting the same fight, you a commander, I in the ranks." Both of them recognized the "continuously creative nature of reality." By reviving the concepts of free will and spiritual belief, Bergson offered a tonic contribution to the anti-positivistic crusade. John Dewey observed in 1912 that no "philosophic problem will ever exhibit just the same face and aspect that it presented before Professor Bergson invited us to look at it in its connections with duration as a real and fundamental fact." Bergson's efforts to "find the reality hidden beneath . . . customary images" and his emphasis on movement rather than rest and the blending of past and present challenged common-sense empiricism and the popularity of literal representation. An art critic announced in 1913 that the primary effect of "Bergsonism" was to legitimize "and glorify the sub-conscious, the impalpable, and the imprecise."[17]

Bergson's notion of reality as a fluid and continuous process shared many affinities with the emerging art of motion pictures. The cinematographer's ability to capture life in motion gave movies a greater sense of actuality than that offered in a photograph, play, novel, or painting. By introducing time into space, films conveyed the interaction of past and present. Cross-cutting enabled directors to show events occurring simultaneously in different places; multiple camera angles offered a many-sided glimpse of characters; close-ups magnified the effect of objects and people. "This world of ours," declared the editors of *The Independent* in 1914, "is a moving world and no static art can adequately represent it. There is no such thing as still life. . . . Everywhere and always there is motion and only motion, and any representation of reality at rest is a barefaced humbug." Motion pictures now offered "for the first time the possibility of representing, however crudely, the essence of reality, that is, motion." The editors based this assertion on their reading of modern philosophy and physics. "Bergson has shown us what a paralyzing influence static conceptions of reality have had upon the history of philosophy and how futile have been all attempts to represent movement by rest. The scientist of today thinks in terms of motion. All modern thought is assuming kinetic forms and we are coming to see the absurdity of the old ideas of immutability and immobility."[18]

The maturation of the cinema as a new art form led many artists and critics to declare that conventional figurative art had become anachronistic—static, stale, dead. "How can Salon painting compete," asked the French painter Ferdinand Léger, "with what is available in every cinema in the world?" Robert MacCameron, an American painter who labeled himself an "academic realist," said that the growing sophistication of the cinema would force "the modern artist to turn back to formula and symbol. What he sees is not as important as what he feels, and art presently will enter the sphere of pure poetry."[19]

The Reality of the Irrational

In his 1895 presidential address to the American Psychological Association, William James declared that recent theories of the subconscious had demolished the conventional belief that the mind was simply a reflexive receiver of external stimuli. Six years later he asserted that the "world of mind" involved "something infinitely more complex than was suspected." Such comments reflected a spreading appreciation of Sigmund Freud, the Viennese father of psychoanalysis. James had become aware of Freud in 1893, but another decade passed before his ideas were widely known in the United States. When Freud gave five lectures at Clark University in Massachusetts in 1909, he was surprised to find himself so famous "even in prudish America." The enthusiastic reception accorded his controversial ideas revealed to him that psychoanalysis "was no longer a product of delusion, it had become a valuable part of reality."[20]

By then Freud had carried analysis deeper into the human psyche than had ever been done before, revealing corners of the personality long hidden or "repressed" from public awareness. Freud claimed that reason offered only a thin coating of protection against the anarchic impulses and atavistic tendencies lurking beneath consciousness. In *The Interpretation of Dreams* (1899), he offered clinical "proof" based on patient interviews that the dreaming mind had a logic of its own. By recalling fragments of dreams, people could fasten upon a new order of communication. The unconscious, he concluded, "is the true psychic reality; in its inner nature it is just as much unknown to us as the reality of the external world, and it is just as imperfectly communicated to us by the data of the consciousness as is the external world by the reports of our sense-organs."[21]

It did not take long for Freud's ideas—or distortions of his ideas—to penetrate society at large. The notion of unruly and unpredictable sexual desires and aggressive instincts at the bottom of the psyche captivated people who were eager to restore a sense of vitalism and uncertainty to theories of human behavior. By 1916 there were some 500 "psychoanalysts" in New York City alone. The Greenwich Village playwright Susan Glaspell remembered that "you could not go out to buy a bun without hearing of someone's complex." People began to talk openly about libido, inhibitions, Oedipus complexes, sublimation, and repression. Artists mined psychoanalysis for symbols and imagery, convinced that the subconscious was rich with drama and hidden meanings. Walter Pach, an early American champion of modern art, noted that much of his distrust of fixity and academic convention came from the discovery of "the role played by the unconscious in our lives."[22]

The Retreat from Likeness

The new physics, philosophy, and psychology trumpeted the same basic message: reality was not what it seemed. This revelation spawned a baffling multiplicity of "modernist" cultural experiments in Europe and the United States designed to give artistic form to hidden realities: unruly states of mind and feeling. In music, composers such as Arnold Schoenberg and Igor Stravinsky disdained conventional melody or harmonics in favor of dissonant and, eventually, atonal works. In the theatre, August Strindberg wrote "anti-realistic" plays in contrast to the middle-class dramas of Henrik Ibsen. In the art world, post-impressionists, fauvists, cubists, futurists, expressionists, and others discarded literal representation of recognizable subjects in favor of vibrant color masses, simplified forms, or geometric shapes. The modern artist, Hutchins Hapgood reported, does "not render objects, as such, at all. He would reproduce . . . the underlying forms of matter, the atoms and molecules of form, rather than the objects such as the camera or the imitative academic painter understands them." The "post-impressionist" painter sought to render "metaphysical form," the "unseen but the deeply real" essence underlying visible objects.[23]

Several women authors saw in literary modernism a means to escape the confining gender stereotypes still governing cultural expression. The British novelist Dorothy Richardson explained in the preface to *Pilgrimage* (1913) that she had attempted to produce a "modern" novel that would be "a feminine equivalent of the current masculine realism." Men, she believed, were so absorbed in the pursuit of material gain and so busy skating along the surfaces of public life and massaging their egos that they were blind to much of life's hidden pleasures and pains. Women, however, led marginalized lives of relative silence and solitude that enabled them to explore the epiphanies of the mind and the elastic qualities of language. Such "contemplated reality," Richardson decided, was richer and deeper than the adventurous public exploits favored by male novelists.[24]

In sum, cultural modernists—both male and female—were less concerned with reforming or even representing society than with exploring the nature of consciousness or with experimenting with purely formal artistic concerns. For writers and artists shorn of religious belief and faith in reason, the sincerity and intensity of personal expression seemed more important than meticulous representations of public life. Walter Pach pointed out in 1914 that the most salient manifestation of the impact of modern science upon modern art was a growing focus on the inner feelings, often irrational or absurd, tucked in the folds of the individual psyche. "As Romanticism tinged the mind and art of the early nine-teenth century, and Realism . . . [tinged] its later years, so every indication to-day points to a deepening interest in the matters which go beyond self-conscious reasoning, and are dealt with by the power of intuition."[25]

American Modernists

Cultural modernism promoted an international sensibility that weakened national allegiances and created a global community of migratory intellectuals, writers, and artists. The experimental American painter Marsden Hartley decided that "the idea of modernity is but a new attachment to things universal." He reported from Paris in 1912 that his reading of William James had prompted him to abandon still-life realism "in favor of intuitive abstractions." He was now painting "very exceptional things of a most abstract psychic nature—things which are in themselves pure expression without subject matter." With a gusto born of fresh discovery, Hartley announced that the "realistic folly of imitation has been exposed" and that he was shedding his "Americanness." Painters must now "express themselves in terms of pure language."[26]

Georgia O'Keeffe also found the attractions of modernist art irresistible. In 1907 she enrolled in the Art Students League in New York City and began study under William Merritt Chase. She liked his use of a "warm, vivid palette" and his "pleinarist approach to the fleeting and transforming quality of light," but she found his impressionism "determinedly masculine" in its broad brush strokes and aggressive literalism. O'Keeffe yearned instead to use intense colors and precise detailing to express the "things I see in my head" with "a woman's feeling." She was determined to "say something that a man can't," and the abstract emphasis of modernism offered her the freedom to do just that.[27]

The expressive abstractions produced by O'Keeffe and other modern artists combined to crack the mirror of representational realism. An art critic reported in 1913 that "realism is receiving its death blow and individualism is asserting itself once more" in an effort to express the "real inwardness of life as opposed to that outward reality glorified by the realists." With each passing year, the most advanced artists of Europe and the United States distorted, dissociated, or dissected identifiable reality so far beyond common recognition that the con-ventional relationship between object and painting disappeared.[28]

The human conduit for modernist painting in the United States was the charismatic photographer and art impresario Alfred Stieglitz. In 1908, at the same time that Robert Henri and the other members of "The Eight" were launching their controversial exhibition at the MacBeth Gallery in New York City, Stieglitz and his associates were showing the even more unsettling works of Henri Matisse at the Photo-Secession gallery at 291 Fifth Avenue. There was little affection between the Stieglitz crowd and the "Ash Can" realists. Although both hated the mossbacks in the Academy, they each claimed to represent the most "progressive" force in American art. Sloan alluded to the tension between the two factions when he noted in his diary that Stieglitz and his supporters were "hot under the collar about our show. The whole curious bunch of 'Matisses' seem to hang about him, and I imagine he thinks we have stolen his thunder in exhibiting 'independent' artists."[29]

In late 1905 Stieglitz had opened an art gallery in the three-room attic of a brownstone at 291 Fifth Avenue. He announced in his magazine *Camera Work* that the new gallery was "neither the servant nor the product of a medium. It is a spirit." Stieglitz pledged his support to any photographers, artists, or sculptors who displayed "honesty of aim, honesty of self-expression, honesty of revolt against the autocracy of convention." He told a New York reporter that he and his comrades were revolting "against all authority in art, in fact against all authority in everything." Within a few months, Stieglitz's "291" gallery had become the center of a self-conscious cultural modernism in the United States. Stieglitz's unstinting enthusiasm for artistic experimentation attracted a galaxy of rebellious young American painters: Hartley, O'Keeffe, Arthur B. Dove, John Marin, Abraham Walkowitz, and Max Weber. Their common denominator, *Camera Work* announced, was a desire to liberate themselves from academic gentility and "from realistic representation."[30]

Stieglitz took a circuitous route in arriving at his modernist perspective. His father, a German immigrant who prospered as a wool merchant, wanted Alfred to become a mechanical engineer. So he sent his seventeen-year-old son to Germany for training in 1881. While enrolled at the Berlin Polytechnic, however, young Stieglitz studied photo-chemistry and soon abandoned his engineering studies. "I went to photography really a free soul," he remembered, "—and loved it at first sight with great passion." Within five years his photographs were winning awards in Germany and England.[31]

In 1890 Stieglitz returned from Europe and began work as a professional photographer in Manhattan. Unlike most of his peers, however, he rejected the use of a studio, costumed models, and "poetic" subjects. Instead he took his hand-held camera outside to capture the pictorial possibilities of his immediate surroundings—the streets, wharves, ferries, railroad yards, back allies, and tenement court yards. "New York, its streets and people," he recalled, ". . . all fascinated me." Like Dreiser and Bellows, Stieglitz roamed the city in search of evocative scenes and picturesque human subjects. He, too, believed that a single moment warranted celebration and preservation. Yet he also maintained that the photographer should integrate his own "feeling" about a scene into the image.[32]

Winter—Fifth Avenue (1893), one of Stieglitz's most famous photographs,

Alfred Steiglitz, *Winter on Fifth Avenue*, 1893. *(George Eastman House, Rochester, New York.)*

illustrates his effort to combine visual fact and emotional feeling. He stood for three hours in a ferocious blizzard at the corner of 35th Street and Fifth Avenue before photographing a lonely teamster and his horses. Not content to use the camera as a passive register of random scenes, Stieglitz waited for "the moment in which everything is in balance" so as to emphasize the important facts of the scene and bring "what is not visible to the surface." The result was not a mere copy or transcript of the scene but an evocative photograph warm with human feeling, not unlike a Bellows painting or a Dreiser story.[33]

Stieglitz urged people to accept the camera for what it was—the most powerful medium for mimetic representation—and pursue "straight" photography; that is, images of "real" life that were "in no way faked, doctored, or retouched." He detested "artistic" camera men who "go against life, against truth, against the reality of experience, against the spontaneous living out of the sense of wonder—of fresh experience, freshly seen and communicated." The "academic" photographers, Stieglitz argued, should quit using quirky camera angles, scratching their negatives, and smearing their printing paper in an effort to produce "painterly" images.[34]

In early 1902 Stieglitz formed a society in New York to promote his belief that photography was a fine art capable of expressing a "deep realism." He named it the "Photo-Secession," choosing the term "secession" because avant-garde painters in Germany and Austria had used it to signify their rebellion from the accepted idea of what constituted a photograph. The Photo-Secession group publicized and exhibited the works of American photographers and published two magazines, first *Camera Notes* and then *Camera Work*. By 1906 one of Stieglitz's associates could announce in *Camera Work* that "the real battle for the recognition of photography is over. The chief purpose for which the Photo-Secession was established has been accomplished—the serious recognition of photography as another medium of pictorial expression."[35]

While promoting photography as a legitimate art form, Stieglitz saw his own artistic outlook transformed. His early commitment to "straight" photography began to give way to a growing interest in aesthetic experimentation. In a 1902 interview with Dreiser, he revealed his willingness to tamper with negatives in order to express his artistic "individualism." When Dreiser countered that such manipulation was "not photography," Stieglitz replied that, as long as the result was "beautiful, call it by any name you please."[36]

Stieglitz's growing interest in abstract form helps explain his excited promotion of non-pictorial painting. Because the camera offered the best medium for recording the face and mood of familiar scenes, he reasoned, avant-garde artists were free to explore the hidden realms of consciousness. When asked about the role of mimetic representation in art, he replied: "My camera can do that. If you are an artist you must do more." Stieglitz thus encouraged the radical subjectivism of "post-impressionist" art, acknowledging that the mind of the painter offered a more powerful source of inspiration than the "real" world. "Art, for me," he explained, "is nothing more than the expression of life. That expression must come from within and not from without."[37]

In 1911 Stieglitz devoted a special issue of *Camera Work* to modern art and invited Gertrude Stein to write essays on Matisse and Picasso. Stein, a transplanted American living in Paris with her brother Leo, an aesthete and art collector, hosted a famous Saturday evening salon that included the most prominent European painters and writers. She also accumulated a vast collection of modern art that attracted a stream of visitors. "I have never in my life seen so many dreadful paintings in one place," Mary Cassatt said after viewing works by Gauguin, Matisse, and Picasso at the Steins, "and I want to be taken home at once." Yet when Stieglitz and his American friends visited the Stein salon, they loved the modernist works hanging on the walls. Arthur Dove told Harriet Monroe that the most exciting "modern" artists in Paris and New York were those "reaching out toward an art of pure color and form dissociated from 'representation.'"[38]

The Armory Show

The crusading efforts of Stieglitz and others to champion a European-inspired "modernism" in the United States culminated in the Armory Show in 1913, the most dramatic event—and publicity stunt—in the history of American art. Plans for the exhibition begin in December 1911, when a group of two dozen "advanced" and "independent" painters, led by Walt Kuhn and Arthur B. Davies, formed the Association of American Painters and Sculptors (AAPS), a group dedicated to "exhibiting the works of progressive and live painters, both American and foreign." The inclusion of "foreign" artists opened a rift between the AAPS and the Henri circle that would widen into a public split at the Armory Show.[39]

Kuhn worried that Henri's chauvinistic devotion to American subjects and artists would impede his own efforts to promote the European avant-garde and an international sensibility. For now, however, Kuhn grudgingly solicited the support of the "soft boys" among the "Ash Can" group. "Henri and the rest will have to be let in," he told a friend, "—but not until things are chained up so that they can't do any monkey business." From the start of the AAPS, the organizers treated the "Ash Can" painters as unwelcome but necessary allies. As Sloan recalled, "Henri and the rest of the crowd around him were rebuffed whenever they asked questions." Yet the two groups temporarily joined forces to organize a massive exhibition of "progressive" art and sculpture, an enterprise so large in scope and ambition that it would dwarf the earlier efforts of "The Eight."[40]

To gather paintings for the Armory Show, Kuhn and Davies took an extended trip to Europe in 1912. What they saw in studios in The Hague, Munich, Berlin, Cologne, Paris, and London convinced them to increase the foreign representation at the exhibition and thereby catapult European "modernism" into the forefront of cultural discussion in the United States. The shift in emphasis surfaced in the Armory Show's official title: the International Exhibition of Modern Art. To house the 1300 works of art they had collected, Kuhn and Davies leased the cavernous 69th Regiment Armory at Lexington Avenue and 25th Street. Two months before the opening, Kuhn sensed its electric significance. "We want this old show of ours," he wrote Pach, "to mark the starting point of a new spirit in art, at least as far as America is concerned."[41]

The Armory Show opened on February 17, 1913, and created an immediate sensation. "A new world has arisen before our eyes," announced *Camera Work*. Although two-thirds of the works were by Americans, it was the Europeans who captured most of the attention. The *New York Times* warned visitors who shared the "old belief in reality" that they would enter "a stark region of abstractions" at the Armory Show that was "hideous to our unaccustomed eyes." The experimentalist artists on display were "in love with science but not with objective reality," and this led them to produce paintings that were "revolting in their inhumanity."[42]

Picasso's cubist works provoked howls of outrage among the Academicians who visited the Armory Show. "The real meaning of the cubist movement," sputtered Kenyon Cox, "is nothing else than the total destruction of the art of painting." Art critic Frank Jewett Mather denounced the European radicals as representing the "harbinger of universal anarchy." The puzzling canvases on display demonstrated that Europe was suffering from "the licentiousness of over-estheticism, the madness of ultra culture." The *New York Times* saw in such "unrealistic" art a threat to Western civilization itself. Cox adopted a similarly apocalyptic tone: "We have reached the edge of a cliff and must turn back or fall into the abyss." The intensity of such criticism revealed that there was more at stake in the Armory Show than art; the "modern" temper itself was on trial.[43]

Yet in addition to provoking shocked indignation and critical ridicule, the Armory Show also generated excited praise. Its riotous energies and courageous experiments fascinated the art critic James Huneker. He warned readers that "to miss modern art is to miss one of the few thrills that life holds." Another critic encouraged people to visit the exhibition so as to enter a "realm where subjectivity reigns supreme." Over 87,000 people flocked to see the controversial show, and it dominated headlines in the New York newspapers for weeks. From New York the Armory Show traveled to Chicago and Boston, and in each city it aroused the same boisterous response and attracted throngs of visitors. A quarter-million people viewed the exhibition in the three cities, and buyers bought some 250 paintings, most of them by the Europeans. As a reporter for the *New York Globe* predicted, "American art will never be the same again."[44]

Stieglitz viewed the Armory Show as both a direct assault on academic gentility and a mortal blow against "realistic" art as well. He was no longer interested in "The Eight" or their fellow travelers. In a full-page advertisement for the Armory Show that he placed in the *New York American*, Stieglitz noted that the painters trained under Chase and Henri would continue "doing business at the old stands. Sometimes the dead don't know when they're dead—and it's folly to expect them to bury each other." Yet he saw no point in "breeding little Chases, little Henris" any longer. The camera and modern art had made their imitative representations of contemporary life obsolete.[45]

Stieglitz's confrontational language set the tone for a public showdown between American "modernists" and "realists." Just before the exhibition opened, Henri visited the Armory and found that the French works were hanging in the most prominent sections of the hall. "I hope that for every French picture that is sold," he told Pach, "you sell an American one." Pach replied with steely certitude: "That's not the proportion of merit." An indignant Henri thereupon issued an empty threat: "If Americans find that they've just been working for the French, they won't be prompted to do this again."[46]

Henri and the other New York Realists were virtually forgotten amid the furor over the European modernists at the Armory Show. "After the Armory Show," said Stuart Davis, "I was impressed by . . . the realization that the Henri school, . . . this American free naturalism, wasn't the answer, that all kinds of new areas were opened up." Ernest Lawson confided to Henri that "we are now old fogies, my dear man, with all this howling dervish mob of cubists, &c." Henri, however, refused to abandon his commitment to humanistic realism. While appreciating the audacity of the "ultra-modern" Europeans, he had no patience with purely formalistic exercises. Art, he continued to insist, must be rooted in humanity rather than the self; it must retain some direct relation to society. Others among the "Ash Can" painters were even more resentful of the insurgent modernism. Glackens, for instance, dismissed much of the "modern" art on display at the Armory Show as "the pouring out through symbols of a half-baked psychology, a suppressed adolescence."[47]

At the next meeting of the AAPS, the simmering controversy boiled over. "Old friends argued and separated," Kuhn reported, "never to speak again." Henri, Luks, Du Bois, Bellows, Myers, and other "New York Realists" resigned from the AAPS. Jerome Myers, who described himself as "an American art patriot who painted ashcans and the little people around him," sadly concluded that "our land of opportunity was thrown wide open to foreign art, unrestricted and triumphant; more than ever before our great country had become a colony;

more than ever before, we had become provincials." George Luks issued a similarly splenetic assessment of the "foreign invasion" at the Armory Show. The vagaries of the abstractionists were so much "wall paper" to him, and he yearned for the day when such "adolescent" experiments would be "swept aside and artists set themselves to record the deep pulse of the national life and capture its wayward rhythms."[48]

John Sloan never shared Luks's "supreme contempt" for the "ultra-modern school" of artists. In his memoirs, he recalled that the Armory Show removed "the blinders from my eyes." Having discovered that "visual verisimilitude was no longer important," he "opened up his palette" to bold new color combinations and experimented with the geometries of form. Bellows responded in a similar manner, turning away from the warm intimacy and vigor of his early work in favor of new color theories designed to make his art more "modern." Yet in trying to adapt to changing currents, both Bellows and Sloan diminished their natural talents and lost the endearing charm and human interest that distinguished their earlier work. By the end of 1914 critic Henry McBride could write that Henri was "almost alone among the American realists in continuing to be in earnest when he paints" in such "conventional" ways. Two years later another art critic delivered a solemn, if premature, obituary: "Realism is dead."[49]

The Revolution of the Word

The modernist impulse displayed at the Armory Show prompted a surge of anarchic experimentation throughout American culture. "Make it new!" urged poet Ezra Pound, and rebels responded with a host of new ideas, causes, and forms of expression. Mabel Dodge, the wealthy hostess of a Fifth Avenue salon for cultural radicals of all sorts, announced that the Armory Show was the most important event in American life since the Revolution. "It seems as though everywhere, in that year of 1913," she wrote, "barriers went down and people reached each other who had never been in touch before." Nearly "every thinking person nowadays is in revolt against something." Bohemians in Greenwich Village promoted free love and women's suffrage, birth control and Freudian psychology. "We are revolutionaries without a revolution," Dell wrote a friend in 1914.[50]

Modernism, declared *The Dial* in 1913, had gained such a secure foothold in the United States that the "tumult and the shoutings of the captains of 'realism' have died away." Avant-garde writers and poets were abandoning the conventions of fact-giving and story-telling in order to convey the bristling complexities of private consciousness. "Realism in literature has had its run," Pound reported from London in 1914. The transplanted American poet had "heard all that the 'realists' have to say." He was now more interested in the "creative

Marsden Hartley (Edmund Hartley), 1877–1943, *Provincetown*, oil on composition
board, 1916, 61.3 × 50.9 cm. (Art Institute of Chicago, Alfred Stieglitz Collection,
1949, 545.)

faculty as opposed to the mimetic." What need, he asked, "was there for a work
of art to resemble something?" The "only measure of truth is our own perception
of truth."[51]

By 1915 the literary critic H. W. Boyton could observe that modernists were
dismissing "the word 'realism' as a shabby counter of speech possessing little or
no intrinsic value." Howells provided a tempting target for those eager to discard

middle-class realism. Literary experimentalists savaged him for having failed to delve beneath the surface of his characters and probe their psyches. He was now an elder to be pulled down, as elders must always be pulled down. The "amazing revelations of Sigmund Freud and his school of psychoanalysis," said a contributor to *Current Opinion*, had discredited "the whole theory of realism in fiction exemplified in the work of William Dean Howells and his disciples." Howells's failure to deal with the subconscious, declared the literary critic Alexander Harvey, "has done an enormous amount of damage to American literature." By refusing to transport the reader into "the heart of life's mystery," his "realism is without reality."[52]

Many modernists also savaged Howells's "sentimental" faith in democratic humanism. Common folk, they contended, were not worthy of great art. Literature no longer needed to represent the contemporary social world; it should instead provide a vehicle for creative genius and for the expression of language as a self-sufficient reality. "There is no respect for mankind" among the avant-garde, Pound claimed, "save in respect for detached individuals." His friend Margaret Anderson, the poet and editor of *The Little Review*, echoed the sentiment: "It's a fact that humanity is the most stupid and degrading thing on the planet."[53]

A chance encounter between Robert Frost and Wallace Stevens suggests the changing literary landscape of the early twentieth century. Stevens reportedly said that the "trouble with your poems, Frost, is that they have subjects and they make the visible world too easy to see." Stevens felt that the modern poet should "escape from Facts" and become "the priest of the invisible." What was most "real" to Stevens was not "a collection of solid, static objects extended in space" but the intangible realm "within or beneath the surface of reality."[54]

By questioning external appearances and by developing idiosyncratic new forms of expression, modernists helped jolt realists out of their complacency about the self-evident nature of the real. Yet like many illuminating new perspectives, modernism often blinded people with the light of its own anarchic vitalism. The tormenting sense of alienation at the heart of the modernist sensibility steered many aesthetes down a solipsistic cul-de-sac. Contemptuous of bourgeois philistinism and the vulgarities of mass society, the most extreme modernists substituted their own aesthetic values and formal concerns for the social content that had long sustained cultural expression and from which democratic realism had always drawn its sustenance. Modernist oracles developed a raging contempt for ordinary men and women that led them to relegate pictorial art and mimetic literature to the mass public while reserving for themselves an abstract realm of pure feeling and unfamiliar forms. As Charles Caffin stressed in 1908, realism in art was now considered "but food for children, fit only for men and women whose minds have not passed beyond the child stage of picture books."[55]

Willard Huntington Wright, a young art critic and raffish editor of *Smart Set*, who became an outspoken proponent of "modern" art, flaunted his indifference to public concerns or popular taste. The "antagonism of the masses to the artist," he wrote in *Modern Painting* (1915), "sprang up simultaneously with the disgust of the artist for the masses." Wright dismissed the very idea of democratic realism. Any effort "to democratise art," he wrote, "results only in the lowering of the artistic standard." For painting to "reach its highest point of artistic creation, its realistic aspect had to go." He thus encouraged artists to spurn "the lovers of reality and sentiment" and indulge in deliberate obscurities.[56]

Why this strenuous effort to sever art from common folk and common experiences? "Were realism the object of art," Wright explained, "painting would always be infinitely inferior to life—a mere simulacrum of our daily existence, ever inadequate in its illusion." Modern painting was instead an exclusive medium for expressing complex emotions. It was meant for the initiated, not for the masses. Only people with highly developed aesthetic sensibilities could appreciate its abstruse spirit. Not content to remain slaves "to the common vision of reality," modern artists, according to Wright, wanted to "sacrifice the fact" in order to express their innermost feelings and produce images that were "incomprehensible to the untutored person who regards art as an imitation of nature."[57]

The "Young Americans"

Many artists and critics who otherwise shared the modernist attack upon the genteel tradition balked at such avant-garde elitism and the promotion of unfettered—and indecipherable—individualism. The modern artistic imagination, they decided, did not have to be hermetic. Rather, it offered powerful new ways of connecting with humanity. "With the increase of democracy," observed James Oppenheim, the former social worker turned experimental poet and the founding editor of *The Seven Arts*, "one would naturally expect art to contact more and more of the majority. The opposite seems to be true."[58]

During the second decade of the twentieth century, Oppenheim and other young intellectuals associated with new magazines such as *The Seven Arts*, *The New Republic*, and *The Masses* promoted a cultural renaissance drawing upon the positive aspects of both twentieth-century modernism and nineteenth-century realism. Van Wyck Brooks labeled them the "Young Americans" and listed Oppenheim, Walter Lippmann, Randolph Bourne, Paul Rosenfeld, and George Soule among the leaders of the group. Devoted to aesthetic experimentalism yet determined to relate the arts to the concerns of mass society, they were "disillusioned optimists" and "subjective realists" who self-consciously identified with the wave of "progressive" reform movements that swept across American politics and society in the first years of the new century. "It is our faith and the faith of many," said the editors of *The Seven Arts* in their inaugural issue in 1916, that the "arts [should] cease to be private matters." Although aware of the significance of "modern developments" in European culture, they refused to sacrifice their national identity for the sake of internationalist modernism. As Americans, "we have our own fields to plough; our own reality to explore and flush with vision."[59]

Like Henri and many of the "Ash Can" painters, these cultural nationalists promoted an American art and literature organically related to the everyday world. They remained confident that there was a discernible reality "out there" independent of subjective perception and that modern forms of artistic representation could help people understand and improve their social order. Any viable new cultural movement must draw its energies from the human hive rather than from the isolated soul, and it must be supple enough to adapt to the shifting "movement of real life."[60]

The "Young Americans" thus refused to stand aloof and apart from the common folk. Their goal was solidarity rather than separatism. Lippmann labeled their program "applied pragmatic realism" because of their faith in the capacity of the scientific spirit to transform facts into dreams. The real movers and shakers of the new century, he predicted, would not be politicians but the "publicists and educators, scientists, preachers and artists," for it "is out of culture that the substance of real revolutions is made."[61]

Implementing such a vague cultural revolution required that "vital and virile" writers and artists adopt a "new realism" that would break through the polite barriers delimiting the fiction of Howells and James while at the same time avoiding the self-referential conceits of modernist art and literature. They "must look to the present," Oppenheim declared, "face reality as it is," and then "recreate reality" by "converting knowledge into vision, and the intangible and abstract into concrete forms." Such a "new realism" must see the relevance of common life and common folk while at the same time use new insights gleaned from modern science and psychology to pierce beneath the shell of facts.[62]

Impatient with the modern artist or writer who cultivated "a little private world of his own," George Soule found "something cleansing" about the "enthusiasm for reality" displayed by the "Young Americans." Their infectious zest for life led them "to look the world in the face and explore every cranny of it." The challenge for the "Young Americans" was to foster a "modernist realism" capable of representing both visible facts and subjective impulses. They thus championed Theodore Dreiser for exploring sensual realities long hidden from view. Harold Stearns charged that the conventional literary realism of the day too often resulted in "mere surface accuracy, in a firm grasp on unessentials." Most "realistic" writers still dealt with the veneer of daily life and ignored its underlying essences and emotional complexities. The characters in Sinclair Lewis's novels, for instance, struck Stearns as "external and a bit unreal." Their "interior lives" were underdeveloped. Yet Stearns maintained that such a "realistic failure was more honorable than a romantic success."[63]

Lippman made perhaps the most fervent plea for a "new realism" combining modern psychology with social engagement. If the purpose of art is to "increase life," he wrote in *The New Republic*, then clearly "art cannot do its work if it remains incommunicable." Most modern artists, he observed, had formed a closed community that admitted only those who could speak the language of abstraction. Their contempt for the representation of contemporary social themes led modernists to favor "symbols of the incommunicable." Lippmann, however, refused to believe that "the subject of art does not matter," and he promised that a "modern realism" drawing upon the recent revelations of physics, philosophy, and psychology could offer a personally meaningful and socially responsible aesthetic. With an almost mystical faith in the wisdom of the literate public, Lippmann concluded that the objective of journalists, writers, and artists must be to provide people "with the sharpest report of reality we can make."[64]

Modern intellectuals and artists, Lippmann added, needed to free themselves from Howells's restrictive Victorianism and learn from the "Freudian school of psychologists" that "crime and civilization, art, vice, insanity, love, lust, and religion" derived from the same sources—the unconscious drives and passions of the psyche. "Real life," he concluded, "is largely agitated by impulses and habits, unconscious needs, faith, hope, and desire."[65]

Oppenheim, too, charged that earlier realists such as Howells focused too much on external details and surface descriptions. The result was "an unreal realism, a realism which left himself and his deep needs out of consideration." Twentieth-century writers, he urged, must recognize that the real exceeded the visual and the logical and included the visionary and passionate. Prejudices, moral judgments, and stereotypes colored or distorted a person's outlook, and any worthwhile political or cultural movement had to take these subjective factors into account.[66]

Among those whom Oppenheim cited as exemplars of the "new realism" was Sherwood Anderson, a veteran of the Spanish-American War from Ohio who quit his job as a paint factory manager in 1912 in order to "find truth" and pursue a literary career in Chicago. Anderson promoted a "democratic literature," but he maintained that it was not enough simply to include a few common folk in fiction. He wanted desperately to "get out of myself [and] truly into others." Literature, he believed, enlarged and enriched existence to the extent that it admitted people into the inner lives of others. Fiction, Anderson wrote, "must constantly feed on reality or starve," yet the writer should aspire to be more than simply a camera with a pencil. The "modern realist" must explore the hidden passions and frustrations within a person's interior life. "Real" authors wrote "out of the people" rather than about the people. This required traveling "on the road of subjective writing" and "crude expression" that "such masters of American prose as James and Howells did not want to take." Howells, Anderson felt, had erred in viewing people primarily as rational, self-controlled beings; he had failed to consider the immense realm of unconscious urges and sexual drives affecting behavior.[67]

In *Winesburg, Ohio* (1919), a collection of twenty-one loosely connected stories about life in a small midwestern town, Anderson revealed a new modernist prose style moored to a representational humanism. Unlike his contemporary, Sinclair Lewis, he gave little sustained attention to the physical description of the town or its residents. His concern was not with surface appearances or with dramatic action but with how psychically bruised people dealt with wayward and often lustful impulses while living within an oppressive village environment. Most of community life, he felt, was actually lived in private, inside the mind and heart, and he viewed it as his special mission to expose such inner realities to his readers. In this regard he followed the advice of one of his characters: "You must not become a mere peddler of words. The thing to learn is to know what people are thinking about, not what they say."[68]

The lonely souls of Winesburg are misunderstood eccentrics who lead stunted and repressed lives, unable to connect or communicate with one another. Kate Swift expresses their common dilemma: "It seemed to her that between herself and all the other people of the world, a wall had been built up." In telling about such folk, Anderson adopts the persona of George Willard, a young newspaper reporter who feels a burning need "to come close to some other human, touch someone with all his hands." The town's solitary grotesques reach out to him because he is uncorrupted, and he, in turn, finds himself reaching out to them. The searing poignancy with which Anderson described the struggles of Winesburg's "grotesques" against personal inhibitions and social conventions endowed them with dignity and revealed how much more truth— and mystery—a modern version of realism could unsheath.[69]

John Dos Passos, a Chicago-born writer fresh out of Harvard, also championed a modern realism capable of representing the nation's inclusive ideal as well as its tapestry of diverse peoples. Instead of "our inane matter-of-factness" or the modernist "cult of the abstract," he appealed for a new style of cultural expression that would restore a sense of "dramatic actuality" to literature and the other arts. "Shall we pick up the glove Walt Whitman threw at the feet of posterity?" To do so, he wrote a friend, required a "frank and sympathetic understanding of reality" and that "is not got by closing doors" to any fact or experience, no matter how distasteful.[70]

Of all the "Young Americans," Van Wyck Brooks became the most vigorous and eloquent proponent of a pluralistic cultural nationalism that would enable people to "see our life as it at present is" and then transform it into a "beloved community." In a series of provocative books and essays written during the second decade of the century, Brooks savaged the "tepidity" and "weightlessness" of an American bourgeoisie so preoccupied with commercial gain and moral prudery that it had yet to discover the depths of its "poetic spirit" amid harsh social realities. At the same time, he charged that the "dazzling sophistication" of modern art depended upon a "deliberate obliviousness to the facts of life." Blessed with a huge reservoir of criticism to dispense, Brooks also flayed the limitations of liberal reformers. They "had no realistic sense of life," he observed. "They ignored the facts of the class struggle" and "accepted illusions like that of the 'melting-pot.'" But "worst of all," they "had no personal psychology."[71]

Brooks asserted that the "real awakeners" of American thought and culture during the previous quarter-century had been sociologists, scientists, and "realistic philosophers" such as William James and John Dewey, who "have not only accepted reality, they have claimed reality." Now it was time for "modern" writers and artists to cultivate a "new and more sophisticated" program of cultural expression designed to produce the "wholesale reconstruction of a social life" in which the emotional and passional needs of the spirit would also be revered and nourished. For too long, realists had exalted the "seeing-relation over the feeling-relation." Political and social reformers had focused so much on transforming institutions that they had ignored the need to revitalize the human spirit. Both the crass materialism of the consumer culture and the individualistic conceits of modern art and literature, Brooks charged, were self-indulgent and ultimately self-defeating. Commercial-obsessed businessmen had lost touch with civic ideals, while the most "abstract" modernists had lost contact with the social scene. Brooks's proposed solution was a cultural flowering that would create a "genial middle tradition" between "vaporous idealism and self-interested practicality," a tradition combining "thought and action" and instilling a sense of reverence for collective experience and "self-subordinating service" that would promote the decentralization of government authority and the nurturing of truly "organic" social ties.[72]

Brooks and the other "Young Americans" injected an invigorating new energy and urgency into intellectual life, but they never explained precisely how such a Whitmanesque cultural revival was going to bring about the salvation of society. Less a program than a plea, their reform crusade blending pragmatic liberalism with cultural modernism was hopelessly ambitious and largely rhetorical. The nostalgic "beloved community" of their dreams encompassed only intellectuals like themselves rather than common folks. Too often, as Casey Blake has recently written, their "generous vision of a democratic culture gave

way to a rhetoric of mystical wholeness, prophetic leadership and organic mutu-
ality that dissolved politics in silent communion and emptied their ideal of
community of the voices and aspirations of the real people in real
communities."[73]

The War to End All Wars

Even more disastrous to the "Young Americans" and their ambitious cultural
ideal was Woodrow Wilson's decision to declare war on Germany in April 1917.
American entry into the three-year-old conflict provoked a fractricidal dispute
among the "progressive" intellectuals that sundered their tenuous union of cultural radicalism with political liberalism. The unexpected turn of events especially devastated Randolph Bourne, who vigorously opposed American intervention. "One has a sense of having come to a sudden, short stop at the end
of an intellectual era," he lamented in the October 1917 issue of *The Seven
Arts*. When his friends Lippman and John Dewey endorsed intervention, he
felt "left in the lurch." The "philosophy upon which I had relied to carry us
through no longer works." Dewey's "realistic" endorsement of war was "a mere
surrender to the actual, an abdication of the ideal." Bourne thereafter became
bitterly skeptical about the possibilities for meaningful cultural change. American intervention in the war had "spiritually wasted" the efforts of the "Young
Americans" to practice a "new realism" intended to reform harsh social facts
and create a "beloved community."[74]

Every war is worse than expected, but no one foresaw the unparalleled
agonies of the protracted conflict in Europe. In just the first five months of
fighting, the Germans suffered a million casualties; French losses totaled
300,000 in two weeks. The average life expectancy of an infantryman in France
was three months. By the end of the war, ten million combatants had been
killed on both sides. And as the American Civil War had demonstrated, such
prolonged slaughter grinds an edge on youthful innocence. The ghastly revelations from the battlefields outraged common sense. Oppenheim observed that
"it is not with the intellect that we can grasp the meaning of this war. It has
assumed proportions of mysterious terror and grandeur."[75]

During the course of the Great War, the same sequence of emotions that
had surfaced during the Civil War reappeared: initial conviction and zeal gave
way to what one writer called numbing "disillusion, cynicism, unbelief in God
and man." To live in rat-infested and mud-carpeted trenches, to experience
constant shelling and sleepless nights, to participate in suicidal frontal assaults
across open land stitched with barbed wire, sown with mines, scented with
mustard gas, and defended with machine guns, was to join in a fraternity of

horror. For many, the absurd devastation of the Great War exposed the inadequacy of conventional forms of artistic representation. "The reality," observed one veteran, "surpasses all literature, all painting, all imagination."[76]

The unparalleled slaughter of the Great War also served to turn the generation gap into a chasm. Ernest Hemingway charged that "worm-eaten old men" and "moralizing hacks" had instigated the war. Dos Passos, who in 1916 worked in France as an ambulance driver and as a hospital volunteer hauling buckets of amputated arms and legs, stressed that "the ridiculous horror of war's actuality is less hateful than the lies" of politicians. He and others who returned to the United States from the Western front felt a disorienting sense of alienation. They talked of hollow men and waste lands, the rupturing of reason and the shattering of old assurances. "The old perspectives are seen to be blurred," said a novelist in 1919, "and we are now in the first stages of the process of readjusting our instruments of social vision."[77]

The Great War caused many sensitive souls to stage a confused retreat from public concerns so as to nurse private chagrins and pursue private certainties. "In literature, in art, in politics, in all departments of life," Floyd Dell observed in 1919, "there has been an alienation of the younger generation from traditional modes of action." The war's ghastly reality had become so simultaneously *real* and *unreal*, that it could neither be condoned nor understood. T. S. Eliot correctly observed that "Human kind/Cannot bear very much reality." The war's concussive impact sent sensitive people careening in every direction in search of new faiths—mechanistic science, nihilism, mysticism, and varied forms of personal preoccupations. The postwar era was, Oppenheim concluded, a period "when everything is possible . . . when all is in flux and the next tide may destroy us."[78]

Especially disheartening for young American intellectuals were the domestic effects of the war: the eruption of savage ethnic prejudices, the suppression of civil liberties and freedom of speech, the collapse of Woodrow Wilson's presidency, and the violent social turmoil accompanying the sharp postwar recession and demobilization. "In our national life today," Harold Stearns lamented in 1921, "the young intellectual speedily finds that *he is not wanted.*" By calling into question commonsense notions of reason and proportion and by raising doubts about the very sanity and fate of Western civilization, the war convinced many that representational art was bankrupt. A "modern" sensibility in pursuit of private truths was the only viable intellectual stance. The "folly of imitation has been exposed," Marsden Hartley announced. Painters sensitive to such disillusioning carnage could not possibly copy its horrors; they must "express themselves in terms of pure language."[79]

Hartley's friend Gertrude Stein called the rudderless artists and intellectuals who had served in the war a "lost generation," and the misleading tag has stuck like a gaudy emblem. Although the extent and significance of the "lost generation" have been greatly exaggerated, it is true that several of the most prominent young artists and thinkers in Europe and America decided in the aftermath of the Great War that social progress was an illusion and that conventional forms of meaning were meaningless. "Literature, like life," wrote the novelist Robert Herrick after the war, "no longer seems to have an unhesitating conviction about anything." In much of the wartime poetry, fiction, art, and criticism, a dark irony displaced the democratic humanism of the "Young Americans." Van Wyck Brooks, for example, abandoned his earlier emphasis on social engagement and called on his disenchanted friends to insulate themselves "against the common life and its common values."[80]

Yet the still popular assumption that the Great War provided the final impetus for a nonrepresentational modernism to supersede referential realism is misleading. Realism in all its diverse manifestations survived the Great War as well as the epistemological challenge of modern science and psychology. "The war is more than anything else for us an awakening to realism," said a writer in *The Nation* in 1919. "The old idealism must pass and a new realism must emerge." At the same time that modernist abstraction witnessed growing popularity among some writers and artists, a "new realism" flourished as well. In 1920 a prominent literary critic seized the occasion of Howells's death to explain that the "father of American realism" had founded a school that had "passed its meridian, but there are other groups, very eager, very vital and industrious, and they may soon achieve" a high place in the cultural affairs of the world.[81]

Waldo Frank spoke on behalf of "modern" realism when he wrote in *Our America* (1919) that "we are the true realists, we who insist that in the essence of all reality lies the ideal. America is for us indeed a promise and a dream." The true modernist, he argued, must "go to the basic materials of life" rather than fly "off from an opaque world into translucent ether." Frank and others still manifested a Howellsian commitment to facing and representing the facts of contemporary life. Their perception of the facts, however, had been deepened, sharpened, and sobered by the insights of modern physicists and psychologists, the challenges posed by abstract modernism, and the shocking effects of the Great War. The war, as the editors of *The Nation* observed, "has set the whole world to questioning what this reality is which is worth such a cataclysm as we are witnessing today." Echoing the cultural nationalists of *The Seven Arts* group, they called for a "new" and "unconventional" realism that would use the insights of both positivistic science and nonrational psychology to come "into close touch with actual life."[82]

Such a "new realism," explained literary critic H. W. Boynton, "interprets, creates, transforms raw fact (if there is such a thing) into something more real than fact, and infinitely more pregnant." After 1920 this effort to represent "both outer and inner realism" found bold expression in the fiction of Anderson, Dos Passos, Ellen Glasgow, Willa Cather, Eudora Welty, Ernest Hemingway, and

William Faulkner, the art of Edward Hopper and Thomas Hart Benton, the music of Aaron Copland, and the plays of Eugene O'Neill. "On the whole, I think this recurrent demand for realism makes for sanity in literature," said a young literary critic. "At the same time, it is obvious that it is not realism, but reality, which is the ultimate excellence in fiction or in drama. Realism is the result of observation; reality is the result of imagination." His distinction reveals how, by 1920, notions of realism had grown more sophisticated, more daring, and, in many eyes, more "real."[83]

EPILOGUE

There is something in staying close to men and women and looking on them,
and in the contact and odor of them, that pleases the soul well. All things
please the soul, but these please the soul well.

—Walt Whitman,
"I Sing the Body Electric"

Realism has not fared well of late within the higher reaches of the
academy and among prominent segments of the artistic community.
Much of the fiction and art produced during the Gilded Age is as
out of fashion as the hansom cab and hooped skirt. Today, it seems,
the gaslit, brownstone world represented by William Dean Howells and John
Sloan evokes the poignance of a bygone era. What was vivid and valid for its
own day has become bland and antique a century later. Many realistic works
of art and literature have lost the sheen of novelty and seem terribly outdated.

But there is more at work than changing cultural fashions. Since the end
of World War II, realism and its accompanying ideology of liberal humanism
have become objects of ferocious contempt. The horrifying revelations of the
Holocaust, the looming threat of nuclear annihilation, and the banal priorities
and conformist imperatives of a runaway consumer culture have led many avant-
garde writers and artists to view contemporary society as a nightmare from which
they are trying to escape. Modern American culture has supposedly become so
suffocating and corrupting, so vacuous and idiotic, that it is no longer worthy
of mimetic representation. In 1961, for instance, novelist Philip Roth noted
how difficult it was "to understand, describe, and then make *credible* much of
American reality. It stupefies, it sickens, it infuriates, and finally it is even a
kind of embarrassment to one's own meager imagination." Many thoughtful
Americans, he reported, were spurning the "grander social and political phenomena of our times" in order to "take the self as their subject": the "sheer fact
of self, the vision of self as inviolate, powerful, and nervy, self as the only real
thing in an unreal environment."[1]

More recently, a burgeoning group of critical theorists—structuralists, post-structuralists, deconstructionists—has assaulted the mimetic tradition from a different angle. Drawing upon an array of reality-defying philosophers—Kant, Nietzsche, Heidegger—and employing an arsenal of opaque jargon and ingenious utterances, they dismiss representational art as a "vulgar illusion" and a "naive subterfuge." In their view, "reality" is a linguistic construct rather than something outside the self. The arts, therefore, can never truly represent or imitate the "actual" world. At best they can only mimic the discourses which fabricate that world. Viewed from this perspective, the methods of conventional realism—visual curiosity, common-sense empiricism, and rational analysis— are utterly inadequate tools for the imaginative artist living in a "postmodern" age. As literary critic Sven Birkerts maintains, the representational imperative championed by Whitman and Howells is now impossible because "our common reality has gradually grown out of reach of the realist's instruments."[2]

Still other "antirealists" argue that the godless world outside the self is not so much elusive and absurd as it is illusory. As a consequence, the best works of art and fiction "deconstruct" themselves. That is, they reveal their own inability to mirror reality. In his brilliantly playful story, "The Death of the Novel," Ronald Sukenick declares: "Reality doesn't exist, time doesn't exist, personality doesn't exist." Sukenick contends that a "post-realist" literature should not try to replicate the "insipid" and "unreal" activities of bourgeois society. The "subject" of art is no longer relevant. Postmodern art and literature should escape from the tyranny of appearances and generate their own reality rooted in the idiosyncratic self, even "at the risk of solipsism." It is time, he asserts, to abandon the false hope of representing a public reality and cultivate an extreme subjectivity that borders on the hermetic and impenetrable.[3]

Novelist William H. Gass said much the same when he announced in 1970 that the postmodern writer must keep readers "kindly imprisoned in his language, [for] there is literally nothing beyond." At its best this anarchic insistence that meaning exists only in the mind of the beholder, that it is something to be created rather than mirrored, opens up infinite possibilities for inventive brilliance and the unfettered expression of things private and inward. At its worst, however, such a purely aesthetic perspective leads to a desiccated self-consciousness and cheerful nihilism detached from the human community. Flannery O'Connor had such a self-referential outlook in mind when she said that the borders of reality for radical modernists were the sides of their skulls.[4]

Some champions of postmodern fabulism have become so zealous in their rejection of the mimetic tradition that they deny *any* legitimacy to realism as a mode of cultural expression. They dismiss representational art as an aesthetic placebo for bourgeois philistines. Realism, claims critic Robert Scholes, is dead, or, at least, irrelevant. Those hopelessly backward writers and artists who still embrace the mimetic fallacy are like "headless chickens unaware of the decapitating axe." Such harsh attacks have led some to question whether realism is worth studying even as an *historical* phenomenon. Literary historian George Levine asks: "Given the disrepute of realism and the exposure of representation [as a false idea], what could be the point in reconsidering their status once again?"[5]

An answer to Levine's question must begin with an obvious correction: reports of the death of realism and reality are greatly exaggerated. Some observers understood this almost a century ago. "Realism," the literary critic Bliss Perry predicted in 1902, "has wrought itself too thoroughly into the picture of the modern world, it is too significant a movement, to allow any doubt as to the permanence of its influence." The realistic impulse has since demonstrated a remarkable perseverance. Today, even in the face of the chic piety that the very notion of representing reality is a "sham," many Americans still savor artistic renderings of the world "out there" beyond the self. Countless writers and artists continue to produce convincing and compelling representations of observed fact that provoke powerful, complicated emotions. They do so without being defensive or apologetic. As artist Richard Estes explains, "I try to work in the tradition of Eakins, Degas, Vermeer, and the many other great realists. I'm . . . trying to paint what I see. . . . Nor do I have any verbal theories behind the work; it's all looking at something and trying to paint it."[6]

Even more than their nineteenth-century predecessors, contemporary realists are diverse in outlook and application. Some, such as writer Tom Wolfe, seek to revive the fact-mongering documentary emphasis of nineteenth-century fiction. He practices and promotes "a highly detailed realism based on reporting, . . . a realism that would portray the individual in intimate and inextricable relation to the society around him." Others profess a new realism informed by the psychological emphases and formal innovations of modernism. As a result, they are more aware of the problematic nature of representing *what is there* than the Dreiser who blithely asserted that "truth is what is." They recognize that seeing is not exactly the same as knowing and that facts are not always unambiguous. Reality to them is not simply what exists; it is fluid and pluralistic, provisional and uncertain, culturally conditioned and filtered through individual perception. Suggestion thus becomes as important a skill as enumeration. Yet for all of their modern sophistication, contemporary realists sustain the essential mission undertaken by Howells and Homer, Jewett and Wharton, Sullivan, Eakins, Henri, and Lippmann: to find an accessible language or form to represent the vital concerns and recognizable features of the everyday world. "Making reality real," as Eudora Welty insists, "is art's responsibility."[7]

Welty posits an essential truth that we are always in danger of forgetting amid the narcissistic cant of self-reflexive modernism: we cannot afford to abandon the larger social scene, nor can we ignore the aesthetic pleasures contained in unadorned fact. Despite the self-contradictory claim that everything is merely fictive and verbal, there is a redemptive public realm "out there." It is not simply a figment of the imagination nor a semantic construct. Our common phenomenal world is a fact of life deserving of respect and recognition. The urge to represent it seems essential to human nature. "Reality is not something I chose," the "New Realist" painter Philip Pearlstein noted in 1973, "but something I found myself in." The ideal of facing and expressing the facts of everyday life and social relations, however imperfect the process might be, persists in large part because of the quiet grandeur inherent in the close observation and faithful rendering of the figures, objects, and concerns outside the self. "All I'm trying to do," Pearlstein remarked, "is to see things as they are."[8]

It is in this context that the nineteenth-century realists retain a vibrant significance. The realistic enterprise, born out of faith in the importance of the social world and nourished by the methods and prestige of empirical science, brought a radiant energy to thought and the arts during the nineteenth century. In the process it helped displace a flaccid idealism. "The illegitimate monopoly which the genteel tradition had established over what ought to be assumed and what ought to be hoped for has been broken down," reported the Harvard philosopher George Santayana in 1913. While using "facts" to assault the patrician elite's stranglehold over high culture, realists helped create a self-conscious American philosophy, literature, art, and architecture. As Saul Bellow has testified, "the development of realism in the nineteenth century is still the major event of modern literature."[9]

Whatever their animating impulse—a simple fascination with the visual world or an unabashed love of life, a curiosity about others, a quest for documentary truth or a commitment to functional design, a pagan acceptance of things as they are or an impassioned protest against injustice—the realists sought to transmute the raw materials of life into works of palpable truth and enduring insight. By rendering the ordinary significant and the hidden visible, by refusing to offer easy consolations or to rest content with cheap ironies, they demonstrated the power of representation to sustain, assure, and enlarge us. The realists, wrote Henry James, provided the pleasure of recognition and the prod of self-examination. In the process they taught the hard but vital lesson that life is often antagonistic to desire and that most people are morally mixed.

In this super-sophisticated postmodern age, the realism of fact spawned during the nineteenth century remains thoroughly unfashionable and fundamentally important. Some quite intelligent and discerning people still long for the sympathetic humanism and trustworthy satisfactions evident in old-fashioned realistic fiction and art. "One of the literary secrets of our time," confessed novelist Cynthia Ozick in 1987, "is that we miss the nineteenth-century novel. We miss it intensely, urgently," and long for "an excursion, however humanly universal, into [its] lost conventions."[10]

Ozick's honest yearning for a humane realism suggests that an essential task of thought and the arts is to help people connect with others. By entering into the emotions, aspirations, and limitations of others, nineteenth-century realists manifested a rejuvenating passion for the strange spectacle of humanity. Although often fraught with class biases, overheated virility, and unknowing condescension, their democratic aesthetic reminded people that they are helplessly interdependent. It also appealed to the unrelenting infatuation people have with themselves, their entanglements with neighbors, and their jarring encounters with the larger society and its ephemera. The literary critic Ludwig Lewisohn recognized in 1921 that it was realism's basic respect for people that made it "so sanative an influence amid the heat, the turmoil, and the moral malignities of society. To it there are no outcasts; none are disinherited, none wholly guilty, none stale or discarded. Thus beauty and truth, art and humanity meet and are one."[11]

Today realism retains its artistic appeal as the United States struggles to embody its ideal as a nation proud of its diversity and just in its social transactions. Painter Willard Midgette stressed in 1978, shortly before his death, that he strove in his art to help "the people looking at the picture to make contact with the people in the picture." He clung tenaciously to the belief that realistic representation possesses a peculiar ability to intensify and extend a sense of social communion. For this reason alone, the creation of an "aesthetic of the common" and a "beloved community" remains a beckoning goal. "The idea of a democratic culture that extends beyond the academy," literary critic Irving Howe observed in 1989, "is one that we cannot afford to surrender, not if we remain attached to the idea of a democratic politics."[12]

The extraordinary resilience of realism as a mode of cultural expression attests to its inherent appeal. In reflecting upon the redemptive ardor of the realistic enterprise, novelist John Updike recently declared that "Howells's agenda remains our agenda: for the American writer to live in America and to mirror it in writing, with 'everything brought out.'" Updike then quoted a letter in which an elderly Howells told Charles Eliot Norton that he was "not sorry for having wrought in common, crude material so much; that is the right American stuff. . . . I was always, as I still am, trying to fashion a piece of literature out of the life next at hand." Updike's conclusion shimmers with conviction: "It is hard to see, more than eight decades later, what else can be done."[13]

LIST OF ABBREVIATIONS

AM	*Atlantic Monthly*	LS	Louis Sullivan
AS	Alfred Stieglitz	NAR	*North American Review*
EW	Edith Wharton	NR	*New Republic*
FLIN	*Frank Leslie's Illustrated News*	OWH	Oliver Wendell Holmes
		PSM	*Popular Science Monthly*
FLW	Frank Lloyd Wright	RHD	Rebecca Harding Davis
FN	Frank Norris	RTC	Rose Terry Cooke
HB	Henri Bergson	RWE	Ralph Waldo Emerson
HDT	Henry David Thoreau	SC	Stephen Crane
HG	Hamlin Garland	SOJ	Sarah Orne Jewett
HJ	Henry James	TD	Theodore Dreiser
HM	*Harper's Monthly*	TE	Thomas Eakins
HW	*Harper's Weekly*	TSP	Thomas Sergeant Perry
JR	Jacob Riis	WDH	William Dean Howells
JRL	James Russell Lowell	WJ	William James
JWD	John W. De Forest	WW	Walt Whitman

NOTES

Introduction

1. Charles G. Leland, "Editor's Easy Talk," *Graham's Magazine* 53 (1858):554; "The Study and Practice of Art in America," *Christian Examiner* 71 (July 1961):80–81; HJ, "The Last French Novel," *Nation* 3 (11 October 1866):286–88.

2. HG, *Crumbling Idols* (Boston, 1894), 141, 143.

3. Lewis Mumford, Vernon Parrington, Oliver Larkin, and Henry Steele Commager were exceptions to this trend, yet their works, though influential and valuable, are dated and in significant ways incomplete. Parrington died before finishing his magisterial trilogy, *Main Currents in American Thought*, and as a result the third volume, entitled *The Beginnings of Critical Realism* (New York, 1930), lacks his characteristic polish and interpretive force. Larkin's *Art and Life in America* (New York, 1949) is much more successful in dealing with developments in the arts than in society, and Commager's *The American Mind* (New York, 1950), while encyclopedic, takes for granted a vast amount of knowledge on the part of readers. Mumford's *The Brown Decades* (New York, 1931) was a truly seminal work that alerted twentieth-century Americans to the underlying vitality of the arts during the "brown decades" known as the Gilded Age. But in the introduction, Mumford acknowledged that he was presenting "only the outline of a book" rather than a comprehensive treatment.

More recently, Miles Orvell has published a sparkling analysis of realism entitled *The Real Thing: Imitation and Authenticity in American Culture, 1880–1940* (Chapel Hill, 1989). The most comprehensive survey of American literary realism remains Warner Berthoff, *The Ferment of Realism, 1884–1919* (New York, 1965) while Amy Kaplan's *The Social Construction of American Realism* (Chicago, 1988) provides fresh interpretive insights into the motives of key literary realists. Equally useful is Michael Davitt Bell's study of the impact of a realistic agenda upon writers. See *The Problem of American Realism: Studies in the Cultural History of a Literary Idea* (Chicago, 1993).

4. "Miss Wilkins — An Idealist in Masquerade," *AM* 83 (May 1899):665–75; Arlo Bates, "Realism and the Art of Fiction," *Scribner's* 2 (August 1887):241.

Theodore Dreiser wanted to straddle both realism and romanticism. "For all my modest repute as a realist," he wrote, "I seem, to my self-analyzing eyes, somewhat more of a romanticist than a realist." He then confused matters by adding that realism constituted "the very substance" of both romance and drama. Likewise, novelist Elizabeth Barstow Stoddard insisted that "I am *not* realistic — I am *romantic*, the very bareness and simplicity of my work is a trap for its romance." TD, *Dawn* (New York, 1931), 198–99; Elizabeth Barstow Stoddard to E. C. Stedman, 21 April 1888, Edmund Clarence Stedman Papers, Rare Book and Manuscript Library, Columbia University.

5. Henry Mills·Alden, "Editor's Study," *HM* 115 (September 1907): 646; Edward Everett Hale, Jr., "Some Further Aspects of Realism," *The Dial* (16 March 1893):170. For a stimulating critique of the use of labels such as realism, see Michael Fried, *Realism, Writing, and Disfiguration: On Thomas Eakins and Stephen Crane* (Chicago, 1988).

6. Leo Stein, "American Optimism," *Seven Arts* 1 (May 1917):83; FN, *Vandover and the Brute* (New York: Library of America Edition, 1986), 185–86, 224; FN, "The Need of a Literary Conscience," in *The Collected Writings of Frank Norris*, 10 vols. (Garden City, N.Y., 1928), 7:38–39.

7. Montrose J. Moses, "Interview with Theodore Dreiser," *New York Times Review of Books*, 23 June 1912, pp. 377–78; Ambrose Bierce, *Devil's Dictionary* (1906; Owings Mills, Maryland, 1978), 209; WDH, *Criticism and Fiction* (New York, 1891), 73.

8. Samuel Osgood, "The Real and the Ideal in New England," NAR 84 (April 1857): 549; HJ to Robert Louis Stevenson, 31 July 1888, Robert Louis Stevenson Collection, Beinecke Rare Book and Manuscript Library, Yale University.

9. Willa Cather, "William H. Crane in *Brother John*," *Nebraska State Journal*, 5 April 1894, 6.

10. Richard Fisguill, "Death and La Mort," NAR 199 (January 1914):96; Countess Du Bury, "Contemporary French Literature," NAR 86 (January 1858):232.

11. WDH, "Editor's Study," *HM* 77 (July 1888):317; WW, *Democratic Vistas* (1871), in *Walt Whitman: Complete Poetry and Collected Prose* (New York: Library of America Edition, 1982), 962.

12. *Letters of Elizabeth Palmer Peabody: American Renaissance Woman*, ed. Bruce A. Ronda (Middletown, Conn., 1984), 49.

13. "Certain Dangerous Tendencies in American Life," AM 42 (October 1878):398; L. Lewis and H. S. Smith, *Oscar Wilde Discovers America: 1882* (New York, 1936), 153. For a useful corrective to the idea that women dominated American cultural life, see Kathleen D. McCarthy, *Women's Culture: American Philanthropy and Art, 1830–1930* (Chicago, 1991). The issue of a gendered culture is explored in Mark Carnes, *Secret Ritual and Manhood in Victorian America* (New Haven, 1989); Mary Ann Clawson, *Constructing Brotherhood* (Princeton, 1989); E. Anthony Rotundo, *American Manhood: Transformations in Masculinity from the Revolution to the Modern Era* (New York, 1993).

14. Willa Cather, "The Passing Show," *The Courier* (23 November 1895):7–8. Edith Wharton agreed that men were more "realistic" than women. As a child she developed a fondness for "little boys and puppies" and a disdain for the "negligible" feminine world of crinoline and dolls as well as the feminine tendency "to make things prettier and prettier." Wharton, "Life and I," unpublished autobiography, pp. 12, 15, 1–2, Edith Wharton Collection, Beinecke Rare Book and Manuscript Library, Yale University.

15. Margaret Fuller, *Woman in the Nineteenth Century* (New York, 1855), 157–58; Fuller, *Literature and Art* (New York, 1832), 2:130–31; Nancy Hale, *Mary Cassatt* (Reading, Mass., 1987), 72–73, 104. See also Lois W. Banner, *American Beauty* (New York, 1983); Frances B. Cogan, *All-American Girl: The Ideal of Real Womanhood in Mid-Nineteenth-Century America* (Athens, Ga., 1989); Catherine Clinton, *The Other Civil War: American Women in the Nineteenth Century* (New York, 1984); Susan Phinney Conrad, *Perish the Thought: Intellectual Women in Romantic America, 1830–1860* (New York, 1976); Mary Ryan, *Womanhood in America: From Colonial Times to the Present* (New York, 1975); Carroll Smith-Rosenberg, *Disorderly Conduct: Visions of Gender in Victorian America* (New York, 1985); Barbara Welter, *Dimity Convictions: The American Woman in the Nineteenth Century* (Athens, Ohio, 1976).

16. HJ, *The Bostonians* (1886; New York: Penguin Edition, 1984), 327. For a recent analysis of this phenomenon, see Elise Miller, "The Feminization of American Realist Theory," *American Literary Realism* 23 (1990):20–39.

17. TD, "The New Humanism," *Thinker* 2 (July 1930):9; TD, "The Literary Shower," *Ev'ry Month* 2 (July 1896):24; Whitman quoted in Lawrence W. Levine, *Highbrow/Lowbrow: The Emergence of Cultural Hierarchy in America* (Cambridge, Mass., 1988), 224.

18. William Roscoe Thayer, "Realism," unpublished manuscript, n.d., pp. 1, 20, in William Roscoe Thayer Papers, Houghton Library, Harvard University.

19. Morton White's *Social Thought in America: The Revolt Against Formalism* (New York, 1949) retains its stature as the best overview of developments in philosophy, law, economics, and history. Trends in the theater are surveyed in Brenda Murphy's *American Realism and American Drama, 1880–1940* (Cambridge, 1987).

20. "Common Sense as the Obstacle to the Progress of Modern Physics," *Current Opinion* 68 (April 1920):501.

21. HJ, "Two Minor French Novelists," *Galaxy* 21 (February 1876):219–33. The text-context debate within the discipline of intellectual history is masterfully addressed in William J. Bouwsma, "Intellectual History in the 1980s: From History of Ideas to History of Meaning," *Journal of Interdisciplinary History* 12 (Autumn 1981):279–90, and John E. Toews, "Intellectual History after the Linguistic Turn: The Autonomy of Meaning and the Irreducibility of Experience," *American Historical Review* 92 (October 1987):879–907.

Part I. Setting the Stage

1. "The Abuse of Realism," *New York Times Saturday Review of Books and Art*, 26 November 1898, p. 785.

2. "Contemporary French Literature," NAR 86 (January 1858):232; Hjalmar H. Boyeson, "The Realism of American Fiction," *Independent* 44 (3 November 1892):3.

Chapter 1. Antebellum Idealism

1. George Santayana, "The Genteel Tradition in American Philosophy," in *Winds of Doctrine: Studies in Contemporary Opinion* (New York, 1913), 186–215; Henry Adams, *History of the United States* (New York, 1889), 1:76–110.

Useful studies of the genteel elite include Lawrence Buell, *New England Literary Culture: From Revolution Through Renaissance* (Cambridge, 1986); Richard Bushman, *The Refinement of America: Persons, Houses, Cities* (New York, 1992); Gordon Milne, *George W. Curtis and the Genteel Tradition* (Bloomington, 1956); Vernon L. Parrington, *Main Currents in American Thought: The Romantic Revolution in America, 1800–1860* (1927; New York, 1954); Stow Persons, *The Decline of American Gentility* (New York, 1973); Bryan Jay Wolf, *Romantic Re-Vision: Culture and Consciousness in Nineteenth-Century Painting and Literature* (Chicago, 1982).

2. *A Philadelphia Perspective: The Diary of Sidney George Fisher*, ed. Nicholas B. Wainwright (Philadelphia, 1967), 104, 150.

3. JRL, [review of Nathaniel Hawthorne, *Notebooks*], NAR 108 (January 1869):323. For a revealing contemporary assessment of the genteel elite, see "The Two Everetts," *Literary World* 1 (24 April 1847):270–71.

4. OWH, Sr., *The Poet at the Breakfast Table* (New York, 1872), 2; JRL, *The Writings of James Russell Lowell* (Boston, 1891), 7:320; JRL, "The Imagination," in *The Function of the Poet and Other Essays* (Boston, 1920), 75, 77, 78; *The Letters of James Russell Lowell*, ed. Charles E. Norton (New York, 1894), 1:377.

5. Larzer Ziff, *Writing in the New Nation: Prose, Print, and Politics in the Early United States* (New Haven, 1991), xi. See also Shirley Samuels, ed., *The Culture of Sentiment: Race, Gender, and Sentimentality in Nineteenth-Century America* (New York, 1992).

6. Martin Duberman, *James Russell Lowell* (Boston, 1966), 252; JRL, *Letters*, 1:377; JRL, "The Function of the Poet," (1855), reprinted in *Century* 47 (January 1894):418, 433–34; Wagenknecht, *James Russell Lowell*, 117; Jared Flagg, ed., *The Life and Letters of Washington Allston* (New York, 1892), 36, 299.

7. John S. C. Abbott, "The Father," *Parent's Magazine* 2 (1842):174; WDH, *A Boy's Town* (New York, 1890), 76. There is a rich body of scholarship illuminating the "cult of domesticity" as a social phenomenon. See especially Ruth Bloch, "American Feminine Ideas in Transition," *Feminist Studies* 4 (1978):101–26; William Bridges, "Family Patterns and Social Values in America," *American Quarterly* 17 (1975):3–11; Nancy F. Cott, *The Bonds of Womanhood: "Women's Sphere" in New England* (New Haven, 1977); Anne L. Kuhn, *The Mother's Role in Childhood Education: New England Concepts, 1830–1860* (New Haven, 1947); Glenna Matthews, *"Just a Housewife": The Rise and Fall of Domesticity in America* (New York, 1987); Mary Ryan, *Cradle of the Middle Class: The Family in Oneida County, New York, 1790–1865* (Cambridge, 1981); Kathryn Kish Sklar, *Catharine Beecher* (New Haven, 1973); Barbara Welter, "The Cult of True Womanhood: 1820–1860," *American Quarterly* 18 (1966):157–74. Joy Kasson provides a captivating study of the sculptural representation of women in *Marble Queens and Captives: Women in Nineteenth-Century American Sculpture* (New Haven, 1991).

8. J. Weiss, "War and Literature," *AM* 9 (June 1862):684.

For thorough treatments of "female fiction" during the antebellum era, see also Barbara Bardes and Suzanne Gossett, *Declarations of Independence: Women and Political Power in Nineteenth Century American Fiction* (New Brunswick, N.J., 1990); Nina Baym, *Women's Fiction: A Guide to Novels by and about Women in America, 1820–1870* (Ithaca, 1978); Ann Douglas, *The Feminization of American Culture* (New York 1977); Fred Pattee, *The Feminine Fifties* (New York, 1940).

9. Louisa May Alcott to Maria S. Porter, 1874, in *The Selected Letters of Louisa May Alcott*, eds. Joel Myerson and Daniel Shealy (Boston, 1988), 190; Joyce W. Warren, *Fanny Fern: An Independent Woman* (New Brunswick, N.J., 1992); Harriet Beecher Stowe to Sara Parton, July 25, 1869, Sara Payson Parton Papers, Sophia Smith Collection, Smith College, Northampton, Massachusetts.

10. Greenwood quoted in Welter, "The Cult of True Womanhood," 160; William G. Simms, *Views and Reviews in American Literature, History and Fiction: First Series*, ed. C. Hugh Holman (Cambridge, Mass., 1962), 259. See also Myra Jehlen, "Archimedes and the Paradox of Feminist Criticism," *Signs* 6 (1981):587, and Nancy Walker, "Wit, Sentimentality and the Image of Women in the Nineteenth Century," *American Studies* 22 (1981):5–21.

11. See Susan K. Harris, "'But is it any good?': Evaluating Nineteenth-Century Women's Fiction," *American Literature* 63 (March 1991):43–61; Sandra M. Gilbert and Susan Gubar, *The Madwoman in the Attic: The Woman Writer and the Nineteenth-Century Literary Imagination* (New Haven, 1979); Fred Kaplan, *Sacred Tears: Sentimentality in Victorian Literature* (1987); Jane Tompkins, *Sensational Designs: The Cultural Work of American Fiction, 1790–1860* (New York, 1985); Laura Wexler, "Tender Violence: Literary Eavesdropping, Domestic Fiction, and Educational Reform," *Yale Journal of Criticism* 5 (Fall 1991):151–87. For a succinct analysis of the interpretive

debate concerning women's fiction, see Mary Kelley, "The Sentimentalists: Promise and Betrayal in the Home," *Signs* 4 (Spring 1979):434–46.

12. The literature on Transcendentalism is extensive and uneven. The best recent treatments are Paul F. Boller, Jr., *American Transcendentalism, 1830–1860: An Intellectual Inquiry* (New York, 1974); Lawrence Buell, *Literary Transcendentalism* (Ithaca, 1973); Anne Rose, *Transcendentalism as a Social Movement* (New Haven, 1981). For a comprehensive bibliography see Joel Myerson, ed., *The Transcendentalist Movement: A Review of Research and Criticism* (New York, 1984).

13. RWE, "The Transcendentalist," (1842), in *Emerson: Essays and Lectures*, ed. Joel Porte (New York: Library of America, 1983), 195.

14. Ibid.; *The Journals and Miscellaneous Notebooks of Ralph Waldo Emerson*, eds. William H. Gilman et al., 16 vols. to date (Cambridge, Mass., 1960–), 5:125 (hereafter cited as RWE, *Journals*); Orestes Brownson, "Synthetic Philosophy," *Works* 1:102.

15. *The Early Lectures of Ralph Waldo Emerson*, eds. Stephen E. Whicher, Robert Spiller, and Wallace E. Williams, 3 vols. (Cambridge, Mass., 1959–72), 3:217; RWE, "American Scholar," *The Complete Works of Ralph Waldo Emerson*, ed. Edward W. Emerson, 12 vols. (Boston, 1903–4), 1:110–11.

16. RWE, "American Scholar," *Works*, 1:110–11. RWE, *Works*, 12:43; RWE, "Plato," *Works*, 4:55. The ideal and the real, Emerson noted, fought for his attention like "two boys pushing each other on the curbstone of the pavement." On Emerson's desire to synthesize the real and ideal, see Sherman Paul, *Emerson's Angle of Vision* (Cambridge, Mass., 1952); Stephen Whicher, *Freedom and Fate: An Inner Life of Ralph Waldo Emerson* (Philadelphia, 1953).

17. *The Heart of Emerson's Journals*, ed. Bliss Perry (Boston, 1909), 21–23, 109, 204; RWE, "The Young American," *Works*, 1:369; RWE, *Journals*, 4:132, 74; RWE, *Works*, 1:39; 4:4. Emerson's attitude toward urban America is ably assessed in Michael H. Cowan, *City of the West: Emerson, America and Urban Metaphor* (New Haven, 1967). The complex theme of masculinity and cultural life during the antebellum period receives extended treatment in David Leverenz, *Manhood and the American Renaissance* (Ithaca, 1989).

18. RWE, "Thoughts on Modern Literature," *Dial* 1 (July 1840):155–56; RWE, *Works*, 2:221.

19. RWE, *Journals*, 13:66.

20. HDT, *Walden*, in *The Writings of Henry David Thoreau*, eds. Bradford Torrey and Francis H. Allen, 20 vols. (Boston, 1906), 2:107–8; HDT, *Journals of Henry David Thoreau*, eds. John C. Broderick et al. (Princeton, 1981–), 1:50–51; HDT, *Writings*, 8:43.

21. On the attitude of the "brooding idealists" toward the Transcendentalists, see Harry Levin, *The Power of Blackness: Hawthorne, Poe, Melville* (New York, 1964); William Braswell, "Melville as a Critic of Emerson," *American Literature* 9 (1937):317–34; Perry Miller, "Melville and Transcendentalism," *Virginia Quarterly Review* 29 (Autumn, 1953):556–75; Henry Bamford Parkes, "Poe, Hawthorne, and Melville: An Essay in Sociological Criticism," *Partisan Review* 16 (February 1949):157–65. More general studies of Poe, Melville, and Hawthorne include: Michael Davitt Bell, *The Development of American Romance: The Sacrifice of Relation* (Chicago, 1980); Richard Brodhead, *Hawthorne, Melville, and the Novel* (Chicago, 1976); Joel Porte, *The Romance in America: Studies in Cooper, Poe, Hawthorne, Melville, and James* (Middletown, Ct., 1969).

22. Newton Arvin, *Hawthorne* (Boston, 1929), 92.

23. John Ostrom, ed., *Letters of Edgar Allan Poe*, 2 vols. (Cambridge, Mass., 1948), 1:256–57; James A. Harrison, ed., *The Complete Works of Edgar Allan Poe*, 17 vols. (New York, 1902), 8:161, 10:37–38; Poe, *Broadway Journal* 1 (1845):281–82; Levin, *The Power of Blackness*, 104; Taylor Stoehr, *Words and Deeds: Essays on the Realistic Imagination* (New York, 1986), 24; Daniel Hoffman, *Poe* (New York, 1985), 202. See also Ziff, *Literary Democracy*, 67–86.

24. Herman Melville to John Murray, 25 March 1848, in *The Letters of Herman Melville*, eds. Merrill R. Davis and William H. Gilman (New Haven, 1960), 70.
A character at the end of *Pierre* (1852) expresses Melville's own impatience with mundane reality: "I must get on some other element than earth. I have sat on earth's saddle till I am weary." The "physical world of solid objects now slidingly displaced itself from around him, and he floated into an ether of visions." Herman Melville, *Pierre, or The Ambiguities* (1852; Evanston and Chicago, 1971), 85.

25. Herman Melville, *Moby-Dick, Or, the Whale* (1851; New York: W. W. Norton, 1976), 164.

26. Nathaniel Hawthorne, *The Scarlet Letter* (Columbus, Ohio: Centenary Edition, 1962), 1:36.

27. WDH, *My Literary Passions* (New York, 1895), 139; *The Memoirs of Julian Hawthorne*, ed. Edith G. Hawthorne (New York, 1938), 127.

28. Anthony Trollope, "The Genius of Nathaniel Hawthorne," NAR 129 (September 1879):204–5, 207.

29. On the prevailing popular sentiment see Lewis O. Saum, *The Popular Mood of Pre-Civil War America* (Westport, Conn., 1980).

Chapter 2. New Paths

1. "Novels: Their Meaning and Mission," *Putnam's* 4 (October 1854):392, 395.

2. *Narrative of the Life of Frederick Douglass*, ed. Benjamin Quarles (Cambridge, Mass., 1960), 9–10. On the slave narratives, see Charles T. Davis and Henry Louis Gates, Jr., *The Slave's Narrative* (New York, 1985); Frances Smith Foster, *Witnessing Slavery: The Development of the Slave Narratives* (1979); Edward Margolies, "Antebellum Slave Narratives: Their Place in American Literary History," *Studies in Black Literature* 4 (Autumn, 1973): 1–8.

3. "Introductory," *Putnam's* 1 (January 1853):1; "Editor's Notes," ibid. 7 (May 1856):546; "The World of New York," ibid. 7 (May 1856):560; "Our Young Authors," ibid. 1 (January 1853):7.

4. "Literary Notices," *Knickerbocker* 57 (February 1861):218.

5. George Foster, *New York Naked* (New York, 1854), 12; Foster, *Fifteen Minutes Around New York* (New York, 1854), 53; Foster, *New York by Gas-light* (1850; Berkeley, 1990), 92.

6. T. C. Clarke, "Architects and Architecture," *Christian Examiner* 49 (September 1850):286.

7. Horatio Greenough, *The Travels, Observations, and Experiences of a Yankee Stonecutter* (New York, 1852), 70, 135, 133; Greenough, *Yankee Stonecutter*, 200, 33. Greenough's vernacular aesthetic is discussed in John A. Kouwenhoven, *The Arts in Modern American Civilization* (New York, 1967), 78–86, 91, 102; Sylvia Crane, "The Aesthetics of Horatio Greenough in Perspective," *Journal of Aesthetics and Art Criticism* 24 (Spring 1966):415–27; Harold A. Small, ed., *Form and Function* (Berkeley, 1957).

8. Horatio Greenough to RWE, 28 December 1851, RWE Papers, Houghton Library, Harvard University; RWE, *The Journals and Miscellaneous Notebooks of Ralph Waldo Emerson*, eds. William H. Gilman et al., 16 vols. (Cambridge, Mass., 1970), 13:85. On Greenough's friendship with Emerson, see Theodore Brown, "Greenough, Paine, Emerson and the Organic Aesthetic," *Journal of Aesthetics and Art Criticism* 14 (March 1956):304–17; Charles Metzger, *Emerson and Greenough* (Berkeley, 1954). Greenough died in 1852, before his writings and theories could make much of an imprint on American cultural expression.

9. RWE, "Experience," in *Emerson: Essays and Lectures* (New York: Library of America Edition, 1983), 491, 478; RWE, "Conduct of Life," ibid., 952, 1025, 1062. See also David Porter, *Emerson and Literary Change* (Cambridge, Mass., 1978).

10. "American Painters: Their Errors as Regards Nationality," *Cosmopolitan Art Journal* 1 (June 1857):116–19. See also Evert Duyckinck, "Bad News for the Transcendentalist Poets," *Literary World* 1 (20 February 1847):53.

11. Charles G. Leland, "Editor's Easy Talk," *Graham's* 53 (1858):553; "To Readers and Correspondents," *Graham's* 51 (July 1857):82; 50 (May 1857):468–69.

12. Eliot quoted in F. O. Matthiessen, *American Renaissance: Art and Expression in the Age of Emerson and Whitman* (New York, 1941), 519; John Updike, "Whitman's Egotheism," in *Hugging the Shore* (New York, 1984), 117.

13. *The Early Lectures of Ralph Waldo Emerson*, eds. Stephen E. Whicher, Robert E. Spiller, and Wallace E. Williams, 3 vols. (Cambridge, Mass., 1959–72), 3:362.

14. Joseph J. Rubin and Charles H. Brown, eds., *Walt Whitman of the New York "Aurora"* (State College, Penn., 1972), 105; RWE to WW, 21 July 1855, in Edwin H. Miller, ed., *Walt Whitman: The Correspondence*, 6 vols. (New York, 1961–77), 1:41.

15. *New York Daily Times*, 22 April 1852.

16. *Uncollected Poetry and Prose of Walt Whitman*, ed. Emory Holloway (New York, 1921), 1:122.

17. WW, *Leaves of Grass*, eds. Harold W. Blodgett and Sculley Bradley (New York: Norton Critical Edition, 1965), 160.

18. Clifton Furness, ed., *Walt Whitman's Workshop* (Cambridge, 1929), 129; Horace Traubel, *With Walt Whitman in Camden*, 3 vols. (New York, 1906), 1:267; "Preface to *Leaves of Grass*," in *Walt Whitman: Complete Poetry and Collected Prose*, ed. Justin Kaplan (New York: Library of America Edition, 1982), 23.

19. Justin Kaplan, *Walt Whitman: A Life* (New York, 1980), 157; Traubel, *With Walt Whitman in Camden*, 3:365; Bayard Still, ed., *Urban America: A History with Documents* (Boston:, 1974), 198.

20. *Brooklyn Advertiser*, 21 December 1850; WW, "An English and American Poet," in *Re Walt Whitman*, eds. Horace L. Traubel, Richard M. Bucke, and Thomas B. Harbed (Philadelphia, 1893), 29, 31; WW, *Leaves of Grass*, 202; Holloway, ed., *Uncollected Poetry and Prose*, 2:69–70.

21. Fanny Fern, "Peeps from Under a Parasol," *New York Ledger*, 19 April and 10 May 1856.

22. *FLIN* 3 (20 December 1856):42; "Studies Among the Leaves," *Crayon* 3 (January 1856):30–32. The scholarly literature analyzing Whitman's poetry is incredibly prolific. I have benefited most from F. O. Matthiessen, *American Renaissance*, 517–625; Floyd Stovall, *The Foreground of Leaves of Grass* (Charlottesville, 1974); Paul Zweig, *Walt Whitman: The Making of a Poet* (New York, 1984). See also Gay Wilson Allen, *The New Walt Whitman Handbook* (New York, 1986) for a complete bibliography of critical studies.

23. *Walt Whitman: Complete Poetry and Collected Prose*, 14.

24. WW, *Leaves of Grass*, 53.

25. Ibid., 63; *Complete Poetry and Collected Prose of Walt Whitman*, 206.

26. WJ, "On a Certain Blindness in Human Beings," in *William James: The Essential Writings*, ed. Bruce W. Wilshire (Albany, N.Y., 1984), 333; Sherman Paul, *Louis Sullivan: An Architect in American Thought* (Englewood Cliffs, N.J., 1962), 1; Howells quoted in "Walt Whitman's Seventieth Birthday," *The Critic* (June 8, 1889):287; HG, "Walt Whitman," p. 4, unpublished manuscript, Hamlin Garland Papers, Doheny Library, University of Southern California.

27. John G. Whittier, "The Nervous Man," in *Whittier on Writers and Writing*, eds. Edwin H. Cady and Harry Hayden Clark (Syracuse, 1950), 99; Whittier, "American Literature," *Whittier on Writers*, 24. For an informative discussion of the changes in short fiction during the 1850s, see Robert F. Marler, "From Tale to Short Story: The Emergence of a New Genre in the 1850s," *American Literature* 46 (May 1974):153–69.

28. Harry Levin, *The Gates of Horn: A Study of Five French Realists* (New York, 1966), 166, 199; HJ, *Literary Criticism: French Writers, Other European Writers* (New York: Library of America Edition, 1984), 50; Levin, *Gates of Horn*, 168. See also Samuel Rogers, *Balzac and the Novel* (Madison, 1953).

29. HJ, "The Present Literary Situation in France," (1899), *Essays*, 111; TD, *A Book About Myself* (New York, 1922), 410–13. See also "Henri [sic] de Balzac," *United States Magazine and Democratic Review* 32 (April 1853):325–29; John L. Motley, "The Novels of Balzac," NAR 65 (July 1847):85–108.

30. Countess Du Bury, "Contemporary French Literature," NAR 86 (January 1858):230. See also "Recent French Literature," NAR 85 (October 1857):532–33; "French Literature," *United States Magazine and Democratic Review* 42 (November 1858):393.

31. See also Ioan Williams, *The Realist Novel in England: A Study in Development* (Pittsburgh, 1974).

32. Nathaniel Hawthorne, *The House of the Seven Gables* (New York: Penguin Edition, 1982), 1–3; Nathaniel Hawthorne to James J. Fields, 11 February 1860, James J. Fields Collection, Huntington Library, San Marino, California.

33. Caroline Kirkland, "Novels and Novelists," NAR 76 (January 1853):105, 110. On Kirkland, see Langley C. Keyes, "Caroline M. Kirkland: A Pioneer in Realism," (Ph.D. diss., Harvard University, 1936); Annette Kolodny, *The Land Before Her* (Chapel Hill, 1984); William S. Osborne, *Caroline Kirkland* (New York, 1972).

34. Alice Cary, "Uncle William's," in *Clovernook; or, Recollections of Our Neighborhood in the West* (New York, 1853), 146, 364.

35. The best source of biographical information about Terry is Jean Downey, "A Biographical and Critical Study of Rose Terry Cooke" (Ph.D. diss., University of Ottawa, 1956). More sources and an introduction are found in Elizabeth Ammons, ed., *How Celia Changed Her Mind and Selected Stories* (New Brunswick, N.J., 1986). See also Katherine Kleitz, "Essence of New England: The Portraits of Rose Terry Cooke," *American Transcendental Quarterly* 47 (1980):127–39; Susan Toth, "Rose Terry Cooke," *American Literary Realism* 42 (Spring 1971):170–76; Perry D. Westbrook, *Acres of Flint: Sarah Orne Jewett and Her Contemporaries* (Metuchen, N.J., 1981), 78–85.

36. RTC, "Uncle Josh," *Putnam's Monthly* 10 (September 1857):347–55. For a historian's perspective on this kind of relationship between women, see Caroll Smith-Rosenberg, "The Female World of Love and Ritual," *Signs* 1 (Autumn 1975):1–29. A

complete listing of Cooke's works can be found in Jean Downey, "Rose Terry Cooke: A Bibliography," *Bulletin of Bibliography* 21 (May–August 1955):159–63.

37. Harriet Prescott Spofford, *A Little Book of Friends* (Boston, 1916), 144; JRL, "Reviews and Literary Notices," *AM* 7 (March 1861):382.

38. See RHD, *Bits of Gossip* (Boston, 1904), and Tillie Olsen, *Silences* (New York, 1978), 47–120.

39. Ibid., 32–33.

40. RHD to James Fields, 26 January 1861, Richard Harding Davis Papers, Barrett Collection, Alderman Library, University of Virginia.

41. RHD, "Life in the Iron-Mills," *AM* 7 (April 1861):430. Emily Dickinson wrote a note to her sister-in-law in 1861: "Will Susan please lend Emily 'Life in the Iron-Mills'?" *Letters of Emily Dickinson*, ed. Thomas H. Johnson, 3 vols. (Cambridge, Mass., 1958), 2:372.

42. RHD, "Life in the Iron-Mills," 430–34.

43. Ibid., 439.

44. Ibid., 431. On Davis's fatalism, see Sharon M. Harris, "Rebecca Harding Davis: From Romanticism to Realism," *American Literary Realism* 21 (1989):4–20. Also useful are: John Conron, "Assailant Landscapes and the Man of Feeling: Rebecca Harding Davis's 'Life in the Iron-Mills,'" *Journal of American Culture* 3 (1980):487–500; Jean Pfaelzer, "Rebecca Harding Davis: Domesticity, Social Order, and the Industrial Novel," *International Journal of Women's Studies* 4 (1981):234–44; Jane Atteridge Rose, "Reading 'Life in the Iron Mills' Contextually: A Key to Rebecca Harding Davis' Fiction," in *Conversations: Contemporary Critical Theory and the Teaching of Literature*, eds. Charles Moran and Elizabeth Penfield (Urbana, 1990), 187–99.

45. On Courbet and the emergence of realism in French painting, see George Boas, *Courbet and the Naturalist Movement* (Baltimore, 1938); Michael Fried, *Courbet's Realism* (Chicago, 1990); Bernard Goldman, "'That Brute' Courbet and Realism," *Criticism* 9 (1967):22–41; Linda Nochlin, *Gustave Courbet: A Study of Style and Society* (Garland, N.Y., 1976); Sarah Faunce and Linda Nochlin, eds., *Courbet Reconsidered* (Brooklyn, 1988).

46. Linda Nochlin, ed., *Realism and Tradition in Art*, 33; Jerrold Siegel, *Bohemian Paris: Culture, Politics and the Boundaries of Bourgeois Life, 1830–1930* (New York, 1986), 84; Traubel, *With Walt Whitman in Camden*, 1:63, 3:89. For an extended discussion of Whitman's enthusiasm for Millet, see Laura Meixner, "'The Best of Democracy': Walt Whitman, Jean-Francois Millet, and Popular Culture in Post-Civil War America," in *Walt Whitman and the Visual Arts*, eds. Geoffrey M. Sill and Roberta K. Tarbell (New Brunswick, N.J., 1992), 28–52. On trends in French art, see Gerald Needham, *19th-Century Realist Art* (New York, 1988); Linda Nochlin, *Realism* (New York, 1971); Gabriel Weisberg, *The Realist Tradition: French Painting and Drawing, 1830–1900* (Bloomington, 1980).

47. Martha A. S. Shannon, *Boston Days of William Morris Hunt* (Boston, 1923), 40; Greta, "Our Boston Letter," *The Art Amateur* 1 (October 1879):96. On Hunt see Helen Knowlton, *Art-Life of William Morris Hunt* (Boston, 1899); Martha Hoppin, "William Morris Hunt and His Critics," *American Art Review* 2 (September 1975):79–91.

48. Ibid.; William M. Hunt, *Talks on Art*, ed. Helen M. Knowlton (London, 1878), 50, 24. Hunt applied such principles in his own portraits and landscapes. In 1856 a critic assessed Hunt's distinctive new style in *Putnam's Monthly*. "In works like these,"

he wrote, "we recognize a future for painting. Realism in art has been pushed to its last term in our day, and we have been summoned to surrender our old-fashioned faith. . . ." The realist, he added, "reproduces with fidelity and feeling what he sees," and, according to such a definition, Hunt deserved the label. See "The World of New York," *Putnam's* 7 (May 1856):560.

49. See Allen Staley, *The Pre-Raphaelite Landscape* (Oxford, 1973); Christopher Wood, *The Pre-Raphaelites* (New York, 1981).

50. Charles Dickens, "Old Lamps for New Ones," *Household Words* (15 June 1850):265–66; *Westminster Review* 63 (January 1855):294. A French critic compared the Pre-Raphaelites to the French realists. See Paul Mantz, "Salon de 1855," *Revue Française* (1 July 1855):125.

51. John Ruskin, *Modern Painters* (New York, 1882), 3:45, 264, 67; 1:423.

52. Ibid., 3:22, 69. For an American assessment of Ruskin, see "Ruskin's Architectural Works," *Southern Quarterly Review* 11 (April 1855):372–423; "The Place of Modern Painters," *Christian Examiner* 70 (January 1861):29–48. On the American admirers of Ruskin and the Pre-Raphaelites, see, in addition to Stein, David H. Dickason, *The Daring Young Men: The Story of the American Pre-Raphaelites* (Bloomington, 1963).

53. "The National Academy of Design," *Putnam's* 5 (May 1855):506; Peter Wight, *The Development of New Phases of the Fine Arts in America* (Chicago, 1884), 22. A richly detailed account of the group associated with *The New Path* can be found in Linda S. Ferber and William H. Gerdts, *The New Path: Ruskin and the American Pre-Raphaelites* (New York, 1985).

54. "Introductory," *New Path* 1 (May 1863):1–2; "Association for the Advancement of Truth in Art," 11; Thomas Farrer, "A Few Questions Answered," *New Path* 1 (June 1863):13–16; Charles H. Moore, "Fallacies of the Present School," *New Path* 1 (October 1863):61; "The Artists Fund Society," *New Path* 1 (December 1863):94.

55. Inness quoted in "Pictures on Exhibition," *New Path* 1 (December 1863):99; "The British Gallery in New York," 502. On Inness, see Nicolai Cikovsky, Jr., and Michael Quick, *George Inness* (New York, 1985). A contemporary assessment of the *New Path* can be found in "Art Criticism," *HW* 9 (15 April 1865):226.

Part II. The Generative Forces

1. WDH to TSP, 28 January 1886, in Mildred Howells, ed., *The Life and Letters of William Dean Howells*, 2 vols. (New York, 1928), 1:378.

2. Sumner quoted in Donald Bellomy, "The Molding of an Iconoclast: William Graham Sumner, 1840–1885" (Ph.D. diss., Harvard University, 1980), 251. On this theme of realism as a reaction to disorienting changes, see Roger B. Salomon, "Realism as Disinheritance: Twain, Howells, and James," *American Quarterly* 16 (Winter 1964):531–44.

Chapter 3. Touched with Fire

1. James Herbert Morse, "The Native Element in American Fiction," *Century* 21 (July 1883):362. The best general accounts of the cultural and intellectual impact of the Civil War are Daniel Aaron, *The Unwritten War: American Writers and the Civil War* (New York, 1971); George Frederickson, *The Inner Civil War: Northern Intellectuals and the Crisis of the Union* (New York, 1965); Anne C. Rose, *Victorian America and*

the Civil War (New York, 1992); Edmund Wilson, *Patriotic Gore* (New York, 1962). A comprehensive collection of primary sources can be found in Richard Rust, ed., *Glory and Pathos: Responses of Nineteenth-Century American Authors to the Civil War* (Boston, 1970).

2. Lars Ahnebrink, *The Beginnings of Naturalism in American Fiction* (New York, 1961), 11.

3. Mark DeWolfe Howe, ed., *Touched with Fire: Civil War Letters and Diary of Oliver Wendell Holmes, Jr.* (Cambridge, Mass., 1946), vii.

4. Brian McGinty, "A Shining Presence: Rebel Poet Sidney Lanier Goes to War," *Civil War Times Illustrated* 19 (May 1980):24–31. See also Wilson, *Patriotic Gore*, 450–66. The mythic qualities of the Civil War can be seen in *The Confederate Image: Prints of the Lost Cause*, eds. Mark E. Neely, Jr., Harold Holzer, and Gabor S. Boritt (Chapel Hill, 1987).

5. As a Wisconsin veteran reflected, the popular image of the Civil War preferred "the fever of romance to the good health of solid facts." Michael H. Fitch, *Echoes of the Civil War as I Hear Them* (New York, 1905), 239.

6. "Editor's Easy Chair," *HM* 23 (November 1861):844.

7. James quoted in *Milwaukee Sentinel*, 2 December 1883; C. Vann Woodward, ed., *Mary Chesnut's Civil War* (New Haven, 1981), 69.

8. OWH, Sr., "The Inevitable Trial," in *The Writings of Oliver Wendell Holmes* (Boston, 1891), 8:86, 117. See also Aaron, *Unwritten War*, 25–34; Fredrickson, *Inner Civil War*, 23–35.

9. *Walt Whitman and the Civil War*, ed. Charles Glicksberg (New York, 1963), 17; Ednah Cheney, *Louisa May Alcott: Her Life, Letters, and Journals* (Boston, 1890), 132, 127.

10. *The Journals and Miscellaneous Notebooks of Ralph Waldo Emerson*, eds. Linda Allardt and David W. Hill (Cambridge, Mass., 1970), 15:299–300, 379–80; *The Complete Works of Ralph Waldo Emerson*, ed. Edward W. Emerson, 12 vols. (Boston, 1903–4), 7:261–63; Nathaniel Hawthorne to Henry Bright, 14 November 1861, in Edward Mather, *Nathaniel Hawthorne: A Modest Man* (New York, 1940), 317. See also Fredrickson, *Inner Civil War*, 176–80.

11. RHD, *Bits of Gossip* (Boston, 1940), 34.

12. Ibid., 35, 36.

13. Nathaniel Hawthorne, "Chiefly About War Matters," *AM* (July 1862):43. See Randall Stewart, "Hawthorne and the Civil War," *Studies in Philology* 34 (January 1937):91–106.

14. The impact of the Civil War on women is treated in Mary E. Massey, *Bonnet Brigades* (1966); George C. Rable, *Civil Wars: Women and the Crisis of Southern Nationalism* (Urbana, Ill., 1989); Ann Douglas Wood, "The War Within a War: Women Nurses in the Union Army," *Civil War History* 18 (September 1972):197–212; Agatha Young, *The Women and the Crisis: Women of the North in the Civil War* (New York, 1959).

15. Rable, *Civil Wars*, 222; Brown quoted in Samuel F. Pickering, Jr., *The Right Distance* (Athens, Ga., 1987), 70. On Emily Dickinson's reaction to the conflict, see Shira Wolosky, *Emily Dickinson: A Voice of War* (New Haven, 1984).

16. R.W.B. Lewis, *The Jameses: A Family Narrative* (New York, 1991), 116; HJ, Sr., to Samuel G. Ward, 1 August 1863, HJ Papers, Houghton Library, Harvard University; HJ to Robert James, 31 August 1864, HJ Papers.

17. WDH, A *Fearful Responsibility and Other Stories* (Boston, 1881), 3; HJ, *Notes of a Son and Brother* (New York, 1914), 379–80; "Athens, or Aesthetic Culture," *Bibliotheca Sacra* 20 (January 1863):171.

18. Howe, ed., *Touched with Fire*, 11–12, 18; Eleanor Tilton, *Amiable Autocrat: A Biography of Dr. Oliver Wendell Holmes* (New York, 1947), 267. On Holmes and the war, see also Aaron, *Unwritten War*, 159–63; Frederickson, *Inner Civil War*, 167–72, 218–22; Wilson, *Patriotic Gore*, 743–96. The best biography is Liva Baker, *The Justice from Beacon Hill: The Life and Times of Oliver Wendell Holmes* (New York, 1991).

19. Ibid., 50–51; Alexander Woollcott, "The Second Hunt After the Captain," *AM* 170 (1942):50. The elder Holmes published a slightly different version of their conversation. He claimed that his son replied "How are you, Dad?" See OWH, Sr., "My Hunt After the Captain," *AM* 10 (1862):754.

20. Mark DeWolfe Howe, *Justice Oliver Wendell Holmes: The Shaping Years, 1841–1870* (Cambridge, Mass., 1957), 94; OWH, Jr., "The Soldier's Faith," in *The Mind and Faith of Justice Holmes: His Speeches, Essays, Letters and Judicial Opinions*, ed. Max Lerner (Boston, 1943), 21; Howe, ed., *Touched with Fire*, 142–43.

21. OWH, Jr., "A Man and the Universe," *The Mind and Faith of Justice Holmes*, 36; Gerald Linderman, *Embattled Courage: The Experience of Combat in the American Civil War* (New York, 1987), 289.

22. Harry C. Shriver, ed., *What Gusto: Stories and Anecdotes About Justice Oliver Wendell Homes* (Potomac, Md., 1970), 5; "Memorial Day" (1884), in Lerner, ed., *Mind and Faith*, 16; "The Soldier's Faith," 23–25.

23. OWH, Jr., "Your Business as Thinkers," in Lerner, ed., *Mind and Faith*, 31; "The Soldier's Faith," ibid., 19.

24. Mark DeWolfe Howe, ed., *Holmes-Laski Letters: The Correspondence of Mr. Justice Holmes and Harold J. Laski, 1916–1935*, 2 vols. (Cambridge, Mass., 1953), 2:36; OWH, Jr., "Parts of the Unimaginable Whole," in Lerner, ed., *Mind and Faith*, 27; Henry S. Commager, *The American Mind: An Interpretation of American Thought and Character Since the 1880s* (New Haven, 1950), 384; "The Soldier's Faith," in Lerner, ed., *Mind and Faith*, 19; Holmes to Frankfurter quoted in Gary J. Aichele, *Oliver Wendell Holmes, Jr.: Soldier, Scholar, Judge* (Boston, 1989), 144; Mark DeWolfe Howe, ed., *Holmes-Pollock Letters*, 2 vols. (Cambridge, Mass., 1941), 1:42.

25. Edwin H. Miller, ed., *Walt Whitman: The Correspondence* (New York, 1961), 1:213; Horace Traubel, *With Walt Whitman in Camden*, 3 vols. (New York, 1906), 2:137. See also Samuel Coale, "Whitman's War: The March of a Poet," *Walt Whitman Review* 21 (September 1975):85–101; Charles Glicksberg, ed., *Walt Whitman and the Civil War: A Collection of Original Articles and Manuscripts* (Philadelphia, 1933); Walter Lowenfels, ed., *Whitman's Civil War* (New York, 1960).

26. *Walt Whitman: Complete Poetry and Collected Prose*, ed. Justin Kaplan (New York: Library of America, 1982), 779, 442; Miller, ed., *Walt Whitman: The Correspondence*, 1:114.

27. Miller, ed., *Correspondence of Walt Whitman*, 1:246–47; *Whitman: Complete Poetry and Collected Prose*, 444, 440, 457.

28. "As I Walk These Broad Majestic Days," *Whitman: Complete Poetry and Collected Prose*, 595–96.

29. Claiborne J. Walton, "One Continued Scene of Carnage," *Civil War Times Illustrated* 15 (August 1976):34–35.

30. RWE, *The Journals and Miscellaneous Notebooks*, 8:126. The history and nature of photography are addressed in Beaumont Newhall, *The History of Photography* (New

York, 1982) and *The Daguerreotype in America* (New York, 1976); Miles Orvell, *The Real Thing: Imitation and Authenticity in American Culture, 1880–1940* (Chapel Hill, 1989), 73–102, 198–239; Floyd and Marion Rinhart, *The American Daguerreotype* (Athens, Ga., 1981); Susan Sontag, *On Photography* (New York, 1977); Robert Taft, *Photography and the American Scene: A Social History, 1838–1889* (New York, 1964); Alan Thomas, *Time in a Frame: Photography and the Nineteenth-Century Mind* (New York, 1977); William Welling, *Photography in America: The Formative Years, 1839–1900* (New York, 1978).

31. "Photographs of War Scenes," *Humphrey's Journal* 13 (1 September 1861):133; "Brady's Photographs," *New York Times*, 20 October 1862, 5; "General Grant's Campaign," *HW* 8 (9 July 1864):442. See also Josephine Cobb, "Matthew B. Brady's Photographic Gallery in Washington," *Columbia Historical Society Records* (1953–54), 28–69; William C. Davis, ed., *The Image of War: 1861–1865*, 6 vols. (Garden City, N.Y., 1981); James D. Horan, *Matthew Brady: Historian with a Camera* (New York, 1955); "Photographers of the Civil War," *Military Affairs* 26 (Fall 1962):127–35; Timothy Sweet, *Traces of War: Poetry, Photography, and the Crisis of the Union* (Baltimore, 1990); Alan Trachtenberg, "Brady's Portraits," *Yale Review* 73 (Winter 1984):230–53, and "Albums of War: On Reading Civil War Photographs," *Representations* 9 (Winter 1985):1–32.

32. *Gardner's Photographic Sketch Book of the Civil War* (1866; New York, 1959), unnumbered. See also Josephine Cobb, "Alexander Gardner," *Image* 7 (1958):124–36; James D. Horan, *Timothy O'Sullivan: America's Forgotten Photographer* (Garden City, N.Y., 1966). On Brady's role see Cobb, "Matthew B. Brady's Photographic Gallery," 28–69; "Photographers of the Civil War," 127–35.

33. *Gardner's Sketch Book*, unnumbered.

34. OWH, Sr., "The Stereoscope and the Stereograph," *AM* 3 (June 1859):744; OWH, Sr., "Doings of the Sunbeam," *AM* 11 (July 1863):11.

35. OWH, Sr., "My Hunt After the Captain," *AM* 10 (December 1862):743–44.

36. Ibid., 749.

37. OWH, Sr., "Doings of the Sunbeam," 11–12; OWH, Sr., "The Inevitable Trial," *Pages from an Old Volume of Life* (Boston, 1899), 116.

38. Samuel Longfellow, ed., *Life of Henry Wadsworth Longfellow; with Extracts from His Journals and Correspondence*, 2 vols. (Boston, 1886), 2:365; *Washington Evening Star*, 16 October 1896. See also J. Cutler Andrews, *The North Reports the Civil War* (Pittsburgh, 1955); Bernard A. Weisberger, *Reporters for the Union* (Boston, 1953).

39. "Our Heroes," *HW* 6 (19 July 1862):450; "To the Bitter End," *HW* 7 (5 September 1863):562–63; Charles G. Leland, "Servile Insurrection," *Knickerbocker* 58 (November 1861):377; "Letter to Editor," ibid. 58 (September 1861):267.

40. William Howe Downes, *The Life and Works of Winslow Homer* (Boston, 1911), xxvii. The best treatment of Homer's Civil War involvement is Marc Simpson, ed., *Winslow Homer: Paintings of the Civil War* (San Francisco, 1988).

41. The Lounger, "The National Academy of Design," *HW* 7 (2 May 1863):274; "The Artists Fund Society," *New Path* 1 (December 1863):95.

42. "Fine Arts," *Nation* 2 (11 May 1866):603; Henry Tuckerman, *Book of the Artists* (New York, 1867), 491.

43. "Art in America," *Godey's Lady's Book* 62 (March 1861):269–70; Fletcher Pratt, *The Civil War in Pictures* (Garden City, N.Y., 1955), 57. Similarly, Confederate General D. H. Hill insisted that he never saw a "Confederate line advancing that was not crooked as a ram's horn." *Battles and Leaders of the Civil War* (New York, 1956), 2:662.

44. Woodward, ed., *Mary Chesnut's Diary*, 581; Albion Tourgée, "The South as a Field for Fiction," *Forum* 6 (December 1888):407, 412. See also Aaron, *The Unwritten War*; Cecil Eby, "The Real War and the Books," *Southwest Review* 47 (Summer 1962):259–65; Irene M. Patten, "Noble Warriors and Maidens Chaste: The Civil War as Romance," *American Heritage* 22 (April 1971):49–53, 109; Rebecca Washington Smith, *The Civil War and Its Aftermath in American Fiction, 1861–1899* (Chicago, 1937); Lawrence S. Thompson, "The Civil War in Fiction," *Civil War History* 2 (1956):83–95.

45. Cooke quoted in George C. Eggleston, *Recollections of a Varied Life* (New York, 1910), 70–71. On Cooke, see Carl Holliday, "John Esten Cooke as a Novelist," *Sewanee Review* 13 (April 1905):216–20; Richard B. Harwell, *Confederate Belles-Lettres* (Hattiesburg, Miss., 1941); Jay Hubbell, *The South in American Literature* (Durham, 1954).

46. John E. Cooke, *Hilt to Hilt; or, Days and Nights on the Banks of the Shenandoah* (New York, 1869), 9; Cooke, *Surry of Eagle's Nest; or, The Memoirs of a Staff Officer* (New York, 1866), 479.

47. John E. Cooke, *Wearing of the Gray* (New York, 1867), xiv; WW, *Complete Poetry and Collected Prose*, 778.

48. Edwin Oviatt, "J. W. De Forest in New Haven," *New York Times*, 17 December 1898, 856; JWD, *Poems, Medley and Palestrina* (New Haven, 1902), ix; JWD to Andrew De Forest, 17 August, 27 November 1863, JWD Papers, Beinecke Library, Yale University. On De Forest see Aaron, *Unwritten War*, 164–80; James Light, *John William De Forest* (New York, 1965); James W. Gargano, ed., *Critical Essays on John W. De Forest* (Boston, 1981); Wilson, *Patriotic Gore*, 669–742.

49. JWD, "Chivalrous and Semi-Chivalrous Southerners," *HM* 38 (January–February 1869):347; JWD, "Our Military Past and Future," *AM* (November 1879):572.

50. JWD, "The Great American Novel," *Nation* 6 (9 January 1868):27–29; Oviatt, "J. W. De Forest in New Haven," 856.

51. Oviatt, "De Forest in New Haven," 856.

52. JWD, *Miss Ravenel's Conversion from Secession to Loyalty* (New York, 1867), 296, 137, 248.

53. Ibid., 279.

54. JWD to WDH, 24 January 1887, WDH Papers; JWD, "The First Time Under Fire," *HM*, 29 (August 1864):480; JWD, *Miss Ravenel's Conversion*, 267, 422. De Forest's description of a division field hospital is even more brutally precise:

> In the centre of this mass of suffering stood several operating tables, each burdened by a grievously wounded man and surrounded by surgeons and their assistants. Underneath were great pools of clotted blood, amidst which lay amputated fingers, hands, arms, feet and legs, only a little more ghastly in color than the faces of those who waited their turn on the table. (269)

55. "Literary" *HW* 11 (8 June 1867):355; *Nation* 4 (20 June 1867):491–92; "Literary Notices," *HM* 35 (August 1867):401.

56. "Reviews," *AM* 20 (July 1867):120–22.

57. WDH, *Some Heroines of Fiction* (New York, 1901), 2:162.

58. Ellen Glasgow, "The Novel in the South," *HM* 157 (December 1928):93–100; Glasgow, "Heroes and Idealism," *Saturday Review of Literature* 12 (4 May 1935):3–4. On the romanticizing of the war in fiction, see Tourgée, "The South as a Field for Fiction," 404–13.

59. "The Nation's Triumphs and Its Sacrifices," *Christian Examiner* 78 (May 1865):439; "Hosmer's Thinking Bayonet," *Nation* 1 (20 July 1865):90; "Literature," *Nation* 1 (6 July 1865):26, 19; "Editor's Easy Chair," *HM* 31 (July 1865):267. See also Samuel Osgood, "New Aspects of the American Mind," *Harper's New Monthly Magazine* 34 (1867):793–801.

Chapter 4. A Mania for Facts

1. James Herbert Morse, "The Native Element in American Fiction," *Century* 21 (July 1883):362; Amelia Gere Mason, "Is Sentiment Declining," *Century* 61 (February 1901):629, 633.

2. Useful studies of the emergence of professional science during this period include Robert V. Bruce, *The Launching of Modern American Science, 1846–1876* (New York, 1987); George H. Daniels, *American Science in the Age of Jackson* (New York, 1968); Sally G. Kohlstedt, *The Formation of the American Scientific Community: The American Association for the Advancement of Science, 1848–1860* (Urbana, Ill., 1976). A good source of primary documents is Nathan Reingold, ed., *Science in Nineteenth Century America: A Documentary History* (New York, 1964).

3. Henry Adams to Charles F. Adams, Jr., 11 April 1862, in *The Letters of Henry Adams*, eds. J. C. Levenson et al. (Cambridge, Mass., 1982), 1:290.

4. Charles G. Leland, "Sunshine in Thought," *Knickerbocker* 59 (April 1862):324; Leland, *Sunshine in Thought* (New York, 1862), 4–5, 69; Leland, "Editor's Easy Talk," *Graham's* 53 (1858):554; Leland, *Sunshine in Thought*, 144–45, 146–47.

5. RWE, "Thoughts on Modern Literature," *Dial* 1 (October 1840):137–58; Hiram M. Stanley, "The Passion for Realism," *Dial* (16 April 1893):239; HJ, [review of John Tyndall, *Hours of Exercise in the Alps*], *AM* 28 (November 1871):634.

6. Andrew Peabody, "The Intellectual Aspects of the Age," *NAR* 64 (April 1847):276.

7. *Philosophical Discussions of Chauncey Wright* (New York, 1877), 407; Chauncey Wright to F. E. Abbot, August 13, 1867, in *Letters of Chauncey Wright, with Some Account of His Life* (Cambridge, 1878), 113; RWE, "Illusions," *AM* 1 (November 1857):59. On Wright, see Edward H. Madden, *Chauncey Wright and the Foundations of Pragmatism* (Seattle, 1963); Daniel J. Wilson, *Science, Community and the Transformation of American Philosophy, 1860–1930* (Chicago, 1990), 18–34.

8. *New York Times*, 23 May 1860, p. 3; Chauncey Wright to Mrs. J. P. Lesley, 12 February 1860, in *The Letters of Chauncey Wright*, 43. See also "Darwin's *Origin of Species*," *CE* 68 (May 1860):449.

9. See Philip Appleman, ed., *Darwin: A Norton Critical Edition* (New York, 1979).

10. H. W. Schneider, "The Influence of Darwin and Spencer on American Philosophical Theology," *Journal of the History of Ideas* 6 (January 1945):3, 17; Titus Coan, "Critic and Artist," *Lippincott's* 13 (March 1874):355. See also Gamaliel Bradford, "The American Out of Doors," *AM* 71 (April 1893):455; "Science and Culture," *Critic* 2 (22 April 1882):118; Charles Sprague, "The Darwinian Theory," *AM* 18 (October 1866):415–25. The literature on Darwinism and its impact in the United States is extensive. The key works include Robert Bannister, *Social Darwinism: Science and Myth in Anglo-American Social Thought* (Philadelphia, 1979); Paul F. Boller, Jr., *American Thought in Transition: The Impact of Evolutionary Naturalism, 1865–1900* (Chicago, 1969); George Daniels, ed., *Darwinism Comes to America* (Waltham, Mass., 1968); Carl

Degler, *In Search of Human Nature: The Decline and Revival of Darwinism in American Social Thought* (New York, 1991); Richard Hofstadter, *Social Darwinism in American Thought* (Boston, 1955); Bert J. Loewenberg, "Darwinism Comes to America, 1859–1900," *Mississippi Valley Historical Review* 28 (December 1941):339–68, and "The Reaction of American Scientists to Darwinism," *American Historical Review* 38 (1933):687–701; Jon H. Roberts, *Darwinism and the Divine in America: Protestant Intellectuals and Organic Evolution, 1859–1900* (Madison, Wis., 1988).

11. WW, *Democratic Vistas* in *Walt Whitman: Complete Poetry and Collected Prose* (New York: Library of America Edition, 1982), 937; WDH, "Certain Dangerous Tendencies in American Life," *AM* 42 (October 1878):394; WDH, *A Modern Instance* (1882; Boston: Riverside Edition, 1957), 18; Clarence Darrow, "Realism in Literature and Art," *Arena* 9 (1894):109. The general state of religion during this period is summarized in Paul Carter, *The Spiritual Crisis of the Gilded Age* (Dekalb, Ill., 1971); Henry May, *Protestant Churches in Industrial America* (New York, 1949); William G. McLoughlin, *The Meaning of Henry Ward Beecher: An Essay on the Shifting Values of Mid-Victorian America* (New York, 1970); James Turner, *Without God, Without Creed: The Origins of Unbelief in America* (Baltimore, 1985).

12. Herbert Spencer, *Principles of Psychology* (1856; New York, 1875), 2:415, 444–503; *First Principles* (1862; London, 1911), 1:122–26. On Spencer's influence in America, see Boller, *American Thought in Transition*, 52–56; Bannister, *Social Darwinism*; Donald C. Bellomy, "Social Darwinism Revisited," *Perspectives in American History* n.s. 1 (1984):1–129; Hofstadter, *Social Darwinism*, 31–50; Josiah Royce, *Herbert Spencer: An Estimate and Review* (New York, 1904).

13. *Autobiography of Andrew Carnegie* (Boston, 1920), 327.

14. E. L. Youmans, "Editor's Table," *PSM* 1 (May 1872):113; "Religion and Science at Vanderbilt," *PSM* 13 (August 1878):492.

15. E. L. Youmans, *The Culture Demanded by Modern Life* (New York, 1867), 2, 27, 34, 375, 1; Burt G. Wilder, "What We Owe to Agassiz," *PSM* (July 1907):12; Eliot quoted in E. L. Youmans, "Science, Literature and Theology," *PSM* 23 (June 1883):268; HJ, *Charles W. Eliot: President of Harvard University, 1869–1909*, 2 vols. (London, 1930), 1:358; John T. Morse, Jr., ed., *The Life and Letters of Oliver Wendell Holmes*, 2 vols. (Boston, 1896), 2:190–91.

16. "The Scientific Attitude of Mind," in *Essays of William Graham Sumner*, eds. Albert Keller and Maurice Davie, 2 vols. (New Haven, 1911), 1:43–54.

17. "Literature," *Nation* 1 (10 August 1865):181; James T. Bixby, "Taine's Philosophy," *NAR* 117 (October 1873):407; Lester Frank Ward, *Dynamic Sociology* (New York, 1883), 1:69–70. On the emergence of the social sciences as organized professions, see Mary O. Furner, *Advocacy and Objectivity: A Crisis in the Professionalization of American Social Science, 1865–1905* (Lexington, Ky., 1975); Thomas Haskell, *The Emergence of Professional Social Science: The American Social Science Association and the Nineteenth Century Crisis of Authority* (Urbana, Ill., 1976); Dorothy Ross, *The Origins of American Social Science* (Cambridge, 1991); *The Organization of Knowledge in Modern America, 1860–1920*, eds. Alexandra Oleson and John Voss (Baltimore, 1979).

18. E. A. Ross, *Social Control* (New York, 1901), 301, viii–ix; Ross, *Seventy Years of It: An Autobiography* (New York, 1936), 7, 33. For an illuminating study of the impact of the social sciences on literature during this period, see Ronald E. Martin, *American Literature and the Universe of Force* (Durham, N.C., 1981).

19. Carroll D. Wright, "The Growth and Purposes of Statistics of Labor," *Journal*

of Social Science (December 1888), 8; "Have We Equality of Opportunity?," *Forum* 19 (May 1895):312. The enthusiasm for statistics during the nineteenth century is assessed in Patricia Cline Cohen, *A Calculating People: The Spread of Numeracy in Early America* (Chicago, 1982); Margo Anderson Conk, *The United States Census and Labor Force Change: A History of Occupation Statistics, 1870–1940* (Ann Arbor, 1987); Daniel Horowitz, *The Morality of Spending: Attitudes Toward the Consumer Society in America, 1875–1940* (Baltimore, 1985); Theodore M. Porter, *The Rise of Statistical Thinking, 1820–1900* (Princeton, 1986).

20. *The Life of Charles Loring Brace, Chiefly Told in His Own Letters* (New York, 1894), 181; E. B. Taylor, "Quetelet on the Science of Man," *PSM* 1 (May 1872):48.

21. Julian Abernethy, "The Invasion of Realism," *Education* 21 (April 1901):469; Francis A. Walker, "Mr. Grote's Theory of Democracy,"*Bibliotheca Sacra* 25 (October 1868):690; Walker, *College Athletics: An Address Before the Phi Beta Kappa Society, Alpha, of Massachusetts at Cambridge, June 23, 1893*, reprinted in *Technology Quarterly* 6 (July 1893):2–3, 6, 13.

22. Keller cited in Robert C. Bannister, *Sociology and Scientism: The American Quest for Objectivity, 1880–1940* (Chapel Hill, 1987), 105; William G. Sumner, "Sociology as a College Subject," in *The Challenge of Facts*, ed. A. G. Keller (New Haven, 1914), 408–10; Sumner, *Earth-Hunger*, ed. A. G. Keller (New Haven, 1913), 73. On the "masculine" tendencies of social science, see Ross, *Origins of American Social Science*, 58–59; Wolf Lepenies, *Between Literature and Science: The Rise of Sociology* (Cambridge, 1988), 8–9, 39, 99–111.

23. Louisa May Alcott, "Fancy's Friend," in *Aunt Jo's Scrap-Bag: My Boys* (Boston, 1878), 218.

24. Bradford, "The American Idealist," 90; William P. Longfellow, "The Architect's Point of View," *Scribner's* 9 (1891):120; Edward J. Harding, "Cutting Adrift," *Critic* 4 (May 17, 1884):229.

25. OWH, Sr., "Mechanism in Thought and Morals" (1870), in John C. Burnham, ed., *Science in America: Historical Selections* (New York, 1971), 203–4; OWH, Sr., to Youmans quoted in Fiske, *Edward L. Youmans*, 315.

26. Mark DeWolfe Howe, ed., *The Holmes-Pollock Letters*, 2 vols. (Cambridge, Mass., 1941), 1:58; F. S. Cohen, "The Holmes-Cohen Correspondence," *Journal of the History of Ideas* 9 (January 1948):14–15; Max Lerner, ed., *The Mind and Faith of Justice Holmes* (Boston, 1943), 388.

27. Mark DeWolfe Howe, *Justice Oliver Wendell Holmes: The Shaping Years, 1841–1870* (Cambridge, Mass., 1957), 17–18, 210; Cushing Strout, "Three Faithful Skeptics at the Gate of Modernity," in Howard H. Quint and Milton Cantor, eds., *Men, Women and Issues in American History* (Homewood, Ill., 1975), 60. See also Philip Weiner, *Evolution and the Founders of Pragmatism* (Cambridge, Mass., 1949).

28. George Santayana, *Persons and Places* (Cambridge, Mass., 1986), 541; F. O. Matthiessen, *The James Family* (New York, 1980), 13; C. Hartley Grattan, *The Three Jameses: A Family of Minds* (New York, 1962), 89–90; HJ, *A Small Boy and Others* (New York, 1913), 220, 216. The best new treatment of the James family is R.W.B. Lewis, *The Jameses: A Family Narrative* (New York, 1991). For a textured account of the intellectual kinship between William and Henry James, see Ross Posnock, *The Trial of Curiosity: Henry James, William James, and the Challenge of Modernity* (New York, 1991).

29. Perry, *The Thought and Character of William James*, 2 vols. (Boston, 1936), 1:282. Henry James developed a similar preference for tangible truths, declaring in 1871

that literature would benefit greatly if writers would adopt science's "more rigorous method." HJ, *AM* 28 (November 1871):634–36.

30. WJ, "What Psychical Research Has Accomplished," *The Will to Believe* (London, 1897), 301; Fred Kaplan, *Henry James: The Imagination of Genius* (New York, 1992), 52.

31. Perry, *Thought and Character of William James*, 2:706, 709–12, 714, 716. See also Howard M. Feinstein, *Becoming William James* (Ithaca, 1984); Gerald E. Myers, *William James: His Life and Thought* (New Haven, 1986).

32. HJ, ed., *The Letters of William James*, 2 vols. (Boston, 1920), 1:152.

33. WJ to HJ, Sr., 18 March 1873, in ibid., 1:171.

34. WJ, *The Will to Believe and Other Essays*, eds. Frederick Burkhardt, Fredson Bowers, and Ignas Skrupskelis (Cambridge, Mass., 1979), 55. See also the probing analysis of James in James T. Kloppenberg, *Uncertain Victory: Social Democracy and Progressivism in European and American Thought, 1870–1920* (New York, 1986).

35. WJ, *Pragmatism* (1907; Cambridge, Mass., 1975), 97; "Professor James Discovers Pragmatism," *New York Times*, 3 November 1907, 5:8.

36. WJ, "Spencer's Definition of Mind," *Journal of Speculative Philosophy* 12 (1878):1–18; WJ, "The Sentiment of Rationality," *Princeton Review* 2 (1882):58–86; WJ, *The Principles of Psychology* in *The Works of William James* (Cambridge, Mass., 1981), 1:191; WJ, "Some Problems in Philosophy" in *William James: The Essential Writings*, ed. Bruce W. Wilshire (Albany, 1984), 3; WJ, *Essays in Radical Empiricism* (New York, 1912), 39–40; WJ, "The Sentiment of Rationality" (1879), in *The Will to Believe* (New York, 1896), 70. By losing contact with "the rich thicket of reality," academic philosophers had grown "stone-blind and insensible to life's more elementary and general goods and joys." WJ, "On a Certain Blindness in Human Beings," in *The Essential Writings*, 339.

37. *Letters of William James*, 2:201.

38. John Dewey, "The Realism of Pragmatism," *Journal of Philosophy, Psychology and Scientific Methods* 2 (1905):324.

39. John Dewey, "Introduction to Philosophy," *John Dewey: The Early, Middle, and Later Works, 1899–1924*, ed. Jo Ann Boydston (Carbondale, Ill., 1969–91), 3:211; Dewey, "The New Psychology," *Early Works*, 1:60; Dewey, "The Need for a Recovery of Philosophy," *Middle Works*, 10:46.

40. John Dewey, "Matthew Arnold and Robert Browning," *Andover Review* 16 (1891):114–16; WJ, "Some Problems of Philosophy," 4; WDH, "Editor's Study," *HM* 83 (July 1891):316.

41. WJ, [review of Herbert Spencer's *Definition of the Mind as a Correspondence* (1878)], in *The Collected Essays and Reviews of William James* (New York, 1920), 67.

Chapter 5. Goods and Surfaces

1. William Graham Sumner, "The Absurd Effort to Make the World Over," *Forum* 17 (March 1894):92–102; HJ, *Partial Portraits* (London, 1905), 206.

2. Lois Dinnerstein, "The Iron Worker and King Solomon: Some Images of Labor in American Art," *Arts Magazine* 54 (1979):115; "Fine Arts," *Nation* 2 (11 May 1866):603. On Weir see Betsy Fahlman, "John Ferguson Weir: Painter of Romantic and Industrial Icons," *Archives of American Art Journal* 20 (1980):1–8; Richard S. Field, "Passion and Industry in the Art of John Ferguson Weir," *Yale University Art Gallery Bulletin* (1991):48–67. On the impact of technology during this period, see John F.

Kasson, *Civilizing the Machine: Technology and Republican Values in America, 1776–1900* (New York, 1976); Robert W. Rydell, *All the World's a Fair: Visions of Empire at American International Exhibitions, 1876–1916* (Chicago, 1984); Cecilia Tichi, *Shifting Gears: Technology, Literature, Culture in Modernist America* (Chapel Hill, 1987).

3. Nathaniel Hawthorne, *The American Note-books* (Cambridge, Mass., 1883), 369, 102–5; Harriet Beecher Stowe, *Oldtown Folks* (New York, 1869), 421. See also Leo Marx, "The Railroad in the Landscape: An Iconological Reading of a Theme in American Art," *Prospects* 10 (1985):77–118; Leo Marx, *The Machine in the Garden: Technology and the Pastoral Ideal in America* (New York, 1964).

4. *New York Times*, 3 May 1864; "Science and Literature," *Dial* 42 (1 May 1907):275. The theme of "incorporation" is explored in Alan Trachtenberg, *The Incorporation of America: Culture and Society in the Gilded Age* (New York, 1982).

5. "Athens, or Aesthetic Culture," *Bibliotheca Sacra* 20 (January 1863):170; E. C. Stedman, *Victorian Poets* (Boston, 1876), 12–13.

6. M. D. Bacon, "Great Cities," *Putnam's* 5 (March 1855):254; David Wasson, "The Modern Type of Oppression," NAR 119 (October 1874):262. Several excellent studies of urbanization and urban culture are available. See Thomas Bender, *Toward an Urban Vision: Ideas and Institutions in Nineteenth-Century America* (Lexington, Ky., 1975); Andrew Lees, *Cities Perceived: Urban Society in European and American Thought, 1820–1940* (New York, 1985); Eric H. Monkkonen, *America Becomes Urban: The Development of U.S. Cities & Towns, 1780–1980* (Berkeley, 1988); Edward K. Spann, *The New Metropolis: New York City, 1840–1857* (New York, 1981).

7. Peter Conrad, *The Art of the City: Views and Versions of New York* (New York, 1984), 3–21; Fanny Bates, "Fiction Reading in the Country," *Dial* 19 (1 July 1895):10.

8. "The Question of Theaters," *Putnam's* 9 (June 1857):642.

9. HG, "Artistic Needs of American Cities," p. 1, undated manuscript, item no. 368, in Hamlin Garland Papers, University of Southern California. See also Michael Denning, *Mechanic Accents: Dime Novels and Working-Class Culture in America* (London and New York, 1987); Adrienne Siegel, *The Image of the City in Popular Literature: 1820–1870* (Port Washington, N.Y., 1981); Janis P. Stout, *Sodoms in Eden: The City in American Fiction Before 1860* (Westport, Conn., 1976).

10. John J. McAleer, *Theodore Dreiser: An Introduction and Interpretation* (New York, 1968), 29–30; TD, *Newspaper Days* (New York, 1922), 100; TD, "Reflections," *Ev'ry Month* 3 (October 1896):6.

11. TD, *Sister Carrie* (1902; New York: Norton Critical Edition, 1970), 47; TD, "Reflections," *Ev'ry Month* 1 (January 1896):4–5.

12. WDH, *The Rise of Silas Lapham* (1885; New York: Norton Critical Edition, 1982), 56; EW, *Novels* (New York: Library of America, 1985), 10; TD, *Sister Carrie*, 74. Several works examine the theme of money in American fiction. See Jan W. Dietrichson, *The Image of Money in the American Novel of the Gilded Age* (New York, 1969); Walter Benn Michaels, *The Gold Standard and the Logic of Naturalism: American Literature at the Turn of the Century* (Berkeley, 1987); John Vernon, *Money and Fiction: Literary Realism in the Nineteenth and Early Twentieth Centuries* (Ithaca, 1984).

13. George Ade, "The Advantage of Being 'Middle Class,'" in *Stories of the Streets and of the Town from the Chicago Record, 1893–1900* (Chicago, 1941), 75. While touring the country at mid-century, the immigrant journalist Carl Schurz saw "the middle-class culture in process of formation." *The Reminiscences of Carl Schurz*, 2 vols. (New York, 1907), 2:158. The emergence of a middle class is treated in Burton Bledstein, *The Culture of Professionalism: The Middle Class and the Development of Higher Edu-*

cation in America (New York, 1976); Stuart Blumin, *The Emergence of the Middle Class: Social Experience in the American City, 1760–1900* (Cambridge, 1989). For a useful discussion of the concept of a middle class, see Peter Stearns, "The Middle Class: Toward a Precise Definition," *Comparative Studies in Society and History* 21 (1979):377–96.

14. WW, *I Sit and Look Out: Editorials from the Brooklyn Daily Times,* eds. Emory Holloway and Vernolian Schwarz (New York, 1932), 145; Richard Ruland, ed., *The Native Muse: Theories of American Literature* (New York, 1972), 1:426–27. A middle-class cultural orientation is examined in Svetlana Alpers, *The Art of Describing: Dutch Art in the Seventeenth Century* (Chicago, 1983); John Michael Montias, *Artists and Artisans in Delft: A Socio-Economic Study of the Seventeenth Century* (Princeton, 1982); Simon Schama, *The Embarrassment of Riches: An Interpretation of Dutch Culture in the Golden Age* (New York, 1987). Also relevant is Gertrude Atherton, "Why Is American Literature Bourgeois," NAR 178 (May 1904):771–81.

15. Henry Tuckerman, "American Society," NAR 81 (July 1855):37–38.

16. Robert Park, "The City: Suggestions for the Investigation of Human Behavior in the Urban Environment," in Richard Sennett, ed., *Classic Essays on the Culture of Cities* (New York, 1969), 126; WDH, *Silas Lapham,* 128, 56.

17. *John Sloan's New York Scene,* ed. Bruce St. John (New York, 1965), 431.

18. HJ, *Autobiography,* ed. Frederick W. Dupee (New York, 1983), 337–38; *Parisian Sketches,* eds. Leon Edel and Ilse Linde (1875–76; London, 1958), 191; HJ, *The American Scene* (1904; Bloomington, 1969), 68, 83; HJ, *Autobiography,* 20. On this theme of urban surveillance, see Rachel Bowlby, *Just Looking: Consumer Culture in Dreiser, Gissing and Zola* (London, 1985); Conrad, *Art of the City,* 178–206; Jonathan Crary, *Techniques of the Observer: On Vision and Modernity in the Nineteenth Century* (Cambridge, Mass., 1990); Guy Debord, *The Society of the Spectacle* (New York, 1990); Michel Foucault, *Discipline and Punish* (New York, 1977); Carolyn Porter, *Seeing and Being: The Plight of the Participant Observer in Emerson, James, Adams, and Faulkner* (Middletown, Conn., 1981); John Rignall, *Realist Fiction and the Strolling Spectator* (London, 1993); Remy G. Saisselin, *The Bourgeois and the Bibelot* (New Brunswick, N.J., 1984). Deborah Epstein Wood examines the gendered nature of urban strolling in "The Urban Peripatetic: Spectator, Streetwalker, Woman Writer," *Nineteenth-Century Literature* 36 (December 1991):351–75.

19. HJ, *The American* (New York,1978), 303; HJ, *The American Scene,* 397.

20. Arthur W. Burks, ed., *Collected Papers of Charles Sanders Pierce* (Cambridge, 1931–58), 7:115; John F. Kasson, *Rudeness and Civility: Manners in Nineteenth-Century Urban America* (New York, 1990), 99.

21. TD, *A Book About Myself* (New York, 1992), 139, 88; TD, *Sister Carrie,* 44, 9, 13; TD, *The Genius* (New York, 1915), 13, 108.

22. William P. Longfellow, "The Architect's Point of View," *Scribner's* 9 (1891):120; George Ade, "From an Office Window," in *Stories of the Streets and of the Town,* 167–68.

23. *John Sloan's New York Scene,* 549, 131, 69.

24. "Broadway," *Crayon* 5 (August 1858):234; WW, "Matters Which Were Seen and Done in an Afternoon Ramble," *Brooklyn Daily Eagle,* 19 November 1846; Stuart Ewen, *All Consuming Images: The Politics of Style in Contemporary Culture* (New York, 1988), 60. The emergence of a consumer culture has received increasing scholarly attention in recent years. It has also provoked growing debate. For some of the key issues, see John Brewer and Roy Porter, eds., *Consumption and the World of Goods* (London,

1993); Simon J. Bronner, ed., *Consuming Visions: Accumulation and Display of Goods in America, 1880–1920* (New York, 1989); Mary Douglas and Christopher Isherwood, *The World of Goods: Toward an Anthropology of Consumption* (New York 1979); Richard W. Fox and T. J. Jackson Lears, eds., *Culture of Consumption: Critical Essays in American History, 1880–1980* (New York, 1983); Neil Harris, "The Drama of Consumer Desire," in his *Cultural Excursions* (Chicago, 1990), 174–97; William Leach, *Land of Desire: Merchants, Money, and the Rise of a New American Culture* (New York, 1994).

25. Gamaliel Bradford, "The American Idealist," *AM* 70 (July 1892):90. See also Thomas W. Higginson, "A Plea for Culture," *Nation* 4 (21 February 1867):151.

26. "Dry Goods," *HW* 11 (4 May 1867):275; "Dry Goods at a Discount," *HW* 1 (24 October 1857):675. See also Clarence Cook, "Some Chapters on Home-Furnishing," *Scribner's* 10 (June 1875):171; Karen Halttunen, *Confidence Men and Painted Women: A Study of Middle-Class Culture in America, 1830–1870* (New Haven, 1982), 33–91.

27. "Regardless of Expense," *HW* 1 (18 April 1857):241. A prominent interior decorator encouraged such clutter when she asserted that "there is hardly likely to be too much in a room." Harriet Spofford, *Art Decoration as Applied to Furniture* (New York, 1877), 222. Two fine studies of nineteenth-century home interiors are Katherine C. Grier, *Culture & Comfort: People, Parlors, and Upholstery, 1850–1930* (Rochester, 1988), and William Seale, *The Tasteful Interlude: American Interiors Through the Camera's Eye, 1860–1917* (Nashville, 1981).

28. Harriet Beecher Stowe, "House and Home Papers," *AM* 13 (January 1864): 40–47.

29. HJ, "Ralph Waldo Emerson," in *Essays on Literature: American Writers, English Writers*, ed. Leon Edel (New York, 1984), 233–71; F. O. Matthiessen and Kenneth Murdock, eds., *The Notebooks of Henry James* (New York, 1947), 47; HJ, *The Bostonians* (1886; New York: Penguin, 1984), 189, 46, 190.

30. Israel L. White, "Childe Hassam — A Puritan," *International Studio* 45 (December 1911):30; TD, *The Financier* (New York, 1912), 187.

31. John Sloan, *The Gist of Art* (New York, 1939), 43; FN, *Vandover and the Brute* (New York: Library of America Edition, 1986), 130–31. "The duty of the subject of realism," as Mark Seltzer has written, "is to fashion a character that corresponds to its representations." Mark Seltzer, "Physical Capital: *The American* and the Realist Body," in Martha Banta, ed., *New Essays on The American* (Cambridge, 1987), 159.

32. EW and Ogden Codman, *The Decoration of Houses* (New York, 1897), 17; EW, *Novels*, 976, 984. On Wharton's emphasis on physical environment, see Judith Fryer, *Felicitous Space: Imaginative Structures of Edith Wharton and Willa Cather* (Chapel Hill, 1986).

33. EW, *Novels*, 27, 41, 313, 316, 336.

34. TD, *Sister Carrie*, 17.

35. Charles Dudley Warner, "Editor's Drawer," *HM* 80 (May 1890):972–73. The surge of women in the work force is assessed in Alice Kessler-Harris, *Out to Work: A History of Wage-Earning Women in the United States* (New York, 1982); Barbara J. Harris, *Beyond Her Sphere: Women and the Professions in American History* (Westport, Conn., 1978); Susan E. Kennedy, *If All We Did Was To Weep at Home: A History of White Working-Class Women in America* (1979); Anne Firor Scott, *The Southern Lady: From Pedestal to Politics* (Chicago, 1970); Martha Vicinus, *Independent Women: Work and Community for Single Women, 1850–1920* (Chicago, 1985); Barbara Wertheimer, *We Were There: The Story of Working Women in America* (New York, 1977). On chang-

ing family and home life, see Dolores Hayden, *The Grand Domestic Revolution* (Cambridge, Mass., 1981); Steven Mintz, *A Prison of Expectations: The Family in Victorian Culture* (1983); Susan Kellogg, *Domestic Revolutions: A Social History of American Family Life* (1988).

36. *Nation* 4 (21 February 1867):155–56.

37. Annie Nathan Meyer, "Letter to Editor," *Critic* (14 September 1889):127. See also Earl Barnes, "The Feminizing of Culture," *AM* 109 (June 1912):770–96; Barbara Solomon, *In the Company of Educated Women: A History of Women and Higher Education in America* (Cambridge, Mass., 1985).

38. E. L. Youmans, "Dr. Dix on the Woman Question," *PSM* 23 (May 1883):122. See also Kate Gannett Wells, "The Transitional American Woman," *AM* 46 (December 1880):817–23.

39. On neurasthenia as a medical phenomenon, see F. G. Gosling, *Before Freud: Neurasthenia and the American Medical Community, 1870–1910* (Urbana, 1987); John S. and Robin M. Haller, *The Physician and Sexuality in Victorian America* (New York, 1974); Charles Rosenberg, "The Place of George Beard in Nineteenth-Century Psychiatry," *Bulletin of the History of Medicine* 26 (1962):245–59.

40. George M. Beard, *American Nervousness: Its Causes and Consequences* (New York, 1881), 207; S. W. Hammond, "Neurasthenia," *Vermont Medical Monthly* 14 (1908):158. For a fascinating analysis of neurasthenia as a cultural force and metaphor, see Tom Lutz, *American Nervousness, 1903: An Anecdotal History* (Ithaca, 1991).

41. Charlotte Perkins Gilman to WDH, 17 October 1919, WDH Papers. See also Ann Douglas, "'The Fashionable Diseases': Women's Complaints and Their Treatment in Nineteenth-Century America," *Journal of Interdisciplinary History* 4 (Summer 1973):25–52; S. Weir Mitchell, *Doctor and Patient* (1887; Philadelphia, 1904), 13, 84, 89, 122.

42. Jane Addams, *Twenty Years at Hull-House* (1910; New York: New American Library, 1960), 59, 216–17, 72.

43. H. H. Gardner, "Immoral Influence of Women in Literature," *Arena* 1 (February 1890):323; HJ quoted in Ralph Barton Perry, *The Thought and Character of William James*, 2 vols. (Boston, 1935), 2:780–81; Charles D. Warner, *As We Go* (New York, 1894), 126–27. See also Margaret Deland, The Change in the Feminine Ideal," *AM* (March 1910):289–302. For a detailed examination of the cultural imaging of women, see Martha Banta, *Imaging American Women: Idea and Ideals in Cultural History* (New York, 1987).

44. Edgar Allan Poe, "The Man in the Crowd," in *The Complete Works of Edgar Allan Poe*, ed. James A. Harrison (New York, 1902), 4:134–45; "Old Things," *Scribner's* 41 (1907):636.

45. George Wakeman, "Advertising," *Galaxy* 3 (15 January 1867):203, 202; George P. Lathrop, "Art and Advertising," *Nation* 20 (20 May 1875):342–43; WDH, *Silas Lapham*, 13; Stuart and Elizabeth Ewen, *Channels of Desire: Mass Images and the Shaping of American Consciousness* (New York, 1982), 82; TD, *Sister Carrie*, 6–7. The phenomenal growth of advertising is explained in T. J. Jackson Lears, "The Rise of American Advertising," *Wilson Quarterly* (Winter 1983):156–67, and "Some Versions of Fantasy: Toward a Cultural History of American Advertising, 1880–1930," *Prospects* 9 (1984):349–405; Roland Marchand, *Advertising the American Dream: Making Way for Modernity, 1920–1940* (Berkeley, 1986); James D. Norris, *Advertising and the Trans-*

formation of American Society, 1865–1920 (Westport, Conn., 1990); Frank Presbrey, *The History and Development of Advertising* (Garden City, N.Y., 1929); Michael Schudson, *Advertising, The Uneasy Persuasion: Its Dubious Impact on American Society* (New York, 1986); Trachtenberg, *Incorporation of America*, 135–39.

46. "The Poetry of Advertising," *HW* 2 (27 March 1858):94–95.

47. Edward E. Hale, Jr., "The Congestion of Cities," *Forum* 4 (January 1888):530. Circulation statistics are included in Alfred Lee, *The Daily Newspaper in America* (New York, 1937), 725–26. Studies of the rise of the penny-press include James L. Crouthamel, *Bennett's New York Herald and the Rise of the Popular Press* (Syracuse, 1990); Hazel Dickens-Garcia, *Journalistic Standards in Nineteenth-Century America* (Madison, Wis., 1989); Dan Schiller, *Objectivity and the News: The Public and the Rise of Commercial Journalism* (Philadelphia, 1981).

48. Isaac C. Pray, *Memoirs of James Gordon Bennett and His Times* (New York, 1855), 216; WW, "The Press—Its Future," *Brooklyn Daily Times*, 31 July 1858, in *I Sit and Look Out: Editorials from the Brooklyn Daily Times by Walt Whitman*, eds. Emory Holloway and Vernolian Schwarz (New York, 1932), 39; *Collected and Other Prose of Walt Whitman*, ed. Floyd Stovall (New York, 1964), 2:367; WW, *Uncollected Poetry and Prose*, ed. Emory Holloway (Garden City, N.Y., 1921), 1:115.

49. TD, "Novels to Reflect Real Life," *New York Sun*, 21 November 1911, 5; George Ade, *Stories of the Streets and of the Town*, xxiii; WDH, *Years of My Youth*, ed. David J. Nordloh (Bloomington, 1975), 122. The relationship between journalism and literature is explored in Shelley Fisher Fishkin, *From Fact to Fiction: Journalism and Imaginative Writing in America* (Baltimore, 1984).

50. "A Journal of Civilization," *HW* 1 (3 January 1857):14; "Pictorial Newspapers in America," *FLIN* 1 (15 December 1855):6; Nathaniel P. Willis, "Frank Leslie: A Life Lengthener," *Home Journal*, reprinted in *FLIN* 11 (15 December 1860):53; Leslie, "Commencement of Our Second Volume," *FLIN* 2 (21 June 1856):18.

51. Marius de Zayas, "Photography," *Camera Work* no. 41 (January 1912):18. Frank Leslie, for instance, assured readers that his magazine's illustrations were taken directly "from daguerreotypes . . . and may be relied upon as correct." "Our Present Number— Triumphant Success," *FLIN* 3 (21 February 1857):189.

See also Beaumont Newhall, *The Daguerreotype in America* (New York, 1976); Richard Rudisill, *The Mirror Image: The Influence of the Daguerreotype on American Society* (Albuquerque, 1971); Martha Sandweiss, ed., *Photography in Nineteenth-Century America* (New York, 1991); Robert Taft, *Photography and the American Scene: A Social History, 1839–1889* (New York, 1938); Alan Trachtenberg, *Reading American Photographs: Images as History, Matthew Brady to Walker Evans* (New York, 1989); John Wood, ed., *America and the Daguerreotype* (Iowa City, 1991).

52. WW quoted in *The Gathering of the Forces*, eds. Cleveland Rodgers and John Black, 2 vols. (New York, 1920), 2:113–17.

53. Robert H. Vance, *Catalogue of Daguerreotype Panoramic Views in California* (New York, 1851), 4. Nineteenth-century Americans displayed a naively exuberant faith in the camera's accuracy and detachment. The Reverend H. J. Morton claimed that the camera "sees everything, and it represents just what it sees. It has an eye that cannot be deceived, and a fidelity that cannot be corrupted." People accepted photographs as mirror images of reality depicted without sentiment, illusion, or decoration, scientific registers of facts and faces which confirmed the existence of the material world. The photograph, wrote an art critic, "cannot deceive; what we see must be TRUE." H. J. Morton,

"Photography as an Authority," *Philadelphia Photographer* 1 (1864):180–81; "America in the Stereoscope," *Art Journal* 6 (July 1860):221.

54. "Development of Nationalism in American Art," *Photographic Art-Journal* 3 (January 1852):39; Maurice Thompson, "The Domain of Romance," *Forum* 8 (November 1889):327; *Photographic Art-Journal* 1 (1851):163. See also Carol Shloss, *In Visible Light: Photography and the American Writer* (New York, 1987).

55. *Crayon* 1 (1855):107. Two excellent studies of the relationship of photography to painting are Van Deren Coke, *The Painter and the Photograph* (Albuquerque, 1964), and Elizabeth Lindquist-Cook, *The Influence of Photography on American Landscape Painting, 1839–1880* (New York, 1977).

56. Mary Cassatt to Mrs. H. O. Havemeyer, 9 January 1912, Mary Cassatt Papers, National Gallery of Art; J. E. Cabot, "On the Relation of Art to Nature," *AM* 13 (February 1864):183; Kenyon Cox, "William M. Chase, Painter," *HM* 78 (March 1889):549, 554.

57. WW, "Notes on the Meaning and Intention of *Leaves of Grass*," in *The Complete Prose Works of Walt Whitman*, ed. Richard Bucke (New York, 1902), 9:21; Clyde E. Henson, *Joseph Kirkland* (New York, 1962), 78–79.

58. "Nebulae," *Galaxy* 4 (1867):367. With these contributions in mind, the literary critic George Lathrop concluded that the new school of realism "must have more than any photograph can give us; with the accuracy of that, should be combined the aesthetic completeness of a picture and a poem in one." George P. Lathrop, "The Novel and Its Future," *AM* 34 (September 1874):324.

59. JRL to Charles E. Norton, 10 December 1869, *The Letters of James Russell Lowell*, ed. Charles E. Norton, 2 vols. (New York, 1894), 2:51.

60. Charles E. Norton, "The Launching of the Magazine," *AM* 100 (November 1907):581.

61. HJ, *Literary Criticism*, ed. Leon Edel (New York: Library of America, 1984), 998; Leon Edel and Lyall Powers, eds., *The Complete Notebooks of Henry James* (New York, 1987), 52; HJ, *Essays on Literature: American Writers, English Writers*, ed. Leon Edel (New York, 1984), 233; HJ, *The Bostonians*, Modern Library Edition, 343; Leon Edel, *Henry James: A Life* (New York, 1985), 95.

62. E. A. Ross, *Social Control* (New York, 1901), 216, 139. For a keen analysis of the democratic objectives of social science, see Edward A. Purcell, Jr., *The Crisis of Democratic Theory: Scientific Naturalism and the Problem of Value* (Lexington, Ky., 1973).

63. Ross, "The Value Rank of the American People," *Independent* 57 (1904):1061–63, and *Social Psychology: An Outline and Source Book* (New York, 1923), 140, 241–43.

64. Ross, *Social Psychology*, 257–61. Additional evidence of this notion can be found in Charles Horton Cooley, *Social Process* (New York, 1918), 410–22.

65. WDH, *The Rise of Silas Lapham* (New York: Norton Critical Edition, 1982), 104; WDH, "Editor's Study," *HM* 74 (February 1887):483; WDH, "Editor's Study," *HM* 75 (September 1887):638–42; WDH, "Editor's Study," *HM* 75 (July 1887):318; WDH, "Two Notable Novels," *Century* n.s. 6 (August 1884):633. See also James Lane Allen, "Local Color," *Critic* (9 January 1886):13.

66. WDH, "Editor's Study," *HM* 83 (July 1891):317; 79 (September 1889):641. As Amy Kaplan has recognized, the essential "project of realism" was "to manage social differences through representation." Kaplan, *The Social Construction of American Realism* (Chicago, 1988), 30.

Part III. Realism Triumphant

1. Joseph Kirkland, "Tolstoi and the Russian Invasion of the Realm of Fiction," *Dial* 7 (August 1886):81. As William Dean Howells admitted, with a nod toward Darwin, "the two kinds do not mingle well, but for a while yet we must have the romantic and the realistic mixed. . . . That is quite inevitable; and it is strictly in accordance with the law of evolution." WDH, "Editor's Study," *HM* 83 (September 1891):478.

Chapter 6. Truth in Fiction

1. "The Minister's Charge," *Critic* (5 February 1887):63; Edmund Gosse, "The Limits of Realism in Fiction," *Forum* 9 (June 1890):392; JWD to WDH, 11 March 1879, JWD Papers, Beinecke Library, Yale University. John Esten Cooke, the unreconstructed Virginia romantic, resigned himself to the realists' success: "They see, as I do, that fiction should faithfully reflect life, and they obey the law while I can not. I was born too soon, and am now too old to learn my trade anew." Cooke quoted in Elisabeth Muhlenfeld, "The Civil War and Authorship," in *The History of Southern Literature*, ed. Louis Rubin, Jr. (Baton Rouge, 1985), 187.

Useful general surveys of American literary realism include: Warner Berthoff, *The Ferment of Realism: American Literature, 1884–1919* (New York, 1965); Edwin H. Cady, *The Light of Common Day: Realism in American Fiction* (Bloomington, 1971); Robert P. Falk, *The Victorian Mode in American Fiction, 1865–1885* (East Lansing, Mich., 1964); Alfred Kazin, *On Native Grounds* (New York, 1942); Harold H. Kolb, Jr., *The Illusion of Life: American Realism as a Literary Form* (Charlottesville, 1969); Jay Martin, *Harvests of Change: American Literature, 1865–1914* (Englewood Cliffs, 1967); Vernon Louis Parrington, *The Beginnings of Critical Realism in America*, vol. 3 of *Main Currents in American Thought* (New York, 1930). See also Hugh Holman, "Notes on How to Read American Realism," *Georgia Review* 18 (Fall 1964):316–24.

The best recent interpretive studies are Michael Davitt Bell, *The Problem of American Realism: Studies in the Cultural History of a Literary Idea* (Chicago, 1993), and Amy Kaplan, *The Social Construction of American Realism* (Chicago, 1988). See also Eric J. Sundquist, ed., *American Realism: New Essays* (Baltimore, 1982).

2. Hjalmar H. Boyeson, "The Realism of the American Fiction," *Independent* 44 (3 November 1892): 4.

3. *Henry James: Autobiography*, ed. Frederick W. Dupee (New York, 1983), 147; HJ quoted in Alfred Habegger, *Henry James and the "Woman Business"* (Cambridge, 1989), 145; WDH, *A Boy's Town* (New York, 1890), 75.

4. *Criticism and Fiction and Other Essays by William D. Howells*, eds. C. M. Kirk and Rudolf Kirk (New York, 1965), 26; WDH, *My Literary Passions* (New York, 1895), 14. On this theme see Michael Davitt Bell, "The Sin of Art and the Problem of American Realism: William Dean Howells,"*Prospects* 9 (1984):115–43; John Crowley, "Howells: The Ever-Womanly," in *American Novelists Revisited: Essays in Feminist Criticism*, ed. Fritz Fleischmann (Boston, 1982); Alfred Habegger, *Gender, Fantasy, and Realism in American Literature* (New York, 1982).

5. HJ, *Autobiography*, 16–17; HJ, *Notes of a Son and Brother* (New York, 1914), 311, 296–97; HJ, *Notes and Reviews* (Cambridge, Mass., 1921), 23; H. A. Huntington, "A Pair of American Novelists," *Dial* 2 (January 1882):214. On the James family, see the fine group biography by R.W.B. Lewis, *The Jameses* (New York, 1991). Henry James's conflict with the Civil War is discussed in Charles and Tess Hoffman, "Henry

James and the Civil War," *New England Quarterly* 62 (December 1989):529–52. On his literary technique, see Peter Buitenhuis, *The Grasping Imagination: The American Writings of Henry James* (Toronto, 1970).

6. HJ, [review of Harriet Prescott, *Azarian*], NAR 100 (January 1865):272–73; HJ, "The Minor French Novelists," *Galaxy* 21 (February 1876):219–33; HJ, "The Last French Novel," *Nation* 3 (11 October 1866):286–88. See also HJ, *Autobiography*, 148–49.

7. HJ to WJ, August 1872, in F. O. Matthiessen, *The James Family* (New York, 1947), 320–21; HJ to TSP, 15 August 1867, in *Henry James: Selected Letters*, ed. Leon Edel (Cambridge, Mass., 1974), 1:71–72; "Alphonse Daudet," *Century* 26 (August 1883):501; Lewis, *The Jameses*, 306. Writing from Paris to his brother William in 1872, Henry James remarked: "The Nortons are excellent, but I feel less and less at home with them." Perry, *Thought and Character of William James*, 1:328.

8. HJ to Charles Eliot Norton, 16 January 1871, Charles Eliot Norton Papers, Houghton Library, Harvard University.

9. WDH, *A Chance Acquaintance* (Boston, 1873), 10.

10. HJ, "The New Novel," *Essays*, 130. For biographical information on Howells see Edwin Cady's two-volume study, *The Road to Realism* and *The Realist at War* (Syracuse, N.Y., 1956, 1958); Kenneth Lynn, *William Dean Howells: An American Life* (New York, 1971); Edward Wagenknecht, *William Dean Howells: The Friendly Eye* (New York, 1969). A penetrating Eriksonian analysis of Howells's youth is found in Rodney Olsen's *Dancing in Chains: The Youth of William Dean Howells* (New York, 1991). Among the many critical studies, see Everett Carter, *Howells and the Age of Realism* (1950; Hamden, Conn., 1966); John W. Crowley, *The Mask of Fiction: Essays on W. D. Howells* (Amherst, 1989); Lionel Trilling, "William Dean Howells and the Roots of Modern Taste," in *The Opposing Self* (New York, 1965).

11. "Real Conversation—A Dialogue Between William Dean Howells and Hjalmar H. Boyeson," *McClure's* 1 (June 1893):3–11; James L. Woodress, Jr., *Howells and Italy* (Durham, N.C., 1952), 52; WDH, *Venetian Life*, 3d ed. (New York, 1867), 10, 15. See also Edd Winfield Parks, "A Realist Avoids Reality: W. D. Howells and the Civil War Years," *South Atlantic Quarterly* 52 (January 1953):93–97.

12. WDH, *Criticism and Fiction*, 66; *Mark Twain-Howells Letters: The Correspondence of Samuel L. Clemens and William D. Howells, 1872–1910*, eds. Henry Nash Smith and William Gibson (Cambridge, Mass., 1960), 1:17; George E. DeMille, "The Infallible Dean," *Sewanee Review* 36 (April 1928):148; WDH, *Criticism and Fiction*, 380.

13. HJ, "A Letter to Mr. Howells," NAR 195 (April 1912):558–62; WDH, "Henry James, Jr.," *Century* 25 (Nov. 1882):28. Robert Louis Stevenson correctly called Howells the "zealot of his school," and Boyeson credited him with assuring the "ultimate triumph of realism." See Robert Louis Stevenson, "A Humble Remonstrance," *Longman's* 5 (December 1884):139–47; H. H. Boyeson, "The Progressive Realism of American Fiction," *Literary and Social Silhouettes* (New York, 1894), 71–2, 78.

14. Albert Bigelow Paine, *Mark Twain: A Biography* (New York, 1912), 1197; *Mark Twain-Howells Letters: The Correspondence of Samuel L. Clemens and William Dean Howells, 1872–1910*, eds. Henry N. Smith and William M. Gibson (Cambridge, Mass., 1960), 534. "At heart," Howells said late in life, "Clemens was romantic, and he would have had the world of fiction stately and handsome and whatever the real world is not, but he was not romanticistic, and he was too helplessly an artist not to wish his own

work to show life as he had seen it." See WDH, *Literary Friends and Acquaintances* (New York, 1900), 286–87.

The most perceptive discussion of Twain's relation to realism is Michael Davitt Bell, "Mark Twain, 'Realism,' and *Huckleberry Finn*," in *New Essays on The Adventures of Huckleberry Finn*, ed. Louis Budd (Cambridge, 1985), 35–60.

15. Mark Twain, *Adventures of Huckleberry Finn* (1884; New York: Norton Critical Edition, 1977), 17; Robert Penn Warren, "Mark Twain," in *New and Selected Essays* (New York, 1989), 118.

16. "Foreign Influences on American Fiction," NAR 149 (1889):118. Kate Chopin recalled being transformed after stumbling upon a volume of Maupassant's stories: "Here was life, not fiction. . . . Here was a man who had escaped from tradition and authority, who had entered into himself and looked out upon life through his own being and with his own eyes; and who, in a direct and simple way, told us what he saw." Kate Chopin, "Confidences," *The Complete Works of Kate Chopin*, ed. Per Seyerstad (Baton Rouge, 1969), 2:700.

17. See "A New Literature," HW 6 (10 May 1862):291; Eugene Benson, "About the Literary Spirit," *Galaxy* 1 (July 1866):487–92. Benson was a transitional figure between romanticism and realism. See Robert J. Scholnick, "Between Realism and Romanticism: The Curious Career of Eugene Benson," *American Literary Realism* 14 (Autumn 1981):242–53.

18. JWD, "The Great American Novel," *Nation* 6 (9 January 1868):27–29.

19. "As Howells Sees Fiction," *New York Sun*, 16 February 1898, 3; Samuel Langhorne Clemens, "What Paul Bourget Thinks of Us," *Literary Essays* (Garden City, N.Y., 1963), 145.

20. Burton Raffel, ed., *The Signet Classic Book of American Short Stories* (New York, 1985), 12. See also James D. Hart, *The Popular Book: A History of America's Literary Taste* (New York, 1930); Michael Spindler, *American Literature and Social Change* (Bloomington, 1983). Useful historical studies of literacy include William J. Gilmore, *Reading Becomes a "Necessity of Life": Material and Cultural Life in Rural New England, 1780–1830* (Knoxville, 1989); David Nasaw, *Schooled to Order: A Social History of Public Schooling in the United States* (1979); Leo Soltow and Edward Stevens, *The Rise of Literacy and the Common School in the United States: A Socioeconomic Analysis to 1870* (Chicago, 1981). On the proliferation of libraries and organized reading, see Esther Jane Carrier, *Fiction in the Public Libraries, 1876–1900* (New York, 1965); Theodora Penny Martin, *The Sound of Our Own Voices: Women's Study Clubs, 1860–1940* (Boston, 1987); Theodore Morrison, *Chautauqua: A Center for Education, Religion, and the Arts in America* (1974).

21. Raffel, ed., *Short Stories*, 12–14. In 1886 De Forest claimed in a letter to Howells that women constituted "four-fifths of our novel-reading public." JWD to WDH, 6 December 1886, JWD Papers, Houghton Library, Harvard University.

22. WDH, "The Man of Letters as a Man of Business," (1893) in *Literature and Life* (New York, 1902), 6–7. On the profession of letters, see Nelson Lichtenstein, "Authorial Professionalism and the Literary Marketplace, 1885–1900," *American Studies* 19 (1978):35–53; Lewis P. Simpson, *The Man of Letters in New England and the South* (Baton Rouge, 1973); Christopher P. Wilson, *The Labor of Words: Literary Professionalism in the Progressive Era* (Athens, 1985).

23. "The Fortunes of Literature Under the American Republic," *Appleton's Journal* 11 (1881):45; Willa Cather, "With Plays and Players," *Nebraska State Journal* (13 March

1894):13; Cather, "As You Like It," *Nebraska State Journal* (16 December 1894):13. The best study of the shifting marketplace's effects on literary realism is Daniel Borus, *Writing Realism: Howells, James, and Norris in the Mass Market* (Chapel Hill, 1989). Borus argues that the "central historical determinant of the realist writing process was the consolidation of the literary marketplace as the locus of literary production, exchange, and circulation" (24). He claims too much, but his book does recognize and document the important role played by the literary marketplace. See also Marcia Jacobson, *Henry James and the Mass Market* (University, Ala., 1983); Henry Nash Smith, *Democracy and the Novel: Popular Resistance to Classic American Writers* (New York, 1978); Christopher P. Wilson, "The Rhetoric of Consumption: Mass-Market Magazines and the Demise of the Gentle Reader, 1880–1920," in *The Culture of Consumption: Critical Essays in American History, 1880–1980*, eds. Richard Wightman Fox and T. J. Jackson Lears (New York, 1983), 39–64.

24. JWD to WDH, 6 December 1886, JWD Papers; WDH, "Editor's Study, *HM* 72 (April 1886):811; WDH, "Novel-Writing and Novel-Reading: An Impersonal Explanation," in *Howells and James: A Double Billing*, ed. William M. Gibson (New York, 1958), 19–20.

25. Constance Fenimore Woolson to WDH, June 28, 1880, WDH Papers.

26. RTC to George Ticknor, 29 October, 1885, 9 January 1886, RTC Papers, Beinecke Library, Yale University; Louisa May Alcott, Journal, May 1864, in Alcott Papers, Houghton Library, Harvard University; Ednah D. Cheney, *Louisa May Alcott: Her Life, Letters, and Journals* (Boston, 1889), 76–77; Louisa May Alcott to Elizabeth Powell, 20 March 1869, in *Selected Letters of Louisa May Alcott*, ed. Joel Myerson (Boston, 1987), 125. During the serial publication of *Little Women* (1868–69), Alcott's editor and readers urged her to have Jo marry the romantic Laurie, but she resisted, determined to demonstrate that unmarried women could live rewarding lives. In the end, however, she succumbed to the pressure and inserted a future husband for Jo.

27. HJ, "The Question of the Opportunities," *Literature* 2 (26 March 1898):356–58; HJ, "Nassau Senior's *Essays on Fiction*," NAR 99 (1 October 1869):580–87; HJ, "The Author of 'Beltraffio,'" in *The Complete Tales of Henry James*, ed. Leon Edel (Philadelphia, 1963), 5:335–36. Howells agreed with James that "the fate of the book is in the hands of the women." On average, he observed, women of means were better-read than men and enjoyed more leisure. "Our men read the newspapers, but our women read the books; the more refined among them read the magazines." As a consequence, the professional author "must resign himself to obscurity unless the ladies recognize him." See WDH, "The Man of Letters as a Man of Business," 21; WDH, "Editor's Study," *HM* 75 (August 1887):477. See also Clarence L. Dean, "Mr. Howells's Female Characters," *Dial* 3 (October 1882):107; Sidney H. Bremer, "Invalids and Actresses: Howells's Duplex Imagery for American Women," *American Literature* 47 (January 1976):599–614.

28. WDH, "Editor's Study," *HM* (Nov. 1889):966; C. Hartley Grattan, "Howells: Ten Years After," *American Mercury* 20 (1930): 43; WDH to Charles Eliot Norton, 1868; WDH, *Life in Letters*, ed. Mildred Howells, 2 vols. (Garden City, N.Y., 1928), 2:132; WDH, *Years of My Youth* (1916; Bloomington, 1975), 88, 42. See also John E. Bassett, "'A Heart of Ideality in My Realism': Howells's Early Criticism," *Papers on Language and Literature* 25 (Winter 1989):67–82. On the role of women in the "rise of realism," see Alfred Habegger, *Gender, Fantasy, and Realism in American Literature* (New York, 1982), and Elizabeth Prioleau, *The Circle of Eros: Sexuality in the Work of William Dean Howells* (Durham, N.C., 1983).

29. Elizabeth Stoddard, Journal, 22 April 1866, Edmund Clarence Stedman Papers, Butler Library, Columbia University; Mary Abigail Dodge [Gail Hamilton], "My Garden," *AM* 9 (May 1862):541.

30. Henry Seidel Canby, "Sentimental America," *AM* 121 (April 1918):500, 505. Recent discussions of sentimentalism include Mark Jefferson, "What Is Wrong with Sentimentality?," *Mind* 92 (1983):519–29; Mary Midgley, "Brutality and Sentimentality," *Philosophy* 54 (1979):385–89; Shirley Samuels, ed., *The Culture of Sentiment: Race, Gender, and Sentimentality in Nineteenth-Century America* (New York, 1992); Robert C. Solomon, "On Kitsch and Sentimentality," *Journal of Aesthetics and Art Criticism* 49 (Winter 1991):1–14; Michael Tanner, "Sentimentality," *Proceedings of the Aristotelian Society* 77 (1976–77):127–47.

31. Ellen Glasgow, "Feminism," *New York Times Book Review*, 30 November 1913, 656–57. "The sentimentalist," explained William James, "is so constructed that 'gushing' is his or her normal mode of expression. Putting a stopper on the 'gush' will only to a limited extent cause more 'real' activities to take its place; in the main it will simply produce listlessness." WJ, *Principles of Psychology* (New York, 1890), 2:467.

32. WDH, "Editor's Study," *HM* 75 (1887):315; WDH, *April Hopes* (New York, 1888), 174; WDH, *Annie Kilburn* (New York, 1889), 8–9; WDH, *A Hazard of New Fortunes* (New York, 1890), 2:177. For similar themes, see also *The Rise of Silas Lapham* (New York: Norton Critical Edition, 1982), 13, 175, 192.

33. Lorne Fienberg, "Mary E. Wilkins Freeman's 'Soft Diurnal Commotion': Women's Work and Strategies of Containment," *New England Quarterly* 62 (December 1989):483–504; Mary Wilkins Freeman, "Old Women Magoun," in *Short Fiction of Sarah Orne Jewett and Mary Wilkins Freeman*, ed. Barbara Solomon (New York, 1979), 485; Clare Benedict, ed., *Constance Fenimore Woolson* (London, 1930), 99, 108; Louisa May Alcott, *An Old-Fashioned Girl* (New York, 1971), 227–28.

34. See Emily Toth, *Kate Chopin* (New York, 1991).

35. Kate Chopin, *The Awakening* (1899; New York: Signet Classic Edition, 1976), 8, 114. To be sure, Edna's "romantic" suicide has raised questions about the psychological costs of such an "awakening" and has led some critics to question its "realism." See also Paul John Eakin, *The New England Girl: Cultural Ideals in Hawthorne, Stowe, Howells and James* (Athens, Ga., 1976); Habegger, *Gender, Fantasy, and Realism*; Mary Suzanne Schriber, *Gender and the Writer's Imagination from Cooper to Wharton* (Lexington, Ky., 1987); Allen F. Stein, *After the Vows Were Spoken: Marriage in American Literary Realism* (Columbus, Ohio, 1987); William Wasserstrom, *Heiress of All the Ages: Sex and Sentiment in the Genteel Tradition* (Minneapolis, 1959).

36. Amelia Barr, "The Modern Novel," *NAR* 159 (November 1894):598; Helen Gray Cone, "Woman in American Literature," *Century* 40 (October 1890):928; George Parsons Lathrop, "Audacity in Woman Novelists," *NAR* 150 (May 1890):616.

37. S. B. Wister to WDH, 26 April 1880, WDH Papers, Houghton Library, Harvard University. Similarly, a reviewer protested after reading James's *Roderick Hudson* (1876) that the "effect of this perpetual analysis is fatiguing; the book never ceases to interest, but it taxes the attention like metaphysics." Another reader said much the same to Kate Chopin. After expressing his regret that her stories too often ended on a bleak note, an attorney acknowledged that such grim conclusions reflected the "school of Realism," and she was "realistic almost to a fault." [Review of HJ, *Roderick Hudson*], NAR 122 (April 1876):424; Per Seyersted, ed., *A Kate Chopin Miscellany* (Natchitoches, La., 1979), 135–36.

38. Florence Jackson, "More of the Strenuous Life," *Overland Monthly* 41 (Feb-

ruary 1903):156; ["A Strong-Minded Woman"], "The Proper Sphere of Men," *Putnam's*
4 (September 1854):306; HJ, "Anthony Trollope," in *Partial Portraits* (London, 1905),
101. For a discussion of the way in which American commentators presumed that women
were peculiarly disposed to romances and sentimental fiction, see Linda K. Kerber,
Women of the Republic: Intellect and Ideology in Revolutionary America (Chapel Hill,
1980), 235–36.

39. Cheney, ed., *Alcott: Letters and Journals*, 36–38; SOJ, *Deephaven* (1877; Boston, 1880), 66. On Jewett see also Richard Cary, "The Rise, Decline, and Rise of Sarah
Orne Jewett," *Colby Library Quarterly* 9 (1972):650–63; Josephine Donovan, *New
England Local Color Literature: A Woman's Tradition* (New York, 1983); Elaine Show-
alter, A *Literature of Their Own* (Princeton, 1977).

40. *Criticism and Fiction and Other Essays by William D. Howells*, 64.

41. SOJ to Annie Fields, 29 January 1891, SOJ Papers.

42. SOJ, 13 July 1872, manuscript diary, SOJ Papers, Houghton Library, Harvard
University; SOJ, "By the Morning Boat," in *Strangers and Wayfarers* (Boston, 1890):197.

43. Willa Cather, preface to *The Country of the Pointed Firs* (1925; New York,
1956), 7–8; Samuel Chapman Armstrong to SOJ, 10 November 1884; Natalie Clark to
SOJ, February 1900, SOJ Papers. For an insightful analysis of such feminine realism,
see Julia Bader, "The Dissolving Vision: Realism in Jewett, Freeman and Gilman," in
Sundquist, ed., *American Realism: New Essays*, 197–211.

44. WDH, "Novel-Writing," 10; WDH, "My First Visit to New England," *HM* 89
(August 1894):447; SOJ to Laura Bellamy, 31 August 1885, *Jewett Letters*, 52; SOJ,
"Looking Back on Girlhood," *Youth's Companion* 65 (7 January 1892):5.

45. E. F. Andrews, "Fashions in Literature," *Cosmopolitan* 9 (May 1890):126; TSP,
"Science and Conscience," *PSM* 23 (May 1883):14, 15; TSP, "Science and the Imagination," *NAR* 137 (July 1883):49–56. In 1884 a Chicago literary critic also attributed
the new realism in American culture as being "largely traceable to a single cause—the
spread of modern scientific ideas, especially those which are associated with the names
of Darwin and Spencer." See Edward J. Harding, "Cutting Adrift," *Critic* 4 (17 May
1884):229.

46. William Morton Payne, "American Literary Criticism and the Doctrine of Evolution," *Dial* 2 (July 1900):42–43; Per Seyerstad, *Kate Chopin: A Critical Biography*
(Baton Rouge, 1980), 84–85. See also "Fiction and Social Science," *Century* 29
(November 1884):153; J. R. Dennett, "The Decline of the Novel," *Nation* 6 (May 14,
1868):390; Albert Tolman, "Natural Science in a Literary Education," *PSM* 49 (May
1896):99. Useful secondary sources on this theme include Harry H. Clark, "The Role
of Science in the Thought of W. D. Howells," *Proceedings of the Wisconsin Academy
of Science, Arts and Letters* 42 (1953):263–303, and Ronald E. Martin, *American Literature and the Universe of Force* (Durham, N.C., 1981).

47. TSP, "The Progress of Literature," in *Gately's World Progress*, ed. Charles E.
Beale (Boston, 1886), 834; WW, *Democratic Vistas*, in *Walt Whitman: Complete Poetry
and Collected Prose* (New York: Library of America Edition, 1982), 949; WDH, "Editor's
Study," *HM* 77 (July 1888):317; HG, "Walt Whitman," Part X of "The Evolution of
Modern Thought," unpublished manuscript, pp. 4, 20, HG Papers.

48. WDH, "Henrik Ibsen," *NAR* 183 (July 1906):12. See also William Alexander,
William Dean Howells: The Realist as Humanist (New York, 1981).

49. WDH, "Editor's Study," *HM* 72 (April 1886):811; WDH to Thomas W. Higginson, 17 September 1879, in Cady, *The Road to Realism*, 189.

50. SOJ to Frederick M. Hopkins, 22 May 1893, *Jewett Letters* (Waterville, Me.,

1967), 83–84; F. O. Matthiessen, *Sarah Orne Jewett* (Cambridge, Mass., 1929), 51; SOJ, "Preface to the 1893 Edition," in *Deephaven and Other Stories*, ed. Richard Cary (New Haven, 1966), 33; SOJ to Imogen Guiney, n.d., SOJ Papers; Louise Collier Willcox, "The Content of the Modern Novel," *NAR* 182 (June 1906):926.

51. SOJ, *Deephaven*, 29, 206–07.

52. Matthiessen, *Sarah Orne Jewett*, 51; Sally Kaign to SOJ, 10 February 1896; John Chadwick to SOJ, 15 July 1900, SOJ Papers.

53. WDH, *HW* 39 (13 April 1895):342; "Fears Realists Must Wait: An Interesting Talk with William Dean Howells," *New York Times*, 28 October 1894, p. 20; WDH, *Their Wedding Journey* (1873; Bloomington, Ind., 1968), 35. Realistic novels, as Amy Kaplan has recently observed, "construct a vision of a social whole, not just as nostalgia for lost unity or as a report of new social diversity, but as an attempt to mediate and negotiate competing claims to social reality by making alternative realities visible while managing their explosive qualities." Kaplan, *Social Construction of American Realism*, 11. On this theme see also Borus, *Writing Realism*, 139–82; George Levine, *The Realistic Imagination* (Chicago, 1981); J. P. Stern, *On Realism* (London, 1973).

54. WDH, *Their Wedding Journey*, 55.

55. HJ to Robert Louis Stevenson, 31 July 1888, in *Henry James: Selected Letters*, ed. Leon Edel (Cambridge, Mass., 1974), 224; HJ, [review of *The Earthly Paradise*, by William Morris], *NAR* (July 1868):324; HJ, "The Art of Fiction," *Longman's Magazine* 4 (September 1884):510–12. The classic accounts of realism as a literary *form* are Erich Auerbach's *Mimesis: The Representation of Reality in Western Literature*, trans. by Willard Trask (New York, 1957), and Ian Watt's *The Rise of the Novel* (Berkeley, 1965).

56. WDH, [review of SOJ, *Deephaven*], *AM* 39 (June 1877):759; WDH, *Criticism and Fiction*, 39. During an interview with Stephen Crane, Howells maintained that it was "the business of the novel to picture the daily life in the most exact terms possible, with an absolute and clear sense of proportion. That is the important matter—proportion." WDH, "Will the Novel Disappear?," *NAR* 175 (September 1902):294.

57. HJ, *The Art of the Novel: Critical Prefaces by Henry James* (New York, 1934), 31–32.

58. Harriet Beecher Stowe, *Oldtown Folks* (New York, 1869), iv.

59. WW, *Democratic Vistas*, 940; WDH, "Emile Zola," in Clara Marburg Kirk and Rudolf Kirk, eds., *William Dean Howells' Representative Selections* (New York, 1961), 381.

60. HJ, [review of Charles Dickens, *Our Mutual Friend*], *Nation* 1 (21 December 1868):786–87; *Literary Criticism: Essays on Literature, American Writers, English Writers*, ed. Leon Edel (New York, 1984), 64; HJ to Violet Paget (Vernon Lee), 10 May 1885, in *Henry James: Selected Letters*, ed. Leon Edel (Cambridge, Mass., 1974), 1:206; HJ, "The Novels of George Eliot," *AM* 18 (October 1866):488.

61. WW, *Complete Prose Works of Walt Whitman* (Philadelphia, 1892), 206, 244, 231; George P. Lathrop, "The Novel and Its Future," *AM* 34 (September 1874):322, 324. On realism's competition with other media, see Kaplan, *Social Construction of Reality*, 12–13, 26–38; Ralph Bogardus, *Pictures and Texts: Henry James, A. L. Coburn and New Ways of Seeing in Literary Culture* (Ann Arbor, 1984), and "A Literary Realist and the Camera: W. D. Howells and the Uses of Photography," *American Literary Realism* 10 (Summer 1977):231–41; Leon Edel, "Novel and Camera," in *The Theory of the Novel: New Essays*, ed. John Halperin (New York, 1974):177–88.

62. Joseph Kirkland, "Realism Versus Other Isms," *Dial* 14 (16 February 1893):99–101.

63. SOJ to Laura E. Bellamy, 31 August 1885, in *Sarah Orne Jewett Letters*, ed. Richard Cary (Waterville, Me., 1967), 51; Benedict, ed., *Constance Fenimore Woolson*, 65; HG, *Crumbling Idols*, 25. As one editor pointed out, "the American middle classes offer a field for achievement to the writer who can understand the life of the every-day people who figure neither in the Four Hundred [social elite] nor in the police court." See "The Abuse of Realism," *New York Times* Saturday Supplement, 26 November 1898, p. 785.

64. HG, "Let the Sunshine in," unpublished manuscript, p. 7, HG Papers. Everyone recognized in Zola an artist of extraordinary energy and honesty. Thomas S. Perry characterized him as "a realist as no man was ever a realist before, — as no man has ever ventured to be." TSP, "Last Excesses of Realism in Fiction," *AM* 39 (May 1877): 610–12.

65. WW, *Democratic Vistas*, 985; HJ, "George Eliot," *AM* (October 1866):374; HJ, "Anthony Trollope," in *Partial Portraits* (London, 1905), 124. See also Herbert Edwards, "Zola and the American Critics,"*American Literature* 4 (May 1932):114–29; William C. Frierson and Herbert Edwards, "The Impact of French Naturalism on American Critical Opinion, 1877–1892," *PMLA* 63 (September 1948):1007–16.

66. HJ to Violet Paget (Vernon Lee), 10 May 1885, HJ Papers, Colby College Library.

67. RHD, *Waiting for the Verdict* (New York, 1867), 324.

68. Ibid., 20; HJ, "Waiting for the Verdict," *Nation* 5 (21 November 1867): 410–11.

69. Stowe quoted from Gerald Langford, *The Richard Harding Davis Years: A Biography of a Mother and Son* (New York, 1961), 54; Leon Edel, *Henry James: A Life* (New York, 1985), 165; HJ quoted in George Merriam Hyde, "Allotrophy of Realism," *Dial* 18 (16 April 1895):232; HJ, "Rudyard Kipling," in Edel, ed., *Literary Criticism*, 1127.

70. EW to Dr. Morgan Dix, 5 December 1905, in *Letters of Edith Wharton*, 99; EW, "Fiction and Criticism," unpublished manuscript, Box 19, Folder 597, EW Papers, Beinecke Library, Yale University.

71. WDH, "Editor's Study," *HM* 81 (September 1890):639–40; *HM* 81 (July 1890):317; SC, "The Function of the Novelist," *New York Times*, 28 October 1894, p. 20; WDH, "Novel-Reading," 20, 24.

72. WDH, "Recent Literature," *AM* 35 (April 1875):490, 494; *Selected Letters of William Dean Howells*, eds. George Arms and Christoph K. Lohmann (Boston, 1979), 2:214. Many readers who otherwise praised James's artistic talents balked at his analytical aloofness. Sarah Butler Wister, for example, acknowledged that James painted "real and living" characters. But she could not identify with them: "we never for a moment cease to be spectators." *NAR* 122 (April 1876):420–24.

73. WDH, "Editor's Study," *HM* 73 (September 1886):641.

74. Ibid.; "Editor's Study," *HM* 74 (August 1887):355. Howells's conservative faith in the virtues of the Protestant work ethic and rugged individualism led him to emphasize in 1884 that "workingmen are in no bad way among us. They have to practice self-denial and to work hard, but both of these things used to be thought good for human nature. When they will not work, they are as bad as club men and ladies of fashion, and perhaps more dangerous." Alexander, *William Dean Howells*, 28.

75. WDH, "Editor's Study," *HM* 73 (September 1886):641; WDH, "Minor Topics," *Nation* 2 (29 March 1866):389.

76. Horace Traubel, *With Walt Whitman in Camden* (New York, 1906), 4:22–23. When criticized for centering her stories on the affluent, Wharton pointed out that the

"assumption that the people I write about are not 'real' because they are not navvies & char-women, makes me feel rather hopeless. I write about what I see, what I happen to be nearest to, which is surely better than doing cowboys de chic." EW to William Crary Brownell, 25 June 1904, in *The Letters of Edith Wharton*, eds. R.W.B. Lewis and Nancy Lewis (New York, 1988), 91.

77. Hamilton Wright Mabie, "A Typical Novel," *Andover Review* 4 (Nov. 1885):417–29. See also George C. Carrington, Jr., *The Immense Complex Drama: The World and Art of the Howells Novel* (Columbus, 1966).

78. Trauble, *With Walt Whitman*, 127; Mary Austin to HG, 1917, in *Tributes to William Dean Howells on the Occasion of His 80th Birthday*, n.p. in WDH Papers.

79. Theodore Roosevelt, "How I Became a Progressive," *The Outlook* 102 (12 October 1912):295; Roosevelt to Edward J. Wheeler, 28 February 1917, WDH Papers.

Chapter 7. Realism on Canvas

1. "Contributor's Club," *AM* 41 (January 1878):130–32; S.G.W. Benjamin, "Tendencies of Art in America," *American Art Review* 1 (1880):105, 107, and *Art in America* (New York, 1880), 174. For examples of the way in which the terms "realism" and idealism" were used in art circles, see Frank Jewett Mather, Jr., "Realism in Art," *Nation* 102 (3 February 1916):129–30; Mary E. Nealy, "The Real and Ideal in Art," *Appleton's Art Journal* 2 (1876):287; William J. Stillman, "Realism and Idealism," *Nation* 41 (31 December 1885):545. H. Wayne Morgan has written a wonderfully textured survey of art criticism during the era. See *New Muses: Art in American Culture, 1865–1920* (Norman, Okla., 1978).

2. Albert P. Ryder, "Paragraphs from the Studio of a Recluse," *Broadway Magazine* 14 (September 1905):10–11. On the various manifestations of the ideal in art during the late nineteenth century, see [The Brooklyn Museum], *The American Renaissance, 1876–1917* (New York, 1979); Wanda M. Corn, *The Color of Mood: American Tonalism, 1880–1910* (San Francisco, 1972); Lois Fink, "The Innovation of Tradition in Late Nineteenth-Century American Art," *American Art Journal* 10 (November 1978):63–71; H. Wayne Morgan, *Keepers of Culture: The Art-Thought of Kenyon Cox, Royal Cortissoz, and Frank Jewett Mather, Jr.* (Kent, Ohio, 1989); Paul D. Schweizer, "Genteel Taste at the National Academy of Design's Annual Exhibitions, 1891–1910," *American Art Review* (September 1975):77–89.

3. Benjamin, *Art in America*, 210; John Ferguson Weir, "American Art: Its Progress and Prospects," *Princeton Review* 8 (May 1878):822; William C. Brownell, "The Younger Painters of America: Part I," *Scribner's* 20 (May 1880):1–15.

4. "Literature," *Nation* 1 (July 1865):19; Sadakichi Hartmann, "A Plea for the Picturesqueness of New York," in Harry Lawton and George Knox, eds., *The Valiant Knights of Daguerre* (Berkeley, 1978), 57. See also "Fine Arts," *Nation* 1 (July 1865):26; "Art and Artists," *Galaxy* 6 (July 1868):141; "About an American School of Art," *Scribner's* 10 (July 1875):381.

5. Anshutz quoted in Harold Rosenburg, "The Problem of Reality," in *American Civilization*, ed. Daniel Boorstin (New York, 1972), 300.

6. Frank Fowler, "The Field of Art," *Scribner's* 30 (August 1901):256. See also Russell Sturgis, Jr., "The Condition of Art in America," *NAR* 102 (January 1866):21. The question of accepting paintings at face value is discussed in two discerning essays: Svetlana Alpers, "Describe or Narrate? A Problem in Realistic Representation," *New*

Literary History 8 (1976):15–41, and Griselda Pollock, "Revising or Reviving Realism?," *Art History* 7 (September 1984):359–68.

For two decades Michael Fried has been in the forefront of those art historians who find the "blandly normalizing" discourse of conventional historians utterly inadequate. All art, Fried insists with characteristic brio, is primarily a representation of the destabilizing internal pressures operating within a painter's psyche rather than a transcription of an external reality. Viewed in this context, art history itself becomes a form of imaginative writing. And Fried's farfetched readings of particular canvases often appear to be the products of a fabulist. Yet although many of his eccentric and tendentious conclusions seem more self-revelatory than credible, they warrant sustained attention. See *Realism, Writing, Disfiguration: On Thomas Eakins and Stephen Crane* (Chicago, 1987).

7. On the still-life tradition, see Alfred Frankenstein, *"After the Hunt"—and After: William Michael Harnett and Other American Still Life Painters, 1870–1900* (Berkeley, 1969); William H. Gerdts, *Painters of the Humble Truth, Masterpieces of American Still Life, 1801–1939* (Columbia, Mo., 1981).

8. Frank Linstow White, "Art Notes," *Epoch* 8 (12 December 1890):300. Of course, art historians have refused to accept this notion of pictorial literalism. On this theme see Barry Maine, "Late Nineteenth-Century *Trompe L'Oeil* and Other Performances of the Real," *Prospects* 16 (1991):281–95.

9. Frankenstein, "After the Hunt," 7.

10. "Science in Its Relations to Art," *New Path* 2 (November 1865):169, 170; W. C. Brownell, "French Art," *Scribner's* 12 (1892):604–7, 618–19. For a probing discussion of the impact of science on nineteenth-century landscape painting, see Barbara Novak, *Nature and Culture* (New York, 1980).

11. Charles H. Moore, "Materials for Landscape Art in America," *AM* 64 (1889):671, 676. Excellent studies of genre painting include Sarah Burns, *Pastoral Inventions: Rural Life in Nineteenth-Century American Art and Culture* (Philadelphia, 1989), and Patricia Hills, *The Painters' America: Rural and Urban Life, 1810–1910* (New York, 1974).

12. HJ quoted in *The Painter's Eye*, ed. John L. Sweeney (Cambridge, Mass., 1956), 140–43.

13. Earl Shinn, "Fine Arts," *Nation* 26 (14 February 1878):120; Kenyon Cox, "The Art of Winslow Homer," *Scribner's* 56 (September 1914):378. Homer represented what one critic called "the new school that is growing up in this country—a school which seeks it subjects from the life, the thoughts and the feelings which are our own." See "The Veteran in a New Field," *FLIN* 24 (13 July 1867):268.

Those interested in learning more about Homer should consult Nicolai Cikovsky, *Winslow Homer* (New York, 1990); Helen Cooper, *Winslow Homer Watercolors* (New Haven, 1986); James T. Flexner, *The World of Winslow Homer, 1836–1910* (New York, 1975); Albert Ten Eyck Gardner, *Winslow Homer* (New York, 1961); Lloyd Goodrich, *Winslow Homer* (New York, 1944).

14. Ernest Knaufft, "Art at the Columbian Exposition," *Review of Reviews* 7 (June 1893):554; "Pictures at the Artists' Fund Society Exhibition," *New York Evening Post*, 23 November 1865; "The Winslow Homer Pictures," *Art Amateur* 24 (February 1891):32.

15. George W. Sheldon, *Hours with Art and Artists* (New York, 1882), 136–41; Lloyd Goodrich, *Winslow Homer* (New York, 1959), 25.

16. "The National Academy of Design," *Art Amateur* 1 (June 1879):4.

17. G. W. Sheldon, "American Painters—Winslow Homer and F. A. Bridgman," *Art Journal* 40 (1878):227.

18. HJ, "On Some Pictures Lately Exhibited," 90, 93, 94.

19. "New England Factory Life," *HW* 11 (25 July 1868):471–72. For another example of Homer's representation of the underside of American life, see "Station-House Lodgers," *HW* 48 (7 February 1874):132.

20. See William H. Gerdts, "Winslow Homer in Cullercoats," *Yale University Art Gallery Bulletin* 36 (Spring 1977):18–35.

21. See Philip Beam, *Winslow Homer at Prout's Neck* (Boston, 1966); Philip Beam, Lois Homer Graham, Patricia Junker, David Tatham, and John Wilmerding, eds., *Winslow Homer in the 1890s: Prout's Neck Observed* (New York, 1990). Once, a New York art collector ventured to Prout's Neck and encountered a rough-looking native with a fishing pole. The visitor offered him a quarter for directions to Homer's cabin. "Where's your quarter?" snapped the fisherman. Upon pocketing the quarter, he curtly said, "I am Winslow Homer."

22. Frederick W. Morton, "The Art of Winslow Homer," *Brush and Pencil* 10 (April 1902): 40–41. Homer's marine paintings are the focus of Bruce Robertson's *Reckoning with Winslow Homer: His Late Paintings and Their Influence* (Cleveland, 1990).

23. Art critic Christian Brinton asserted in 1911 that it "is always reality that Homer offers us." Brinton, "Winslow Homer," *Scribner's* 49 (January 1911):23.

24. *The Philadelphia Press*, 22 February 1914.

25. Catalogue of the Thomas Eakins exhibition (Philadelphia, 1982), xvi.

26. Gordon Hendricks, *The Life and Works of Thomas Eakins* (New York, 1974), 271.

27. Elizabeth Johns has focused on this theme as the major interpretive factor in Eakins scholarship. See her illuminating study, *Thomas Eakins: The Heroism of Modern Life* (Princeton, 1983).

28. TE to Benjamin Eakins, 16 January 1867, Charles Bregler's Thomas Eakins Collection, Pennsylvania Academy of the Fine Arts; TE to Frances Eakins, 1 April 1869, Archives of American Art, Washington, D.C. On Eakins in Paris, see Gerald Ackerman, "Thomas Eakins and His Parisian Masters Gérôme and Bonnat," *Gazette des Beaux-Arts* 73 (April 1969):235–56, and Elizabeth LaMotte Cates Milroy, "Thomas Eakins' Artistic Training" (Ph.D. diss., University of Pennsylvania, 1986).

29. TE to Benjamin Eakins, 6 March 1868, Bregler Collection.

30. Ibid.

31. TE to Benjamin Eakins, 29 October 1868, Bregler Collection.

32. S.G.W. Benjamin, "Society of American Artists," *American Art Review* 2 (1881):72–73. After alluding to such carping about Eakins's "hard and narrow realism," a New York critic concluded that no one could deny "that his realism is powerful; impressive everything he does certainly is." (Review of *At the Piano*, *New York World*, n.d., in Henry McBride Papers, Beinecke Library, Yale University.)

33. "The Third Reception at the Union League," *Philadelphia Evening Bulletin*, 28 April 1871.

34. Edward Strahan, "Pennsylvania Academy Exhibition," *Art Amateur* 4 (November 1880):115; [Earl Shinn], "Notes," *Nation* 18 (12 March 1874):172.

35. Eakins's other famous surgical painting, *The Agnew Clinic* (1889), reveals the progress in medical knowledge over the previous decade. The physician and attending nurses all wear white surgical gowns.

36. [William Clark], *Philadelphia Evening Telegraph*, 28 April 1876. Michael Fried provides an imaginative revisionist reading of *The Gross Clinic* in *Realism, Writing, Disfiguration: On Thomas Eakins and Stephen Crane* (Chicago, 1987). For a stimulating, if often ridiculous, psychosexual reading of Eakins, see David M. Lubin, *Act of Portrayal: Eakins, Sargent, James* (New Haven, 1985), 27–82.

37. Brownell, "Younger Painters in America, Part I," *Scribner's* 20 (May 1880): 1–15.

38. *New York Daily Tribune*, 8 March 1879; *New York Times*, 8 March 1879. A writer in *The Art Interchange* found it "revolting in the last degree." See *Art Interchange* (19 March 1879):42.

39. Charles Bregler, "Thomas Eakins as a Teacher," *The Arts* (March 1931):383–85; William C. Brownell, "The Art Schools of Philadelphia," *Scribner's* (September 1889):737–50.

40. See Gordon Hendricks, "A May Morning in the Park," *Pennsylvania Museum of Art Bulletin* (Spring 1965):48–64; William I. Homer, "Eakins, Muybridge and the Motion Picture Process," *Art Quarterly* (Summer 1963):194–216.

41. This letter is quoted in full in Sylvan Schendler, *Eakins* (Boston, 1967), 90–92.

42. See Goodrich, *Thomas Eakins*, 2:90–95. Such testimony raises an awkward question: Was Eakins's interest in the nude solely aesthetic or was it in part that of a voyeur? His famous painting, *The Swimming Hole* (1885), supports either possibility. It portrays five of his male students, all naked, diving, swimming, and reclining on a stone jetty. On one level, the finely sculptured and solidly modeled figures testify to Eakins's superb ability to make forms painted on a flat surface seem three-dimensional. On another level, however, its message is more problematic. Eakins inserted himself as a figure in the painting, breaststroking toward the others, gazing at the nude bodies. His intention could have been solely artistic, but neither his role nor his motive is clear.

43. Horace Traubel, ed., *With Walt Whitman in Camden* (Boston, 1906), 3:135. See Cheryl Liebold, "Thomas Eakins in the Badlands," *Archives of American Art Journal* 28 (1988):2–15.

44. Lloyd Goodrich, Interview with Weda Cook Addicks, Lloyd and Edith Havens Goodrich—Thomas Eakins Archive, Philadelphia Museum of Art. Discussions of the relationship between Whitman and Eakins appear in Goodrich, *Thomas Eakins*, 2:28–38; F. O. Matthiessen, *American Renaissance: Art and Expression in the Age of Emerson and Whitman* (New York, 1941), 604–10; Henry B. Rule, "Walt Whitman and Thomas Eakins: Variations on Some Common Themes," *Texas Quarterly* 17 (Winter 1974):7–57. The most recent analysis is William Innes Homer, "New Light on Thomas Eakins and Walt Whitman in Camden," in *Walt Whitman and the Visual Arts*, eds. Geoffrey M. Sill and Roberta K. Tarbell (New Brunswick, N.J., 1992), 85–98.

45. *Walt Whitman: The Correspondence*, ed. by Edwin Haviland Miller (New York, 1961–69), 4:133, 135, 143, 154, 163, 160; Traubel, ed., *With Walt Whitman in Camden*, 1:39.

46. See Johns, *Thomas Eakins*, 115–70.

47. "John Sargent, Painter," *Nation* 77 (26 November 1903):428. A Sargent portrait, announced another critic, was a "perfect piece of characterization. Realism in the best sense of the word. But there is more than characterization and realism,—there is decoration. Every object,—hat, cravat, coat, riding whip, boots, columns, and dado in the background, makes a pattern that forms a decoration." Ernest Knaufft, "American Painting To-day," *Review of Reviews* 36 (December 1908):690.

48. Abbey quoted in Elizabeth McCausland, *George Inness and Thomas Eakins*

(New York, 1940), 11. As a Philadelphia art critic recognized, few Americans could appreciate Eakins's "unflattering fidelity." His "severely literal works" would "excite the cordial admiration" only of "those who are capable of appreciating genuine artistic realism." *Philadelphia Telegraph*, 10 December 1877.

49. Traubel, ed., *With Walt Whitman in Camden*, 1:284.

50. The standard treatment of French Impressionism remains John Rewald, *History of Impressionism*, 4th rev. ed. (New York, 1973). Also of considerable value is Charles S. Moffitt, *The New Painting: Impressionism, 1874–1886* (Geneva, 1986). Two more recent works highlight the ideological implications and social context of Impressionism: T. J. Clark, *The Painting of Modern Life: Paris in the Art of Manet and His Followers* (Princeton, 1985), and Robert L. Herbert, *Impressionism: Art, Leisure, and Parisian Society* (New Haven, 1989).

51. Nancy Hale, *Mary Cassatt* (New York, 1975), 72–73; Gemma Newman, "The Greatness of Mary Cassatt," *American Artist* 30 (February 1966):45.

52. HJ, "Parisian Festivity," *New York Tribune*, 13 May 1876.

53. See Meyer Schapiro, "The Nature of Abstract Art," *Marxist Quarterly* (January–March 1937):83; Arnold Hauser, *The Social History of Art*, 4 vols. (New York, 1958), 4:166–225.

54. "Exhibition of the Plein-Airists," *Nation* 42 (15 April 1886):328; Henry G. Stephens, "Impressionism: The Nineteenth Century's Distinctive Contribution to Art," *Brush and Pencil* 11 (January 1903):279–80. General surveys of impressionism in the United States include Richard J. Boyle, *American Impressionism* (Boston, 1974); Wanda Corn, *The Color of Mood: American Tonalism, 1880–1910* (San Francisco, 1972); Moussa M. Domit, *American Impressionist Painting* (Washington, D.C., 1973); William Gerdts, *American Impressionism* (New York, 1984); Donelson F. Hoopes, *The American Impressionists* (New York, 1972). An informative article is Hans Huth, "Impressionism Comes to America," *Gazette des Beaux-Arts* 29 (April 1946):225–52.

55. Childe Hassam to Miss Farmer, 22 February 1933, Childe Hassam Papers, Archives of American Art. For background on Hassam, see Donelson F. Hoopes, *Childe Hassam* (New York, 1979).

56. See Jennifer A. Martin Bienenstock, "Childe Hassam's Early Boston Cityscapes," *Arts Magazine* 55 (November 1980):168.

57. There is an obvious resemblance between Hassam's *Rainy Day, Boston* and Gustave Caillebotte's *Rue de Paris, temps de pluie* (1877), but it is unlikely that Hassam had an opportunity to see the French painting.

58. WDH, *A Hazard of New Fortunes* (1890; New York: New American Library, 1983), 217.

59. Adams, *Childe Hassam*, 52.

60. A. E. Ives, "Mr. Childe Hassam on Painting Street Scenes," *Art Amateur* 27 (1892): 116–17.

61. "New York the Beauty City," *New York Sun*, 23 February 1913; Ives, "Hassam on Painting Street Scenes," 116.

62. Clark Eliot, "Childe Hassam," *Art in America* 8 (June 1920):172–80.

63. "Hassam Speaks for Our Art," *New York Times*, 14 October 1934; Frederick W. Morton, "Childe Hassam, Impressionist," *Brush and Pencil* 8 (June 1901):147. A Chicago art critic concluded that Hassam "has transformed for us the most commonplace of daily scenes and taught us to find beauty in many a jagged vista." See "Water Colors at the Institute," *Chicago Tribune*, 24 March 1891.

64. *The Art Amateur* 16 (December 1886):7.

Chapter 8. Form Follows Function

1. Henry Van Brunt, "Henry Hobson Richardson, Architect," *AM* 58 (November 1886):685. For general surveys of American architecture, see John Burchard and Albert Bush-Brown, *The Architecture of America: A Social and Cultural History* (Boston, 1961); James Fitch, *American Building: The Historical Forces That Shaped It* (Boston, 1966); William Jordy and William Pierson, *American Buildings and Their Architects* (New York, 1970); Thomas E. Talmadge, *The Story of Architecture in America* (New York, 1927).

2. James J. Jarves, *The Art-Idea: Sculpture, Painting and Architecture in America* (New York, 1864), 286–90.

3. The World's Fair receives extended treatment in Reid Badger, *The Great American Fair: The World's Columbian Exposition & American Culture* (Chicago, 1979); David F. Burg, *Chicago's White City of 1893* (Lexington, Ky., 1976). An innovative treatment of the cultural context within which the fair developed can be found in James Gilbert, *Perfect Cities: Chicago's Utopias of 1893* (Chicago, 1991).

4. Charles Moore, *Daniel H. Burnham, Architect, Planner of Cities* (Boston, 1927), 2:147; Montgomery Schuyler, "Last Words About the World's Fair," *Architectural Record* 3 (January–March 1894):291–301. On Burnham see Thomas S. Hines, *Burnham of Chicago: Architect and Planner* (New York, 1974).

5. Henry Van Brunt, "Cast Iron in Decorative Architecture," *Crayon* 6 (January 1859):15–20. "The great problem before the American architect," declared Barr Ferree, "is to mold architectural ideas and forms to the varied conditions of our national life and situation." Ferree, "An American Style of Architecture," *Architectural Digest* 1 (July 1891):45.

6. William P. Longfellow, "The Architect's Point of View," *Scribner's* 9 (January 1891):120, 128; H. Allen Brooks, ed., *Prairie School Architecture: Studies from "The Western Architect"* (Toronto, 1975), 85.

7. Schuyler, "Last Words About the World's Fair," 301.

8. Henry B. Fuller, *With the Procession: A Novel* (New York, 1895), 248. See also Carl S. Smith, *Chicago and the American Literary Imagination, 1880–1920* (Chicago, 1984).

9. Peter B. Wight, "On the Present Condition of Architectural Art in the Western States," *American Art Review* 1 (1880):138. On the "Chicago School," see Siegfried Giedion, *Space, Time, and Architecture* (New York, 1941); Carl W. Condit, *The Chicago School of Architecture* (Chicago, 1964); Wichit Charernbhak, *Chicago School Architects and Their Critics* (Ann Arbor, 1981); Daniel Bluestone, *Constructing Chicago* (New Haven, 1991).

10. On Richardson, see Mariana Griswold Schuyler Van Rensselaer, *Henry Hobson Richardson and His Works* (Boston, 1888); Henry Russell Hitchcock, *The Architecture of H. H. Richardson and His Times* (New York, 1936). The best recent account is James F. O'Gorman, *H. H. Richardson: Architectural Forms for an American Society* (Chicago, 1987).

11. James F. O'Gorman, *Three American Architects: Richardson, Sullivan, and Wright, 1865–1915* (Chicago, 1991), 49–50.

12. LS, *Kindergarten Chats and Other Writings* (New York, 1947), 28–30.

13. Frederick Baumann, "Thoughts on Style," *Inland Architect* 20 (November 1892):36; Allen Pond, "The Evolution of An American Style," *Inland Architect* 10 (January 1888):98.

14. Editorial, *Inland Architect* 16 (January 1891).

15. Peter B. Wight, "John W. Root as a Draftsman," *Inland Architect* 16 (January 1891):88.

16. John W. Root, "A Great Architectural Problem," *Inland Architect* 15 (June 1890):67–68; Root, "Architectural Freedom," *Inland Architect* 8 (December 1886):64–65.

17. John W. Root, "Broad Art Criticism," *Inland Architect* 11 (February 1888):3; Harriet Monroe, *John Wellborn Root: A Study of His Life and Work* (Boston, 1896), 51, 62.

18. John W. Root, "Architectural Ornamentation," *Inland Architect* 5 (April 1885):54–55; Monroe, *John Wellborn Root*, 188, 65.

19. John W. Root, "Architectural Ornamentation," 55; Root, "Style," *Inland Architect* 8 (January 1887):99–101; Root, "Broad Art Criticism," *Inland Architect* 11 (February 1888):5; Root, "A Great Architectural Problem," 69–71.

20. Peter Brooks to Owen Aldis, 25 March 1881, Aldis & Company Archives, Chicago, Illinois.

21. Peter Brooks to Owen Aldis, 15 April, 6 May 1884, Aldis & Company Archives; John W. Root, "The City House in the West," *Scribner's* 8 (October 1890):430.

22. Barr Ferree, "The High Building and Its Art," *Scribner's* 15 (March 1899):312; Robert D. Andrews, "The Broadest Use of Precedent," *Architectural Review* 2 (May 15, 1893):34–35; Harriet Monroe, *A Poet's Life: Seventy Years in a Changing World* (New York, 1938), 114. The Monadnock Block impressed Louis Sullivan as "an amazing cliff of brickwork, rising sheer and stark with the subtlety of line and surface, a direct singleness of purpose, that gave one the thrill of romance." LS, *The Autobiography of an Idea* (New York, 1926), 309. See Also Donald Hoffmann, "The Monadnock Building," *Journal of the Society of Architectural Historians* 26 (1967):269–77.

23. Fiske Kimball, *Architectural Record* 57 (April 1925):289; LS, "Reply to Mr. Frederick Stymetz Lamb," *The Craftsman* 8 (June 1905):336; LS, "Emotional and Intellectual Architecture," *Inland Architect* 24 (November 1894):33; LS, *The Autobiography of an Idea* (1934; New York, 1956), 275, 257, 258. For a survey of Sullivan's constrasting historical reputations, see Rochelle Berger Elstein, "Enigma of Modern Architecture: An Introduction to the Critics," in *Louis Sullivan: The Function of Ornament*, ed. Wim de Wit (New York, 1986), 199–211.

24. LS, *The Autobiography of an Idea*, 85. Among the vast literature on Sullivan, the following were especially useful: Albert Bush-Brown, *Louis Sullivan* (New York, 1960); Willard Connelly, *Louis Sullivan as He Lived* (New York, 1960); Narciso Menocal, *Architecture as Nature: The Transcendentalist Idea of Louis Sullivan* (Madison, 1982); Hugh Morrison, *Louis Sullivan: Prophet of Modern Architecture* (New York, 1935); Sherman Paul, *Louis Sullivan: An Architect in American Thought* (Englewood Cliffs, N.J., 1962); Robert Twombly, *Louis Sullivan: His Life and Work* (New York, 1986).

25. LS, *The Autobiography of an Idea* (New York, 1926), 199.

26. Ibid., 200.

27. Ibid., 213–16, 254–55.

28. LS, *Kindergarten Chats*, 140; LS, "What is Architecture," *The Craftsman* 10 (May 1906):145–49.

29. LS to WW, 3 February 1887, American Institute of Architects Archives, Washington, D.C.; LS, "Characteristics and Tendencies of American Architecture," *The Inland Architect* 6 (1885):58.

30. FLW, *An Autobiography* (New York, 1932), 95.

31. LS, *Kindergarten Chats*, 49. One story has it that John Root, upon examining some of Sullivan's early sketches for the auditorium, lamented that the young genius was going to "smear another facade with ornament." An indignant Sullivan thereupon stripped the exterior of all decoration. See Morrison, *Louis Sullivan*, 88.

32. LS, "Reply to Mr. Frederick Lamb," *Craftsman* 8 (1905):337; LS, *Kindergarten Chats*, Chat 15.

33. Morrison, *Louis Sullivan*, 77.

34. LS, *Kindergarten Chats*, 199, 134.

35. Ibid., 206; LS, "Characteristics and Tendencies of American Architecture," 58–62. Robert Twombly has provided the most substantial argument about Sullivan's supposed homosexuality. Twombly, *Louis Sullivan*, 399–400. See also Narciso Menocal, *Architecture as Nature: The Transcendentalist Idea of Louis Sullivan* (Madison, 1982).

36. LS, "Form and Function Artistically Considered," *Craftsman* 8 (1905):456.

37. Paul Goldberger, *The Skyscraper* (New York, 1981), 19; Bragdon, "Letters from Louis Sullivan," 9.

38. LS, "What Is the Just Subordination, in Architectural Design, of Details to Mass?," *Inland Architect* 9 (April 1887):51–54; Twombly, *Louis Sullivan*, 401.

39. Paul, *Louis Sullivan*, 32; "Reality in Architecture," *Chicago Tribune*, August 1900, reprinted in *Inland Architect* 36 (September 1900):116; LS, "Form and Function Artistically Considered," 458.

40. LS, "Emotional Architecture as Compared with Classical," in *Kindergarten Chats*, 191–201; Lewis Mumford, *The Brown Decades: A Study of the Arts in America, 1865–1895* (New York, 1931), 69–70. For insights into what it was like to work in the early skyscrapers, see Lisa M. Fine, *The Souls of the Skyscraper: Female Clerical Workers in Chicago, 1870–1930* (Philadelphia, 1990). David Andrews has provided an overheated indictment of the disparity between Sullivan's rhetoric and his practice in *Louis Sullivan and the Polemics of Modern Architecture* (Urbana, Ill., 1985).

41. FLW, *Genius and the Mobocracy* (New York, 1949), 72; FLW, *Autobiography*, 266–80; [Van H.] Higgins and [Henry J.] Furber, *The Columbus Memorial Building* (Chicago, 1893). So many workers were injured and killed building the auditorium that the contractor took the unprecedented step of taking out $2000 life insurance policies on the laborers to assuage their mortal fears.

42. Henry B. Fuller, *The Cliff-Dwellers* (New York, 1893), 4, 95–96, 97. Several architects and critics felt the same way. "First, last, and all the time," Barr Ferree asserted, the skyscraper "is a commercial building, erected under commercial impulses, answering to commercial needs, and fulfilling a commercial purpose in supplying its owner with a definite income." By 1903 Montgomery Schuyler was forced to admit that skyscrapers had created more civic problems—congestion, bureaucratization, alienation—than they had solved: "Like Frankenstein, we stand appalled before the monster of our own creation." See Ferree, "The High Building and Its Art," 303; LS, *Autobiography of an Idea*, 289, 291; Schuyler quoted in Alan Trachtenberg, *Reading American Photographs: Images as History* (New York, 1989), 212.

43. LS, *Autobiography of an Idea*, 289–91.

44. *People's Bank: The First Seventy-Five Years* (Cedar Rapids, Iowa, 1978), 18–19.

45. LS, "Lighting the People's Savings Bank," *Illuminating Engineer* 6 (February 1912):631–35; LS, "The Modern Phase of Architecture," *Inland Architect* 33 (June 1899):40. See also Andrew Rebori, "An Architecture of Democracy: Three Recent Examples from the Work of Louis H. Sullivan," *Architectural Record* 39 (1916):437–65; Lauren S. Weingarden, *Louis Sullivan: The Banks* (Cambridge, Mass., 1987); Wim De

Wit, "The Banks and the Image of Progressive Banking," in Wit, ed., *Louis Sullivan: The Function of Ornament*, 159–97.

46. Meryle Secrest, *Frank Lloyd Wright* (New York, 1992), 77.

47. FLW, "In the Cause of Architecture," *Architectural Record* 23 (1908):155.

48. FLW, *The Future of Architecture* (New York, 1953), 228; FLW, "The Architect," *The Brickbuilder* 9 (June 1900):126; FLW, *Autobiography*, 139; FLW, *The Natural House* (New York, 1954), 14–15; FLW, *Autobiography*, 136. Wright has benefited from several excellent scholarly studies. See especially H. Allen Brooks, *The Prairie School: Frank Lloyd Wright and His Midwest Contemporaries* (Toronto, 1972); Henry-Russell Hitchcock, *In the Nature of Materials: The Buildings of Frank Lloyd Wright, 1887–1941* (New York, 1942); Robert Twombly, *Frank Lloyd Wright: His Life and His Architecture* (New York, 1979). For a thorough account of the efforts to reform home design in Chicago, see Gwendolyn Wright, *Moralism and the Model Home: Domestic Architecture and Cultural Conflict in Chicago, 1873–1913* (Chicago, 1980). Two important surveys of residential design in the United States are Clifford E. Clark, Jr., *The American Family Home, 1800–1960* (Chapel Hill, 1986), and David Handlin, *The American Home: Architecture and Society, 1815–1915* (Boston, 1979).

49. FLW, *An Organic Architecture* (Cambridge, Mass., 1939), 14.

50. Ibid., 40; FLW, *The Natural House*, 31.

51. FLW, *The Natural House*, 16.

52. FLW's two house plans were "A Home in a Prairie Town," *Ladies' Home Journal* 18 (February 1901):17, and "A Small House with 'Lots of room in It,'" *Ladies' Home Journal* 18 (July 1901):15. For information about Bok, see David E. Shi, "Edward Bok and the Simple Life," *American Heritage* 36 (December 1984):100–109.

53. Edward Bok, "In an Editorial Way," *Ladies' Home Journal* 24 (May 1907):6; Bok, *The Americanization of Edward Bok* (New York, 1921), 238, 243, 250; FLW, "In the Cause of Architecture," *Architectural Record* 23 (March 1908):12; FLW, *Autobiography*, 163.

54. Edward Bok, "Is It Worth While?," *Ladies' Home Journal* 17 (November 1900):18; Edward P. Pressey, "Village Industries: Home Making," *Country Time and Tide* 1 (June 1902):209. General studies of the transatlantic Arts and Crafts movement include Isabelle Anscombe and Charlotte Gere, *Arts and Crafts in Britain and America* (London, 1967); Robert Judson Clark, ed., *The Arts and Crafts in America, 1876–1916* (Princeton, 1972); Peter Davey, *Architecture of the Arts and Crafts Movement* (New York, 1980); Lionel Lambourne, *Utopian Craftsmen: The Arts and Crafts Movement from the Cotswolds to Chicago* (Salt Lake City, 1980). The best treatment of the peculiar quest for reality expressed by the Arts and Crafts movement is T. J. Jackson Lears, *No Place of Grace: Antimodernism and the Transformation of American Culture, 1880–1920* (New York, 1981), 59–96.

55. See Mary Ann Smith, *Gustav Stickley, the Craftsman* (Syracuse, N.Y., 1983).

56. LS, "The Architectural Discussion: Form and Function Artistically Considered," *The Craftsman* 8 (July 1905):453; Gustav Stickley, "The Craftsman Idea," in *The Best of Craftsman Homes*, ed. Barry Sanders (Salt Lake City, 1979), 234–45; Stickley, "A Word About Craftsman Architecture," 6–9, and "The Living Room," 167, in ibid.

57. Charles Keeler, *The Simple Home* (1904; Salt Lake City, 1979), xlv, 54; Irving Gill, "The Home of the Future: The New Architecture of the West," *Craftsman* 30 (May 1916): 140–51. See also Leslie Freudenheim and Elisabeth Sussman, *Building with Nature: Roots of the San Francisco Bay Region Tradition* (Salt Lake City, 1974).

58. "California's Contribution to a National Architecture: Its Significance and

Beauty as Shown in the Work of Greene and Greene, Architects," *The Craftsman* 22 (August 1912):536; Randell L. Makinson, *Greene & Greene: Architecture as Fine Art* (Salt Lake City, 1977), 160.

59. Keeler, *Simple Home*, 4, 54.

60. Wright, *Moralism and the Model Home*, 131; Robert J. Spencer, Jr., "Brick Architecture in and about Chicago," *The Brickbuilder* 12 (November 1903):222; Herbert Croly, "The Contemporary Suburban Residence," *Architectural Record* 11 (January 1902):79.

Chapter 9. Realism and the Social Question

1. Henry Cabot Lodge, *Congressional Record*, 54 Cong., 1 Sess., 2819. On the nativist phenomenon, see John Higham, *Strangers in the Land: Patterns of American Nativism, 1860–1925* (New York, 1973). Peter Conn examines the theme of clashing social visions in *The Divided Mind: Ideology and Imagination in America, 1898–1917* (Cambridge, 1984).

2. Austin Phelps, Introduction to Josiah Strong, *Our Country: Its Possible Future and Present Crisis* (New York, 1885); WDH, *Letters of an Altrurian Traveler* (1894; Gainesville, Fla., 1961), 16; WDH, "Editor's Easy Chair," *HM* 104 (January 1902):334. For a concise overview of the major cultural tendencies of the era, see John Higham, "The Reorientation of American Culture in the 1890s," in *Writing American History: Essays on Modern Scholarship* (Bloomington, Ind., 1970), 73–102.

3. "The People's Party Platform" *National Economist* 6 (1892):396. Two fine treatments of social and economic developments during the Gilded Age are Paul Boyer, *Urban Masses and Moral Order in America, 1820–1920* (Cambridge, Mass., 1978), and Nell Painter, *Standing at Armageddon: The United States, 1877–1919* (New York, 1987). On the farm problem and its "conspiratorial" elements, see Richard Hofstadter, *The Age of Reform* (New York, 1955), chapters 1–3.

4. The working class receives detailed treatment in Herbert G. Gutman, *Work, Culture and Society in Industrializing America* (New York, 1976); Daniel Nelson, *Managers and Workers: Origins of the New Factory System in the United States, 1880–1920* (1975); Daniel Rodgers, *The Work Ethic in Industrial America* (Chicago, 1978).

5. Josiah Strong, *Our Country: Its Possible Future and Its Present Crisis* (New York, 1885), 129. In his 1904 analysis entitled *Poverty*, Robert Hunter estimated that 10 million of the nation's 82 million residents lived in destitution. See also Robert Bremner, *From the Depths: The Discovery of Poverty in the United States* (New York, 1956).

6. Child quoted in Stuart Blumin, *The Emergence of the Middle Class: Social Experience in the American City, 1760–1900* (Cambridge, 1989), 288. In 1869 the *Nation* reported "the rapidly growing separation between extremes of poverty and wealth" which "every political party and school of economists takes notice of. . . ." See "The Rich Richer—The Poor Poorer," *Nation* 8 (15 April 1869):290. Also of relevance is "Certain Dangerous Tendencies in American Life," *AM* 42 (October 1878):393.

7. TD, *The Titan* (New York, 1914), 186.

8. Ida Tarbell, *All in the Day's Work: An Autobiography* (New York, 1939), 82.

9. Wendell quoted in Van Wyck Brooks, *New England: Indian Summer* (New York, 1940), 409.

10. WJ to WDH, 13 November 1907, WDH Papers. The "realistic" qualities of

utopian novels are discussed in Stanley Cooperman, "Utopian Realism: The Futurist Novels of Bellamy and Howells," *College English* 24 (March 1963):464–67; Ellene Ransom, "Utopus Discovers America; or Critical Realism in American Utopian Fiction, 1798–1900" (Ph.D. diss., Vanderbilt University, 1946); Charles Rooney, "Utopian Literature as a Reflection of Social Forces in America, 1865–1917" (Ph.D. diss., George Washington University, 1968).

11. N. P. Gilman quoted in Laurence Gronlund, "Nationalism," *Arena* 1 (January 1890):153.

12. Edward Bellamy, *Looking Backward, 2000–1887* (1887; Boston: Riverside Edition, 1926), 196, 61.

13. Edward Bellamy, "How I Came to Write Looking Backward," *Nationalist* 1 (May 1889):1–4; Bellamy, Postscript, *Looking Backward*, 336.

14. Walter Blackburn Harte, "Books of the Day," *Arena* 10 (August 1894):vi–vii.

15. Albion Small, "The Meaning of the 'Social Movement,'" *Review of Reviews* 17 (January 1898):92–94.

16. Gertrude Himmelfarb, *Poverty and Compassion: The Moral Imagination of the Late Victorians* (New York, 1991), 5–6.

17. Starr quoted in Anne Firor Scott, "Jane Addams," in Edward T. James et al., eds., *Notable American Women* (1971), 1:17. See also Mina Carson, *Settlement Folk: Social Thought and the American Settlement Movement, 1885–1930* (Chicago, 1990).

18. Leila Bedell, "A Chicago Toynbee Hall," *Women's Journal* 20 (May 25, 1889).

19. Jane Addams, *Twenty Years at Hull House* (New York, 1910), 73, 64. The plague of neurasthenia among women is analyzed in Graham Barker-Benfield, *The Horrors of the Half-Known Life: Male Attitudes Toward Women and Sexuality in Nineteenth-Century America* (New York, 1976); Ann Douglas, "The Fashionable Diseases: Women's Complaints and Their Treatment in Nineteenth-Century America," *Journal of Interdisciplinary History* 4 (Summer 1973):25–52; Carroll Smith-Rosenberg, *Disorderly Conduct: Visions of Gender in Victorian America* (New York, 1985).

20. Allen F. Davis, *American Heroine: The Life and Legend of Jane Addams* (New York, 1973), 65; Jane Addams, "A New Impulse to an Old Gospel," *Forum* 14 (1892):348, 356; Hutchins Hapgood, "Seen from the Outside: An Aspect of Hull House," *New York Evening Post*, 17 March 1905.

21. JR, *Children of the Poor* (New York, 1892), v.

22. JR, *The Making of an American* (1901; New York, 1937), 43. On Riis see Peter B. Hales, *Silver Cities: The Photography of American Urbanization, 1839–1915* (Philadelphia, 1984), 161–218; Carol Schloss, *In Visible Light: Photography and the American Writer: 1840–1940* (New York, 1987), 118–29; Ferenc M. Szasz and Ralph Bogardus, "The Camera and the American Social Conscience: The Documentary Photography of Jacob Riis," *New York History* 60 (October 1974):409–36. The best available biography is James B. Lane, *Jacob A. Riis and the American City* (Port Washington, N.Y., 1974).

23. JR, *Making of an American*, 132, 62; JR, "The Battle with the Slum," AM 83 (1899):626.

24. JR, *The Making of an American*, 267; *New York Sun*, 12 February 1888.

25. *New Haven Palladium*, 20 July 1891; "How the Other Half Lives," *Coup d'Etat* (April 1893), in Scrapbook, p. 81, Box 10, JR Collection, Library of Congress; Riis quoted in *New York Morning Journal*, 12 December 1888.

26. JR, "How the Other Half Lives: Studies Among the Tenements," *Scribner's* 6 (December 1889):643–62.

27. "New Books," *Dial* 11 (April 1891):364. Riis shared many of the ethnic prejudices of the day. He repeatedly used derogatory language and ugly stereotypes to describe Jews, Irish, Italians, Chinese, and African Americans.

28. JR, *How the Other Half Lives: Studies Among the Tenements of New York* (1890; New York, 1937), 22, 282, 3; Francesco Cordasco, ed., *Jacob Riis Revisited* (Garden City, N.Y., 1968), 4; JR, *Theodore Roosevelt: The Citizen* (New York, 1904), 131.

29. JR, *The Making of an American*, 267–68, 176.

30. JR, *How the Other Half Lives*, 282, 265; James Russell Lowell to JR, 21 November 1890, JR Collection, Library of Congress; JR, *Making of an American*, 234; Ernest Poole, *The Bridge; My Own Story* (New York, 1940), 66.

31. Newell Dwight Hillis, "Poets and Essayists as Prophets of a New Era," *Critic* 35 (September 1899):828.

32. Fred Lewis Pattee, "Constance Fenimore Woolson and the South," *South Atlantic Quarterly* (April 1939):132; Charles W. Chesnutt, Journals and Notebook, 25 June 1880, p. 161, 198, Charles Chesnutt Papers, Fisk University, Nashville, Tennessee.

33. Cahan quoted in Hutchins Hapgood, "A Live Bunch," *New York Globe*, undated article in Hutchins Hapgood Papers, Beinecke Library, Yale University.

34. H. H. Boyeson, "Why We Have No Great Novelists," *Forum* 2 (December 1886): 617–18; Boyeson, *Literary and Social Silhouettes* (New York, 1894), 46.

35. WDH, "Editor's Study," *HM* 73 (September 1886):642; WDH to TSP, 14 April 1888, in *The Life in Letters of William Dean Howells*, ed. Mildred Howells (Garden City, N.Y., 1928), 1:413; WDH to HG, 15 January 1888, ibid., 1:407. On Howells and the Haymarket Affair, see Walter Fuller Taylor, "On the Origin of Howells' Interest in Economic Reform," *American Literature* 2 (1931):3–14.

36. WDH, "Editor's Study," *HM* 73 (September 1886):641; WDH, "Editor's Study," *HM* 77 (June 1888):154; WDH, "Editor's Study," *HM* 79 (August 1889):480.

37. WDH, *My Literary Passions* (New York, 1895), 257; WDH, "Editor's Study," *HM* 78 (December 1888):159–60.

38. WDH to Edward Everett Hale, 1888, *Life in Letters*, 1:418–19; WDH to HJ, 10 October 1888, ibid., 1:417. A few days later Howells confessed to his father that he and his friend Twain were "theoretical socialists, and practical aristocrats. But it is a comfort to be right theoretically, and to be ashamed of one's self practically." WDH to William Cooper Howells, ibid., 2:1.

39. WDH, "Editor's Study," *HM* 78 (January 1889):319; WDH, *Impressions and Experiences* (New York, 1896), 163, 45; WDH, "Editor's Study," *HM* 83 (November 1891):965.

40. WDH, "Novel-Writing and Novel-Reading: An Impersonal Explanation," in *Howells and James: A Double Billing*, ed. William M. Gibson (New York, 1958), 20. "The object," as one of the organizers explains, "is to show them that the best people in the community have their interests at heart, and wish to get on common ground with them." WDH, *Annie Kilburn*, in WDH, *Novels 1886–1888* (1888; New York: Library of America), 672.

41. WDH, *Annie Kilburn*, Ibid., 804–5.

42. *Boston Evening Transcript*, 27 December 1888, 6; *The Standard*, 3 November 1888, 8; WDH to Edward Everett Hale, August 1888, *Life in Letters*, 1:416, 419.

43. WDH to Edward E. Hale, 5 April 1889, *Life in Letters*, 1:424–25.

44. WDH, *A Hazard of New Fortunes* (1890; Bloomington, 1976), 90.

45. Ibid., 76.

46. Ibid., 27, 25, 55, 296–97.

47. Ibid., 69, 147.

48. Ibid., 306, 437.

49. Ibid., 486, 412, 436.

50. Mark Twain to WDH, 11 February 1890, WDH Papers, Houghton Library, Harvard University; WJ to WDH, *Letters of William James*, ed. Henry James (Boston, 1920), 1:298.

51. Mary Hill, *Charlotte Perkins Gilman: The Making of a Radical Feminist* (Philadelphia, 1980), 176.

52. Amy Kaplan, *The Social Construction of American Realism* (Chicago, 1988), 63.

53. WDH, *Hazard of New Fortunes*, 183.

54. WDH, *An Imperative Duty*, ed. Martha Banta (Bloomington, 1970), 58.

55. *The Critic* (16 January 1892):34; Julia Cooper, *A Voice from the South* (Xenia, Ohio, 1892), 201–9. On the depiction of blacks in fiction, see Benjamin Brawley, "The Negro in American Fiction," *Dial* 60 (May 11, 1916):445–49; J. Saunders Redding, "American Negro Literature," *American Scholar* 18 (Spring 1949):137–48.

56. B. O. Flower, *The New Time: A Plea for the Union of the Moral Forces for Practical Progress* (Boston, 1894), 152.

57. B. O. Flower, *Civilization's Inferno, Or, Studies in the Social Cellar* (Boston, 1893), 152; Flower, *Persons, Places and Ideas: Miscellaneous Essays* (Boston, 1894), 20; "Editorial Notes," *Arena* 3 (April 1891):632.

58. "Prospectus of 'The Arena,'" *Arena* 1 (January 1890):i.

59. Flower, *Civilization's Inferno*, 13–14. On Flower, see H. F. Cline, "Benjamin Orange Flower and the *Arena*, 1889–1909" and "Flower and the *Arena*: Purpose and Content," in *Journalism Quarterly* 17 (June and September 1940):139–50, 247–57; David H. Dickason, "Benjamin Orange Flower, Patron of the Realists," *American Literature* 14 (May 1942):148–56; Roy P. Fairfield, "Benjamin Orange Flower: Father of the Muckrakers," *American Literature* 22 (November 1950):272–82; Allen J. Matusow, "The Mind of B. O. Flower," *New England Quarterly* 34 (December 1961):492–509.

60. B. O. Flower, "The Present," *Arena* 1 (January 1890):242; Flower, *The New Time: A Plea for the Union of the Moral Forces for Practical Progress* (Boston, 1894), 45–46; "Deplorable Social Conditions," *Arena* 3 (February 1891):378; "Editorial Notes," *Arena* 3 (February 1891):384.

61. B. O. Flower, "Cancer Spots in Metropolitan Life," *Arena* 4 (November 1891):762; Flower, *Civilization's Inferno*, 47.

62. "Editorial Notes," *Arena* 4 (June 1891):127; Flower, "Editorial Notes," *Arena* 4 (July 1891):247; Clarence Darrow, "Realism in Literature and Art," *Arena* 9 (1894):98–113.

63. B. O. Flower, "Books in Review," *Arena* 5 (February 1892):xxix; HG, *A Son of the Middle Border* (1917; New York, 1962), 73. On Garland, see Jean Holloway, *Hamlin Garland: A Biography* (Austin, Texas, 1960); Donald Pizer, *Hamlin Garland's Early Work and Career* (Berkeley, 1960).

64. HG, *Son of the Middle Border*, 210, 186.

65. HG, *Roadside Meetings* (New York, 1930), 7; HG, *Son of the Middle Border*, 321. Howells, who became Garland's hero and mentor, recalled their first meeting: "He was as poor as he was young, but he was so rich in purpose of high economic and social import that he did not know that he was poor." WDH, "Mr. Garland's Books," *NAR* 196 (October 1912):523–28.

66. HG, "Walt Whitman" (unpublished lecture, n.d.), 20, 15, in HG Papers, Do-

heny Library, University of Southern California; Horace Traubel, *With Walt Whitman in Camden* (New York, 1906), 437.

67. HG, *Main-Travelled Roads* (1891; New York, 1962), ix–x.

68. Joseph Kirkland to Family, 4 July 1887, Joseph Kirkland Papers, Newberry Library; HG, *Main-Travelled Roads*, xi; *The Standard*, 3 February 1892, 7; HG, *Son of the Middle Border*, 371, 374, 356. See Clyde Henson, "Joseph Kirkland's Influence on Hamlin Garland," *American Literature* 23 (1952):458–63.

69. HG quoted in John Perry, *James A. Herne: The American Ibsen* (Chicago, 1978), 107; B. O. Flower to HG, 30 April 1890, HG Papers; HG, *Crumbling Idols* (New York, 1894), 51–53.

70. HG, *Son of the Middle Border*, 374; HG, *Crumbling Idols*, 133; HG, "Productive Conditions of American Literature," *Forum* 17 (August 1894):690–94; HG, *Crumbling Idols*, 133.

71. HG, *Main-Travelled Roads* (Boston, 1891), Preface.

72. HG, "Lucretia Burns," in *Prairie Folks* (New York, 1899), 89–90.

73. HG, "Under the Lion's Paw," *Main-Travelled Roads*, 214.

74. HG, "A Branch Road," *Main-Travelled Roads*, 45, 62, 58.

75. WDH, "Editor's Study," *HM* 83 (September 1891):639.

76. HG, *Son of the Middle Border*, 416–17. Garland's fiction, Benjamin Flower observed, presented the "real with all its vivid power, reinforced and vitalized by realistic or truthful idealism." See "Mask or Mirror," *Arena* 8 (1893):305.

77. HG, *Son of the Middle Border*, 418; HG, *My Friendly Contemporaries* (New York, 1932), 395.

78. WDH, "Mr. Garland's Books," 523–28; HG, *Son of the Middle Border*, 412; John T. Flanagan, "Hamlin Garland Writes to His Chicago Publisher," *American Literature* 23 (January 1952):448–53; HG, *Roadside Meetings* (New York, 1930), 187. See also Bernard Duffey, "Hamlin Garland's 'Decline' from Realism," *American Literature* 25 (March 1953):69–74; James D. Koerner, "Comment on Hamlin Garland's 'Decline' from Realism," *American Literature* 26 (November 1954):427–32; Pizer, *Hamlin Garland's Early Work*, 110.

79. HG, *A Daughter of the Middle Border* (New York, 1921), 35; HG, *Back-Trailers from the Middle Border* (New York, 1928), 40.

80. HG, *Roadside Meetings*, 187; JR, *The Battle with the Slum* (1902; New York, 1912), 48.

81. *The Literary World* 25 (22 September 1894):299; "A Spiritual Reaction," *Outlook* (9 December 1899):867–68.

82. Maurice Thompson, "The Prospect in Fiction," *Independent* 52 (17 May 1900):1182; "The Revival of Romance," *Dial* 25 (1 December 1898):387–89. Howells complained in 1900 that "nothing of late has been heard but the din of arms, the horrid tumult of the swashbuckler swashing on his buckler." WDH, "The New Historical Romances," *NAR* 171 (December 1900):935–36. On the battle between romantics and realists, see Grant C. Knight, *The Critical Period in American Literature* (Chapel Hill, 1951). For additional evidence from the period, see Margaret Anderson, "A New Ideal in American Fiction," *Dial* 23 (1897):269–71; Stockton Ashton, "The New Romanticism," *Citizen* 1 (1895):60–63; William Roscoe Thayer, "The New Story-Tellers and the Doom of Realism," *Forum* 18 (1894):470–74.

83. F. Marion Crawford, *The Novel: What It Is* (New York, 1893), 23–44; "What Is a Novel?," *Forum* 14 (January 1893):597.

84. "More Remarks on Realism," *AM* 62 (December 1888):846; HJ to SOJ, 5 October 1901, HJ Papers, Houghton Library, Harvard University.

85. SC, "Fears Realists Must Wait," *New York Times*, 28 October 1894, p. 20.

Chapter 10. Savage Realism

1. HG, "The Work of Frank Norris," *Critic* 42 (March 1903):218; George Wharton James, "A Study of Jack London in His Prime," *Overland Monthly* 69 (May 1917):382; FN, *The Responsibilities of the Novelist* (New York, 1901). 280. The phrase "ultra realism" comes from an anonymous review of Frank Norris's *A Man's Woman* entitled "Mr. Norris's Ultra Realism," *New York Times: Saturday Review of Books and Art* 49 (10 February 1900):82.

2. H. L. Mencken, "Introduction," *Major Conflicts*, Vol. 10 of *The Works of Stephen Crane*, ed. Wilson Follett (New York, 1926), x; FN, "An American School of Fiction," *Boston Evening Transcript*, 22 January 1902, p. 17; Ellen Glasgow, *A Certain Measure* (New York, 1943), 14, 113–14. See also Rodrigue E. Labrie, "The Howells-Norris Relationship and the Growth of Naturalism," *Discourse* 11 (1968):363–71.

3. TD, "The Real Howells," *Ainslee's* 5 (March 1900):137–42; TD, *A Book About Myself* (New York, 1922), 488, 499.

4. WDH, "Novel-Writing and Novel-Reading: An Impersonal Explanation," in *Howells and James: A Double Billing*, ed. William M. Gibson (New York, 1958), 21; WDH, "A Case in Point," *Literature* 1 (24 March 1899), 241–42; WDH to Robert Herrick, 16 February 1909, WDH to Henry Blake Fuller, 10 March 1907, in *Selected Letters of W. D. Howells*, ed. William C. Fisher (Boston, 1983), 5:267, 213.

5. TD, "The Literary Shower," *Ev'ry Month* 2 (July 1896):24; TD, "Reflections," *Ev'ry Month* 2 (August 1896):6; FN, "Why Women Should Write the Best Novels: And Why They Don't," *Boston Evening Transcript*, 13 November 1901, p. 20. Jack London loved to write stories about "adventures I had actually gone through, things I had done with my own hands and head." Jack London, "Stranger Than Fiction," *Critic* 43 (August 1903):124.

6. David G. Phillips, *George Helm* (New York, 1912), 17; Henry Childs Merwin, "On Being Civilized Too Much," *AM* 79 (June 1897):839; James Weir, Jr., M.D., "The Methods of the Rioting Striker as Evidence of Degeneration," *Century* 48 (October 1894):952–53.

7. Robert Grant, "Search-Light Letters: Letter to a Modern Woman with Social Ambitions," *Scribner's* 25 (March 1899):378; Ernest Earnest, *Silas Weir Mitchell, Novelist and Physician* (Philadelphia, 1950), 174; Carl D. Case, *The Masculine in Religion* (Philadelphia, 1906), 21, 111.

8. G. Stanley Hall, "Feminization in School and Home," *World's Work* (May 1908):1037–44. On the theme of besieged masculinity, see Joseph Kett, *Rites of Passage: Adolescence in America* (New York, 1977), 173–214; J. A. Mangan and James Walvin, eds., *Manliness and Morality: Middle Class Masculinity in Britain and America, 1800–1940* (New York, 1987); David Pugh, *Sons of Liberty: The Masculine Mind in Nineteenth-Century America* (Westport, Conn., 1983); Anthony Rotundo, *American Manhood: Transformations in Masculinity from the Revolution to the Modern Era* (New York, 1993).

9. Luther Gulick, *The Efficient Life* (New York, 1907), 182, 62, 35, 46.

10. Theodore Roosevelt, *The Strenuous Life: Essays and Addresses* (New York, 1904), 1, 6–8; Roosevelt, "Professionalism in Sports," NAR 15 (August 1890):187.

11. Three superb secondary studies of the nature cult are Roderick Nash, *Wilderness and the American Mind* (New Haven, 1973); Peter Schmitt, *Back to Nature: The Arcadian Myth in Urban America* (New York, 1969); G. Edward White, *The Eastern Establishment and the Western Experience: The West of Frederic Remington, Theodore Roosevelt, and Owen Wister* (New Haven, 1968).

12. FN, "The 'Nature' Revival in Literature" (1903), in *The Literary Criticism of Frank Norris*, ed. Donald Pizer (New York, 1964), 40–43. See also Richard Slotkin, *Regeneration Through Violence: The Mythology of the American Frontier, 1600–1860* (Middletown, Conn., 1973). Both the creation of "Fresh Air Funds" to send urban kids to the countryside during the summer and the founding of the Boy Scouts of America (BSA) reflected growing fears that city boys were becoming soft and effeminate. "The REAL Boy Scout is not a 'sissy,'" stressed James West, the first Chief Scout Executive. "He is not a hothouse plant, like Little Lord Fauntleroy. . . . He is not hitched to his mother's apronstrings. While he adores his mother . . . he is self-reliant, sturdy, and full of vim." Ernest Thompson Seton, the BSA's Chief Scout, said that the scouting movement was needed to combat the "*city rot*" that "has turned such a large proportion of our robust, manly, self-reliant boyhood into a lot of flat-chested cigarette smokers." Ernest Thompson Seton, *Boy Scouts of America: A Handbook of Woodcraft, Scouting, and Life-craft* (New York, 1910), xi, xii. See also Jefferey P. Hantover, "The Boy Scouts and the Validation of Masculinity," *Journal of Social Issues* 34 (1978):184–95; David E. Shi, "Ernest Thompson Seton and the Boy Scouts: A Moral Equivalent of War?," *South Atlantic Quarterly* 84 (Autumn 1985):379–91; David Mcleod, *Building Character in the American Boy: The Boy Scouts, YMCA, and Their Forerunners, 1870–1920* (Madison, Wis., 1983).

13. TD, *A Book About Myself* (New York, 1922), 107; LS, *Kindergarten Chats* (New York, 1918), 113; Bernarr A. McFadden, *The Virile Powers of Superb Manhood; How Developed, How Lost; How Regained* (New York, 1900), 198. See also Caspar Whitney, "Outdoor Sports—What They Are Doing for Us," *Independent* 52 (1900):1361–63. For thorough scholarly analyses of these trends, see Dominick Cavallo, *Muscles and Morals: Organized Playgrounds and Urban Reform, 1880–1920* (Philadelphia, 1981), and James C. Whorton, *Crusaders for Fitness: The History of American Health Reformers* (Princeton, 1982). Fitness reform was not for men only. See Regina Morantz, "Making Women Modern: Middle Class Women and Health Reform in Nineteenth-Century America," *Journal of Social History* 10 (1977):493–97, and Roberta J. Park, "'Embodied Selves': The Rise and Development of Concern for Physical Education, Active Games and Recreation for American Women," *Journal of Sport History* 5 (1978):5–41.

14. Edgar M. Robinson, *Reaching the Boys of an Entire Community* (New York, 1909), 66–67; Henry S. Curtis, "The Proper Relation of Organized Sports on Public Playgrounds and in Public Service," *Playground* 3 (September 1909):14; Francis A. Walker, *College Athletics: An Address Before the Phi Beta Kappa Society, Alpha, of Massachusetts at Cambridge, June 29, 1893*, in *Technology Quarterly* 6 (July 1893):2–3, 6, 13. On the evolution of organized sports, see Donald J. Mrozek, *Sport and American Mentality: 1880–1910* (Knoxville, 1983).

15. Willa Cather, "As You Like It," *Nebraska State Journal*, 2 December 1894, p. 13; Charles F. Thwing, "The Ethical Functions of Foot-Ball," NAR 173 (November 1901):628; "Football as a Training for Life," *Independent* 59 (30 November 1905):1294.

16. OWH, Jr., "The Soldiers' Faith," in Max Lerner, ed., *The Mind and Faith of Justice Holmes* (Boston, 1943), 18–21.

17. G. Stanley Hall, *Life and Confessions of a Psychologist* (New York, 1923), 578–79. On Eakins's boxing paintings, see Carl S. Smith, "The Boxing Paintings of Thomas Eakins," *Prospects* 4 (1979):403–19. For a wonderfully textured survey of the cultural implications of prize fighting, see Elliot Gorn, *The Manly Art: Bare-Knuckle Prize Fighting in America* (Ithaca, 1986).

18. Price Collier, *England and the English from an American Point of View* (New York, 1916); Alfred T. Mahan, *The Interest of America in Sea Power* (New York, 1897), 18. On this theme, see Amy Kaplan, "Romancing the Empire: The Embodiment of American Masculinity in the Popular Historical Novel of the 1890s," *American Literary History* 2 (Winter 1990):659–90; Gerald Linderman, *The Mirror of War: American Society and the Spanish-American War* (Ann Arbor, 1974).

19. Sarah Grand, "The Man of the Moment," NAR 158 (May 1894):626; Maurice Thompson, "Vigorous Men, a Vigorous Nation," *Independent* (1 September 1898), 610.

20. The term "effeminate realists" comes from Bliss Carman, "Mr. Gilbert Parker," *Chap-Book* 1 (1 November 1894):339–43; Robert Bridges, "Kipling," *Outlook* 61 (4 February 1899):281; Willa Cather, "Scottish Novelists," *Courier*, 30 November 1895, pp. 6–7, and "The Passing Show," *Courier*, 2 November 1895, p. 6. Such chauvinism, however, seemed to heighten Kipling's appeal to women readers. Kipling, explained one critic, had discovered that "to compel a woman reader an author need not be nice to her." In fact, he could "ride over her rough-shod; and the more he does this the more readily will she turn to his next epistle." Bailey Millard, "Why Women Dislike Kipling," *Bookman* 40 (November 1914):328.

21. Grace Isabel Colbron, "The Eternal Masculine," *Bookman* 32 (1910):157. See also Joseph J. Kwiat, "The Newspaper Experience: Crane, Norris, Dreiser," *Nineteenth-Century Fiction* 8 (September 1953):99–117.

22. "Stephen Crane Says: Edwin Markham Is His First Choice for the American Academy," *New York Journal*, 31 March 1900; FN, *A Man's Woman* (New York, 1900), 71; FN, *The Collected Writings of Frank Norris*, 10 vols. (Garden City, N.Y., 1928), 4:114.

23. Jack London, *War of the Classes*, "How I Became a Socialist," in *American Issues* (Chicago, 1955), 724; Jack London to Anna Strunsky, 15 March 1900, in *The Letters of Jack London*, eds. Earle Labor, Robert C. Leitz, and I. Milo Shepard, 3 vols. (Stanford, 1988), 1:60, 366, 173.

24. Jack London, *The Sea-Wolf* (New York, 1904), 42, 121.

25. E. Preston Dargan, "Jack London in Chancery," *New Republic* (21 April 1917):7; TD, "Reflections," *Ev'ry Month* 2 (August 1896):6–7; TD, *The "Genius"* (New York, 1915), 394; TD, *Jennie Gerhardt* (New York, 1911), 135, 414; "A Note on Theodore Dreiser by a Man," *International* (August 1914):249.

26. Barrett Clark, *Eugene O'Neill: The Man and His Plays* (New York, 1929), 130. Among the dozens of works dealing with literary naturalism, see especially Lars Ahnebrink, *The Beginnings of Naturalism in American Fiction* (Cambridge, Mass., 1950); John J. Conder, *Naturalism in American Fiction: The Classic Phase* (Lexington, Ky., 1984); Don Graham, "Naturalism in American Fiction: A Status Report," *Studies in American Fiction* 10 (1982):1–16; Harold Kaplan, *Power and Order: Henry Adams and the Naturalist Tradition in American Fiction* (Chicago, 1981); Ronald Martin, *American Literature and the Universe of Force* (Durham, N.C., 1981); Walter Benn Michaels, *The Gold Standard and the Logic of Naturalism: American Literature at the*

Turn of the Century (Berkeley, 1987); Lee Mitchell, *Determined Fictions: American Literary Naturalism* (New York, 1989); Donald Pizer, *Realism and Naturalism in American Fiction* (1966); Charles C. Walcutt, *American Literary Naturalism, A Divided Stream* (New York, 1956); Christopher P. Wilson, "American Naturalism and the Problem of Sincerity," *American Literature* 54 (1982):511–27.

27. FN, "Fiction Is Selection," *Wave* 16 (11 September 1897):3. Dreiser was equally committed to a mimetic approach liberated from genteel prohibitions: "The extent of all reality is the realm of the author's pen, and a true picture of life . . . is both moral and artistic whether it offends conventions or not." TD, "True Art Speaks Plainly," *The Booklovers' Magazine* 1 (1903):129.

28. FN, "Zola as a Romantic Writer," *Wave* 15 (27 June 1896):3; FN, "Weekly Letter," *Chicago American Literary Review*, 3 August 1901; FN, "Zola as a Romantic Writer," 3; FN, "A Plea for Romantic Fiction," *Boston Evening Transcript*, 18 December 1901, p. 14.

29. Emile Zola, *Thérèse Raquin*, trans. Leonard Tancock (Harmondsworth, 1962), 22; Jack London to Cloudesley Johns, 6 January 1902, in *Letters of Jack London*, 1:270; TD, *Jennie Gerhardt*, 426; Lee Mitchell, *Determined Fictions: American Literary Naturalism* (New York, 1989), 3. See also Gordon O. Taylor, *The Passages of Thought: Psychological Representation in the American Novel, 1870–1900* (New York, 1969). For other analyses of naturalism, see Robert M. Figg, "Naturalism as a Literary Form," *Georgia Review* 18 (1964):308–16; Donald Pizer, "Nineteenth-Century American Naturalism: An Essay in Definition," *Bucknell Review* (December 1965):1–18, and *Realism and Naturalism in Nineteenth-Century American Literature* (Carbondale, Ill., 1984), 9–30.

30. Jack London, *The Call of the Wild* (New York: Penguin Classic), 51.

31. TD, *The Financier* (New York, 1912), 13–14. London's Wolf Larsen expressed the savage reality of primal competition in *The Sea-Wolf*: "The big eat the little that they may continue to move, the strong eat the weak that they may retain their strength. The lucky eat the most and move the longest, that is all." Norris repeated this motif in *Vandover and the Brute* when he described Charlie Geary's brutish philosophy of survival: "Every man for himself, that was what he said. It might be damned selfish, but it was human nature." London, *The Sea-Wolf*, 50; FN, *Vandover and the Brute* (Garden City, N.Y., 1928), 288.

32. Edward A. Ross, *Social Control* (Cleveland, 1901), 196; SC, *George's Mother* (1896; Greenwich, Conn., 1960), 97; Jack London, [review of Upton Sinclair, *The Jungle*], *Wilshire's Magazine* 10 (August 1906):10; TD, *The Titan* (New York, 1914), 516.

Chapter 11. A World Full of Fists

1. SC to Nellie Crouse, 31 December 1895, 5 February 1896, in *The Correspondence of Stephen Crane*, eds., Stanley Wertheim and Paul Sorrentino, 2 vols. (New York, 1988), 1:163, 197–98; SC to WDH, 27 January 1894, WDH Papers, Houghton Library, Harvard University.

2. SC to Nellie Crouse, 5 February 1896, in *Correspondence of Stephen Crane*, 1:198; *The Works of Stephen Crane*, ed. Fredson Bowers, 10 vols. (Charlottesville, 1969–75), 8:364. Crane once told Howells that "'I always thank God that I can have the strongest admiration for the work of a man who has been so much to me personally.'" SC to WDH, 15 August 1893, WDH Papers. See also Thomas Gullason, "New Light

on the Crane-Howells Relationship," *New England Quarterly* 30 (September 1957):389–92.

3. SC to John Northern Hilliard, 2 January 1896, in *Correspondence of Stephen Crane*, 1:166. The best sources of biographical information about Crane are Christopher Benfey, *The Double Life of Stephen Crane* (New York, 1992), and R. W. Stallman, *Stephen Crane: A Biography* (New York, 1968).

4. SC to a Claverack College Schoolmate, November 1890; SC to John Northern Hilliard, 2 January 1896, in *Correspondence of Stephen Crane*, 1:35, 166; Michael Robertson, *Stephen Crane at Lafayette* (Easton, Pa., 1990), 8; John Berryman, *Stephen Crane* (New York, 1950), 22.

5. SC to John Northern Hilliard, February 1895; SC to the Editor of *Demorest's Family Magazine*, April-May 1896; SC to Lily Brandon Munroe, March–April 1894, in *Correspondence of Stephen Crane*, 1:99, 230, 63.

6. SC, "On the Jersey Coast," *New York Tribune*, 24 July 1892, p. 22; Edwin H. Cady, *Stephen Crane* (New York, 1962), 31–32. In July 1892, Crane wrote an article for the *New York Tribune* about an illustrated lecture by Jacob Riis concerning the "unfortunates" living in "crowded tenements." Riis's impassioned presentation inspired Crane to read *How the Other Half Lives* and later introduce himself to the author.

7. SC to J. Herbert Welch, April–May 1896, in *Correspondence of Stephen Crane*, 1:232.

8. SC to John Northern Hilliard, January 1896, in ibid., 1:195.

9. Stallman, *Stephen Crane*, 70. Crane used a small inheritance and funds borrowed from his brother to pay the $869 printing bill.

10. E. J. Edwards, *Philadelphia Press*, 22 April 1894; Edwards, "Realism and a New Realist," *New York Times*, 4 November 1894; WDH, *New York Press*, 15 April 1894.

11. SC, *Maggie: A Girl of the Streets* (New York: Norton Critical Edition, 1979), 17–18, 20, 19.

12. Ibid., 39, 50.

13. H. F. West, *A Stephen Crane Collection* (Hanover, N.H., 1948), x–xi; SC to the Editor of *Demorest's Family Magazine*, April–May 1896; Thomas Beer, *Hanna, Crane, and the Mauve Decade* (New York, 1937), 312; SC to HG, in *Correspondence of Stephen Crane*, 1:230, 53.

14. SC to Catharine Harris, 12 November 1896, in R. W. Stallman and Lillian Gilkes, eds., *Stephen Crane: Letters* (London, 1960):133. See also David Fitelson, "Stephen Crane's *Maggie* and Darwinism," *American Quarterly* 16 (1964):182–94; Donald Pizer, "Stephen Crane's 'Maggie' and American Naturalism," *Criticism* 7 (Spring 1965):168–75.

15. SC, *Maggie*, 19, 28; Cady, *Stephen Crane*, 104. See also Pizer, "Stephen Crane's 'Maggie' and American Naturalism"; 75; SC, *The Third Violet*, in *Stephen Crane: Prose and Poetry* (New York: Library of America Edition, 1984), 320.

16. SC, *Maggie*, 6; "A Work by Stephen Crane," *New York Times*, 31 May 1896, p. 31. See also Milne Holton, *Cylinder of Vision: The Fiction and Journalistic Writing of Stephen Crane* (Baton Rouge, 1972); Arno Karlen, "The Craft of Stephen Crane," *Georgia Review* 28 (Fall 1974):473–77; James Nagel, *Stephen Crane and Literary Impressionism* (University Park, Pa., 1980), and Sergio Perosa, "Naturalism and Impressionism in Stephen Crane's Fiction," in *Stephen Crane: A Collection of Critical Essays*, ed. Maurice Bassan (Englewood Cliffs, N.J., 1967), 81–85; Allan Gardner Smith, "Stephen Crane, Impressionism and William James," *Revue française d'etudes americaines* 17 (May 1983):237–49.

17. Bowers, ed., *Works of Stephen Crane*, 5:49; Corwin K. Linson, *My Stephen Crane*, ed. E. H. Cady (New York, 1958), 37.

18. SC, *Maggie*, 9. On this theme see Eric Solomon, *Stephen Crane: From Parody to Realism* (Cambridge, Mass., 1966), 35–44; Jay Martin, *Harvests of Change: American Literature, 1865–1914* (Englewood Cliffs, N.J., 1967), 57–59.

19. *Nashville Banner*, 15 August 1896; Rupert Hughes, "The Justification of Slum Stories," *Godey's Magazine* 131 (October 1895):431–32; *Boston Beacon*, 27 June 1896; *Bookman* 10 (October 1896):19–20.

20. SC to Nellie Crouse, 26 January 1896, in *Correspondence of Stephen Crane*, 1:186–87; SC, "War Memories," *Works of Stephen Crane*, ed. Wilson Follett (New York, 1925–26), 9:238. For recent scholarship on *The Red Badge of Courage*, see Lee Clark Mitchell, ed., *New Essays on The Red Badge of Courage* (Cambridge, England, 1986).

21. FN, "Stephen Crane's Stories of Life in the Slums," *The Wave* (San Francisco) 15 (4 July 1896):13. Norris and Crane met in 1898 as rival correspondents covering the Spanish-American War.

22. FN, "The True Reward of the Novelist," *World's Work* 2 (October 1901):1339; Berryman, *Stephen Crane*, 288. For astute insights into Norris's attitude toward "style," see Michael Davitt Bell, "Frank Norris, Style, and the Problem of American Naturalism," *Studies in the Literary Imagination* 16 (Fall 1983):93–106; Larzer Ziff, *The American 1890s: Life and Times of a Lost Generation* (New York, 1966), 270.

23. "Books of the Week," *New York Herald*, 10 January 1903, p. 12.

24. On Norris's background see Franklin Walker, *Frank Norris: A Biography* (Garden City, N.Y., 1932); Ernest Marchand, *Frank Norris: A Study* (Stanford, 1942); Warren French, *Frank Norris* (New York, 1962). For a succinct treatment of Norris and his career, see H. Wayne Morgan, "Frank Norris: The Romantic as Naturalist," in *American Writers in Rebellion* (New York, 1965), 104–45.

25. Wallace Everett, "Frank Norris in His Chapter," *Phi Gamma Delta* 52 (April 1930):561; FN, *Vandover and the Brute* in *The Collected Writings of Frank Norris*, 10 vols. (Garden City, N.Y., 1928), 5:24; Franklin Walker, *Frank Norris: A Biography* (Garden City, N.Y., 1932), 65.

26. Donald Pizer, *The Novels of Frank Norris* (Bloomington, 1966), 16.

27. FN, *Moran of the Lady Letty* (Garden City, N.Y., 1928), 214; Walker, *Frank Norris*, 88; FN, "Ethics of the Freshman Rush," *Wave* 16 (4 September 1897):2.

28. Franklin Walker, "An Early Frank Norris Item," *Book Club of California Quarterly News Letter* 25 (Falls 1960):83–86; *Frank Norris of "The Wave"* (San Francisco, 1931), 38; Frank Norris, "A Plea for Romantic Fiction," *Boston Evening Transcript*, 18 December 1901, p. 14.

29. Lewis E. Gates, *Studies and Appreciations* (1900), 41, 59.

30. FN to Mrs. Elizabeth H. Davenport, 12 March 1898, in *The Letters of Frank Norris*, ed. Franklin Walker (San Francisco, 1956), 6; FN, "A Plea for Romantic Fiction," 14; FN, "True Reward," 1338; FN, "The House with Blinds," in *The Collected Writings of Frank Norris*, 4:11–18; FN, "Zola as a Romantic Writer," 3. For collections of Norris's essays and journalism, see FN, *The Responsibilities of the Novelist* (New York, 1903), and Donald Pizer, ed., *The Literary Criticism of Frank Norris* (New York, 1964).

31. FN, "The Country Club at Del Monte," *Wave* 14 (31 August 1895):7; FN, "Novelists of the Future: The Training They Need," *Boston Evening Transcript*, 27 November 1901, p. 14. For provocative psychological explanations of Norris's preoccupation with masculinity, see Maxwell Geismar, *Rebels and Ancestors: The American*

Novel, 1890–1915 (New York, 1953), 3–66; Kenneth Lynn, "Frank Norris: Mama's Boy," in *The Dream of Success: A Study of the Modern American Imagination* (Westport, Conn., 1955), 158–207; Sherwood Williams, "The Rise of the New Degeneration: Decadence and Atavism in *Vandover and the Brute*," *ELH* 57 (Fall 1990):709–36.

32. FN, *A Man's Woman*, in *Collected Writings of Frank Norris*, 6:209; Geraldine Bonner, "A Californian's Novel," *The Argonaut* 43 (21 November 1898):7; "Books of the Week," *Providence Journal*, 11 February 1900, p. 15. A useful compendium of reviews of Norris's novels is Joseph R. McElrath, Jr., and Katherine Knight, eds., *Frank Norris: The Critical Reception* (New York, 1981).

33. FN, *A Man's Woman*, 6:24, 38.

34. At times, as Jack London pointed out, Norris went overboard in his quest for verisimilitude. He *saw* better than he *thought*, but "never mind the realism, the unimportant detail, minute description," for Norris "*has* produced results, Titanic results." Jack London, "The Octopus," *Impressions Quarterly* 2 (June 1901):45–47.

35. FN, "A Problem in Fiction: Truth Versus Accuracy," *Boston Evening Transcript*, 6 November 1901. See also George M. Spangler, "The Idea of Degeneration in American Fiction, 1880–1940," *English Studies* 70 (October 1989):407–35.

36. FN, *McTeague: A Story of San Francisco* (New York: Norton Critical Edition, 1977), 2, 108. The best analysis of *McTeague* remains Donald Pizer, *The Novels of Frank Norris*, 63–79. For the background behind the novel, see Robert D. Lundy, "The Making of *McTeague* and *The Octopus*" (Ph.D. diss., University of California, 1956), 145–53; Jesse S. Crisler, "A Critical and Textual Study of Frank Norris's *McTeague*" (Ph.D. diss., University of South Carolina, 1973).

37. FN, *McTeague*, 16, 18.

38. Ibid., 19, 48, 50, 51.

39. Ibid., 119.

40. Ibid., 171, 174.

41. Ibid., 249; "Aims and Autographs of Authors," *Book News* 17 (May 1899):486.

42. John D. Barry, "New York Letter," *Literary World* 30 (18 March 1899):88–89; Isaac F. Marcosson, "The Story of McTeague," *Louisville Times*, 13 March 1899, p. 3; "A Western Realist," *Washington Times*, 23 April 1899, p. 20.

43. WDH, "A Case in Point," *Literature* n.s. 11 (24 March 1899):241–42.

44. FN to WDH, undated (1899), WDH Papers; Walker, ed., *Letters of Frank Norris*, 22–23.

45. FN, *The Octopus* (New York: Penguin Classics Edition, 1986), 298, 650.

46. Ibid., 12, 382.

47. Ibid., 634.

48. Ibid., 651–52, 332–23.

49. WDH, "Editor's Easy Chair," *HM* (October 1901):824–25; WDH, "Frank Norris," *NAR* 175 (December 1902):776; B. O. Flower, "The Trust in Fiction: A Remarkable Social Novel," *Arena* (May 1902):234.

50. "The Degradation of a Soul," *The Independent* 79 (1914):173.

51. FN to TD, May 1900, in *Letters of Theodore Dreiser*, ed. Robert Elias (Philadelphia, 1959), 1:50–52; FN quoted in Ziff, *The American 1890s*, 344.

52. "The Eighteen-Eighties," *New York Tribune*, 23 May 1914, p. 11. For analyses of Dreiser's fiction, see Richard Lehan, *Theodore Dreiser: His World and His Novels* (Carbondale, Ill., 1969), and Donald Pizer, *The Novels of Theodore Dreiser* (Minneapolis, 1976).

53. Lillian Rosenthal to TD, 25 January 1911, Theodore Dreiser Papers, Van Pelt

Library, University of Pennsylvania. Ellen Moers offered a convincing defense of Dreiser's literary style in "The Finesse of Dreiser," *American Scholar* 33 (Winter 1963):109–14.

54. H. L. Mencken, "Introduction," *An American Tragedy* (Cleveland, 1946), ix–xvi.

55. Good biographies of Dreiser abound. See Robert Elias, *Theodore Dreiser: Apostle of Nature* (New York, 1949); Richard Lingeman, *Theodore Dreiser: At the Gates of the City, 1871–1907* (New York, 1986), and *Theodore Dreiser: An American Journey, 1908–1945* (New York, 1991); W. A. Swanberg, *Theodore Dreiser* (New York, 1965).

56. TD, *Dawn* (London, 1931), 156. In his autobiography Dreiser used the pseudonyms Amy and Janet to refer to Emma and Sylvia.

57. Ibid., 305, 166, 553, 160.

58. Ibid., 569, 343; TD, *An American Tragedy* (New York: Signet Classic edition, 1967), 220; TD, *A Book About Myself* (New York, 1922), 33; TD, *Dawn*, 541, 580; TD, *A Book About Myself*, 326; Robert L. Duffus, "Dreiser," *American Mercury* 7 (January 1926):71.

59. TD, *A Book About Myself*, 373, 132; TD to William C. Lengel, 15 October 1911, Dreiser Papers.

60. TD, *A Book About Myself*, 382, 412, 411; TD, "Literary Notes," *Ev'ry Month* 2 (May 1896):12; TD, *A Book About Myself*, 126.

61. Frank Harris, Contemporary Portraits (New York, 1919), 91; TD, "At the Sign of the Lead Pencil," *Bohemian* 17 (October 1909):431. See also TD, *A Book About Myself*, 457–58.

62. Swanberg, *Theodore Dreiser*, 177; TD, *The Financier* (New York, 1912), 177.

63. TD, *A Book About Myself*, 482; TD, "Reflections," *Ev'ry Month* 3 (February 1897):4–5; TD, *A Book About Myself*, 106–07; TD, "Reflections," *Ev'ry Month* 2 (June 1896):5.

64. TD to George Ade, 16 March 1942, in *The Letters of Theodore Dreiser*, 3:949.

65. TD, *A Book About Myself*, 490–91; TD, "True Art Speaks Plainly," *Booklovers' Magazine* 1 (February 1903):129; Harper's to TD, 2 May 1900, Dreiser Papers.

66. Isaac Marcosson, "Sister Carrie," *Louisville Times*, 20 November 1900.

67. "Sister Carrie," *Toledo Blade*, 8 December 1900; "Sister Carrie," *Chicago Advance*, 27 June 1907; Little [William Marion Reedy], "Sister Carrie: A Strangely Strong Novel in a Queer Milieu," *St. Louis Mirror* 10 (3 January 1901):6. TD referred to the "tea and toast crowd" in "The New Humanism," *Thinker* 2 (July 1930):9.

68. For Dreiser's account of his prolonged depression, see *An Amateur Laborer*, eds. Richard R. Dowell, James L. West III, and Neda Westlake (Philadelphia, 1983), 77, 79, 92, xi. A lengthy treatment of Dreiser's battle with neurasthenia is contained in Tom Lutz, *American Nervousness, 1903: An Anecdotal History* (Ithaca, 1991).

69. "Recent Publications: Sister Carrie," *New Orleans Daily Picayune*, 1 July 1907, p. 14; M. K. Ford, "The Editor's Clearing House," *Putnam's Monthly* 1 (October 1906):123–24.

70. TD, *Sister Carrie* (1900; New York: Norton Critical Edition, 1970), 8.

71. Ibid., 13, 75. Dreiser shared Carrie's obsession with material standards of value. He agreed with Balzac that "human beings are innately greedy—avaricious. Do you know that they dream of fine clothes and fine houses and of rolling about luxuriously in carriages while others beg along their pathway?" Humankind's animating hope and ultimate folly, he testified, was in assuming that happiness was a commodity to be bought. TD, "Reflections," *Ev'ry Month* 2 (September 1896):2.

72. TD, *Sister Carrie*, 17, 51; Walter Benn Michaels, "*Sister Carrie's* Popular Economy," *Critical Inquiry* 7 (Winter 1980):390. For an especially lucid analysis of Dreiser's ideas about material desire, see Amy Kaplan, "The Sentimental Revolt of *Sister Carrie*," in *The Social Construction of American Realism* (Chicago, 1988), 140–60.

73. See Philip Fisher, "Acting, Reading, Fortune's Wheel: *Sister Carrie* and the Life History of Objects," in *American Realism: New Essays*, ed. Eric J. Sunquist (Baltimore, 1982):278–95; Robert Shulman, "Dreiser and the Dynamics of American Capitalism," in *Social Criticism and Nineteenth-Century Fiction* (Columbia, Mo., 1987), 284–316.

74. TD, *Sister Carrie*, 47, 73.

75. Ibid., 226, 229, 318.

76. Ibid., 366.

77. Ibid., 369.

78. Leslie Fiedler went so far as to place Dreiser within the tradition of sentimental domestic fiction. See Fiedler, *Love and Death in the American Novel* (New York, 1960), 241–48.

Those who believe that Dreiser's purpose was to parody rather than reinforce the stereotypical assumptions of "feminine fiction" and the cult of domesticity note that he contrived maudlin chapter titles drawn from the idealizing clichés of popular ballads or melodramas—"The Magnet Attracting: A Waif Amid Forces," "Ashes of Tinder: The Loosing of Stays," and "The Way of the Beaten: A Harp in the Wind"—not as literal introductions but as satirical weapons against conventional pieties and ideals. The title of Chapter 27, "When Waters Engulf Us We Reach for a Star," evokes images of spiritual consolation during times of stress. Yet the chapter actually describes Hurstwood's theft of $10,000 from the saloon safe. On Dreiser's chapter titles, see Phillip Williams, "The Chapter Titles of *Sister Carrie*," *American Literature* 36 (1964):359–65; Sandy Petrey, "The Language of Realism, the Language of False Consciousness: A Reading of *Sister Carrie*," *Novel* (Winter 1977):101–13. For differing analyses of Dreiser's relationship to sentimentalism, see Michael Davitt Bell, "Fine Styles of Sympathy: Theodore Dreiser's *Sister Carrie*," in *The Problem of American Realism* (Chicago, 1993), 149–67; Cathy N. Davidson and Arnold E. Davidson, "Carrie's Sisters: The Popular Prototypes for Dreiser's Heroines," *Modern Fiction Studies* 23 (1977):395–407; Daryl Dance, "Sentimentalism in Dreiser's Heroines, Carrie and Jennie," *CLA Journal* 14 (1970):127–42.

79. "Recent Publications," *New Orleans Picayune*, 14.

80. TD, *Sister Carrie*, 56, 57. See also Yoshinobu Hakutani, "*Sister Carrie* and the Problem of Literary Naturalism," *Twentieth-Century Literature* 13 (April 1967):3–17.

81. TD, *Jennie Gerhardt* (New York, 1911), 401.

82. RB, "Desire as Hero," *NR* 5 (20 November 1915):5; TD, *The Financier*, 244; TD, *The Genius* (New York, 1915), 726.

83. TD, *The Financier*, 409; "Talks with Four Novelists," 393. In an essay entitled "The Dilemma of Determinism," William James asserted that if one truly believes that people behave *solely* according to overpowering forces and uncontrolled instincts, then there is no room for the element of chance nor is there any reason for feeling regret at the personal tragedies or societal ills haunting human existence. This led him to suggest the concept of "soft determinism," a perspective "which allows considerations of good and bad to mingle with those of cause and effect in deciding what kind of universe this may rationally be held to be." WJ, "The Dilemma of Determinism," in *The Will to Believe and Other Essays in Popular Philosophy* (London, 1897), 175, 164, 166.

84. TD, *The Financier*, 261.

85. TD, "Life, Art, and America," *Seven Arts* (February 1917):10.

86. H. L. Mencken to TD, 23 April 1911, in *Letters of Theodore Dreiser*, 1:115.

87. Lewis quoted in Jack Salzman, ed., *Theodore Dreiser: The Critical Reception* (New York, 1972), xv; Sinclair Lewis, "The Passing of Capitalism," *Bookman* 40 (November 1914):280; Randolph Bourne, "Theodore Dreiser," *NR* 2 (17 April 1915):8. Others acknowledged the tonic force of Dreiser's creative power as well as his strenuous efforts to extend and deepen the reach of realism. John Dos Passos, for example, confided to Dreiser that if it had not been "for your pioneer work none of us would have gotten our stuff written or published." See Lehan, *Theodore Dreiser*, xiii.

88. FN, "A Plea for Romantic Fiction," 14.

Chapter 12. Ash Can Realism

1. TD, *The "Genius"* (New York, 1915), 221, 89, 232, 231.

2. Marianne Doezema, *American Realism and the Industrial Age* (Cleveland, 1980), 57; TD, "The Month in New York," *Broadway Magazine* 17 (March 1907):589. For a thorough study of the relationship between Dreiser and Shinn, see Joseph J. Kwiat, "Dreiser's *The "Genius"* and Everett Shinn, the 'Ash-Can' Painter," *PMLA* 67 (March 1952):15–31. Kwiat's dissertation remains a treasure-trove of information about the "Ash Can" painters, for he was able to conduct numerous personal interviews with Sloan and gained access to the private papers of Henri. See Joseph Kwiat, "The Revolt in American Painting and Literature, 1890–1915" (Ph.D. diss., University of Minnesota, 1950).

Other useful studies of the "Ash Can" school include Milton W. Brown, "The Ash Can School," *American Quarterly* 1 (1949):127–34; Peter Conn, *The Divided Mind: Ideology and American Imagination, 1898–1917* (Cambridge, 1983), 259–78; Robert Crunden, *Ministers of Reform: The Progressives' Achievement in American Civilization, 1889–1920* (1982), 102–116; Ira Glackens, *William Glackens and the Ashcan Group* (New York, 1957); William Innes Homer, *Robert Henri and His Circle* (Ithaca, 1969); James T. Wall, "The Ashcan School: Transition in American Art," *South Atlantic Quarterly* 69 (1970):317–29; Rebecca Zurier, "Picturing the City: New York in the Press and the Art of the Ashcan School, 1890–1917" (Ph.D. diss., Yale University, 1988).

3. Leslie Katz, "The Breakthrough of Anshutz," *Arts Magazine* (March 1963):28; Thomas Anshutz to Edward H. Coates, 15 May 1893, Pennsylvania Academy of Fine Arts Archives. A full-length study of Anshutz remains to be published. The most authoritative work remains Sandra Denney (Heard), "Thomas Anshutz: His Life, Art, and Teaching" (M.A. thesis, University of Delaware, 1969). See also her exhibition catalogue; *Thomas Pollock Anshutz* (Philadelphia, 1973).

4. Homer, *Henri and His Circle*, 25.

5. Ibid., 32; Robert Henri, *The Art Spirit* (New York, 1913), 86–87, 125–26.

6. Homer, *Henri and His Circle*, 38; Henri, *Art Spirit*, 240–42; Homer, *Henri and His Circle*, 62.

7. Homer, *Henri and His Circle*, 65, 131. Henri's rebelliousness spilled over to his social views as well. As a philosophical anarchist, he enjoyed taking "the wrong side of things." In a letter to his parents during the violent labor dispute at the Homestead Steel plant, Henri expressed his support for the strikers, and on other occasions he spoke out in behalf of Indians, blacks, and women's rights. Henri, *Art Spirit*, 127.

8. Homer, *Henri and His Circle*, 76, 79.

9. Robert Henri, "The New York Exhibition of Independent Artists," *Craftsman* 18 (1910):161; Henri, *Art Spirit*, 127, 240–42, 141; Sloan quoted in Kwiat, "Revolt in

American Painting," 175. On Henri as a teacher, see Helen Goodman, "Robert Henri, Teacher," *Arts Magazine* 53 (September 1978):158–60. For Henri's interest in Whitman, see Joseph Kwiat, "Robert Henri and the Emerson-Whitman Tradition," *PMLA* 71 (1956):617–36.

10. Everett Shinn, "Life on the Press," in *Artists of the Philadelphia Press* (Philadelphia, 1945), 12. See also Joseph J. Kwiat, "The Newspaper Experience: The 'Ash-Can' Painters," *Western Humanities Review* 6 (1952):337–38. Luks echoed the sentiment: "I think the newspaper experience is the most valuable school for art students I know." "George Luks," *Art Digest* 8 (15 November 1933):5.

11. Homer, *Henri and His Circle*, 82. On the influence of Homer upon Henri and his followers, see Bruce Robertson, *Reckoning with Winslow Homer: His Late Paintings and Their Influence* (Cleveland, 1990), 81–136.

12. Everett Shinn, "Recollections of the Eight," in *The Eight* (Brooklyn, 1943), 11; "John Sloan," *Antiques* 133 (April 1988):778.

13. Bruce St. John, ed., *John Sloan's New York Scene* (New York, 1965), 316; Sloan quoted in Homer, *Henri and His Circle*, 84; Ira Glackens, *William Glackens and the Ashcan Group: The Emergence of Realism in American Art* (New York, 1957), vii. See also Matthew Baigell, "Notes on Realistic Painting and Photography, c. 1900–10," *Arts Magazine* 54 (November 1979):141–43; Mary Liasson, "The Eight and 291: Radical Art in the First Two Decades of the 20th Century," *American Art Review* 2 (September–October 1975):91–101.

14. Sadakichi Hartmann, "A Plea for the Picturesqueness of New York" (1900), in Harry Lawton and George Knox, eds., *The Valiant Knights of Daguerre* (Berkeley, 1978), 57.

15. Homer, *Henri and His Circle*, 107.

16. Henri, *Art Spirit*, 117, 173. See also Judith K. Zilczer, "Anti-Realism in the Ashcan School," *Artforum* 17 (March 1979):44–49.

17. Kwiat, "The Revolt in American Painting," 25; Henri, *Art Spirit*, 41.

18. Robert Henri to his parents, 7 April 1902, Robert Henri Papers, Beinecke Library, Yale University.

19. Robert Henri, "The 'Big Exhibition,' The Artist, and the Public," *Touchstone* 1 (June 1917):174; Homer, *Henri and His Circle*, 133. See also Guy Pène du Bois, *Artists Say the Silliest Things* (New York, 1940), 82.

20. Du Bois, *Artists Say the Silliest Things*, 86, 84, and "Robert Henri: The Man," *Arts and Decoration* 14 (1920):36; Henri, *Art Spirit*, 71.

21. *Robert Henri and Five of His Pupils* (exhibition catalogue; New York, 1946), n.p.

22. Bennard B. Perlman, *The Immortal Eight* (New York, 1962), 118; Du Bois, *Artists Say the Silliest Things*, 83, 88.

23. Margaret Denton, "Just One Rampart Left for Masculinity Says Everett Shinn, — Beards," *Des Moines Register*, 21 January 1927; Guy Pène du Bois, "Exhibition by Independent Artists Attracts Immense Throngs," *New York American*, 4 April 1910, p. 6; "George Bellows, an Artist with 'Red Blood,'" *Current Literature* 53 (1912): 342.

24. Samuel Swift, "Revolutionary Figures in American Art," *HW* 51 (13 April 1907):534. Sloan once wrote in his diary that he relished associating with painters who were "real men like Henri, Luks, Glackens and such." St. John, ed., *Sloan's New York Scene*, 184. See also [Mary Fanton Roberts], "Foremost American Illustrators: Vital Significance of Their Work," *Craftsman* 17 (December 1909):266–84.

25. "An Artist of the New York Underworld," *Current Literature* 48 (March 1910):329; Shinn, "Recollections of the Eight," 13.

26. [Mary Fanton Roberts], "Foremost American Illustrators," 267.

27. Luks quoted in Louis Baury, "The Message of Proletaire," *Bookman* 34 (December 1911):400–401.

28. Perlman, *The Immortal Eight*, 164; Kenyon Cox, *The Classic Point of View* (New York, 1911), 27. On Cox, see H. Wayne Morgan's *Keepers of Culture: The Art-Thought of Kenyon Cox, Royal Cortissoz, and Frank Jewett Mather, Jr.* (Kent, Ohio, 1989).

29. St. John, ed., *Sloan's New York Scene*, 118; Glackens, *William Glackens*, 86.

30. St. John, ed., *Sloan's New York Scene*, 194; *Town Topic* quoted in Glackens, *William Glackens*, 89; unidentified newspaper clipping quoted in *The Eight*, 3.

31. James Huneker, *New York Sun*, 9 February 1908, p. 8. On the Macbeth show, see William Innes Homer, "The Exhibition of 'The Eight': Its History and Significance," *American Art Journal* 1 (Spring 1969):53–64; Bennard Perlman, "Rebels with a Cause — The Eight," *Whitney Museum of Art News* 2 (January 1983):1–8.

32. Glackens, *William Glackens*, 105; Giles Edgerton (Mary Fanton Roberts), "The Younger American Painters: Are They Creating a National Art?," *Craftsman* 13 (February 1908):512–32.

33. John Sloan, *The Gist of Art* (New York, 1939), 189; Mary Fanton Roberts, "John Sloan: His Art and Its Inspiration," *Touchstone* 4 (February 1909):362; Sloan, *Gist of Art*, 41; Kwiat, "The Revolt of American Painting," 277. On Sloan see Bruce St. John, *John Sloan* (New York, 1971). At times in his diary, Sloan fretted over the ethics of his "peeping" at people, but he decided that it was "no insult to the people you are watching to do so unseen." See John Baker, "Voyeurism in the Art of John Sloan: The Psychodynamics of a 'Naturalistic' Motif," *Art Quarterly* 1 (Autumn 1978):379–95.

34. St. John, ed., *Sloan's New York Scene*, 185; Sloan, *Gist of Art*, 41.

35. St. John, ed., *Sloan's New York Scene*, 133; John Butler Yeats, "The Work of John Sloan," *HW* 58 (22 November 1913):21, 20.

36. Sloan, *Gist of Art*, 88.

37. Ibid., 19; St. John, ed., *Sloan's New York Scene*, 13. On Sloan's evolving political consciousness, see Joseph Kwiat, "John Sloan: An American Artist as Social Critic, 1900–1917," *Arizona Quarterly* 10 (Spring 1954):52–64.

38. Sloan, *Gist of Art*, 9; Sloan quoted in Peter Morse, *John Sloan's Prints* (New Haven, 1969), 192.

39. St. John, ed., *Sloan's New York Scene*, 308–9. See Patricia Hills, "John Sloan's Images of Working-Class Women: A Case Study of the Roles and Interrelationships of Politics, Personality, and Patrons in the Development of Sloan's Art, 1905–1916," *Prospects* 5 (1980):157–96.

40. Peter Morse, *John Sloan's Prints* (New Haven, 1969), 141.

41. St. John, ed., *Sloan's New York Scene*, 554, 305–6, xx; Brooks, *John Sloan, a Painter's Life*, (New York, 1955) 56n.

42. Charles W. Barrell, "The Real Drama of the Slums as Told in John Sloan's Etchings," *Craftsman* 15 (February 1909):559; Roberts, "John Sloan: His Art and Its Inspiration," 363.

43. "American Life by American Painters," *Craftsman* 23 (December 1912):371.

44. Conn, *The Divided Mind*, 261.

45. Henri quoted in Homer, "The Exhibition of 'The Eight,'" 54. On Bellows, see

Marianne Doezema, *George Bellows and Urban America, 1905–1913* (New Haven, 1992); Charles H. Morgan, *George Bellows: Painter of America* (New York, 1965).

46. Henry McBride, "George Bellows Paints 'The Call of the City,'" *New York Sun*, 26 December 1915; George Bellows, "'The Art Spirit,' by Robert Henri," *Arts and Decoration* 20 (December 1923):26, 87; "The Big Idea: George Bellows Talks About Patriotism for Beauty," *Touchstone* 1 (July 1917):275.

47. Henry McBride, "Art News and Reviews," *New York Sun*, 20 November 1921; [W.H.D.], "Huns Depicted Vividly," *Boston Evening Transcript*, 13 January 1919.

48. "George Bellows: An Artist with 'Red Blood,'" *Current Literature* 53 (1912):342; McBride, "George Bellows Paints 'The Call of the City.'"

49. Bellows's fascination with boxing is discussed in Doezema, *George Bellows*, 67–122; Robert Haywood, "George Bellows's *Stag at Sharkey's*: Boxing, Violence, and the Male Identity," *Smithsonian Studies in American Art* 2 (Spring 1988):3–15.

50. James Huneker, "Seen in the World of Art," *New York Sun*, 5 March 1911, p. 4; McBride, "George Bellows Paints 'The Call of the City'"; Ameen Rihani, "American Painting, Part III, Luks and Bellows," *International Studio* 71 (August 1920):281. "Bellows loved force and forceful doings," noted McBride, "and even in his portraits he meant to hit you hard." Henry McBride, "Bellows Lithographs on View," *New York Sun*, 28 March 1925.

51. Bellows quoted in Morgan, *George Bellows*, 208.

52. "Annual Exhibition at Carnegie Institute," *Fine Arts Journal* 30 (June 1914):292; unidentified newspaper clipping, *Cliff Dweller* file, Los Angeles County Museum of Art; "Honor George Bellows, Artist Who Blasts Traditions," *New York Herald Magazine*, 29 September 1912, p. 9. Based on his own reading of *Cliff Dwellers*, Theodore Dreiser was convinced that Bellows had no political purpose. He characterized the painter's outlook as: "What will be, will be." See TD, "Cliff Dwellers," *Vanity Fair* 25 (December 1925):55. Huneker praised Bellows for not singing "the songs of the oppressed. No pitying socialistic note spoils his virile art." Bellows showed "the working man as he is." See Morgan, *George Bellows*, 149.

53. Florence Barlow Rithraff, "His Art Shows 'The Big Intention,'" *New York Telegraph*, 2 April 1911, magazine section, p. 2.

54. James Huneker, "Around the Galleries," *New York Sun*, 7 April 1910, p. 6; "What Is the Most Beautiful Spot in the City?," *New York Times*, 18 June 1911.

55. "Passing Up a Few Facts: Everett Shinn's Original Pastels," *Town Topics*, undated clipping in Everett Shinn Papers, Reel D179, Archives of American Art, Washington, D.C.

56. "Sloan in Luks Job," *Art Digest* 8 (December 1933):24; Sloan, *Gist of Art*, 40–41, 189.

57. Estelle H. Ries, "The Relation of Art to Everyday Things," *Arts and Decoration* 15 (July 1921):158; Sloan, *Gist of Art*, 51; [Roberts], "Foremost American Illustrators," 268; *John Sloan: Contemporary Work in Painting* (exhibition catalogue; University of Chicago, 1942), foreword.

58. *New York Sun*, 9 February 1908.

Part V. An Epoch of Confusion

1. Edward Lindeman, ed., *Basic Selections from Emerson* (New York, 1954), 215; *Hamlin Garland's Diaries*, ed. Donald Pizer (San Marino, Calif., 1968), 19; Henry

McBride, "The Growth of Cubism," *New York Sun*, 8 February 1914; Gauguin quoted in Arthur Jerome Eddy, *Cubists and Post-Impressionism* (Chicago, 1914), 41.

Chapter 13. The Modernist Revolt

1. Henry Poincaré, "The Value of Science" (1904), in *The Foundations of Science: Science and Hypothesis, the Value of Science, Science and Method*, trans. G. B. Halsted (New York, 1913), 355. There is no better analysis of the cultural effects of changing concepts of time and space than Stephen Kern's *The Culture of Time and Space, 1880–1918* (Cambridge, Mass., 1983).

2. Ernst Mach, *The Science of Mechanics* (1883; New York, 1919), 223; James Byrnie Shaw, "The Real in Science," *Scientific Monthly* 4 (1917):514, 532; H. Wildon Carr, "The Metaphysical Implications of the Principle of Relativity," *Philosophical Review* 24 (January 1915):10. The impact of "modern" science at the turn of the century is assessed in David Albert, *Quantum Mechanics and Experience* (Cambridge, Mass., 1993); Jerome Bruner, *Actual Minds, Possible Worlds* (Cambridge, Mass., 1986); P. M. Harman, *Energy, Force and Matter: The Conceptual Development of Nineteenth-Century Physics* (Cambridge, 1982); Floyd W. Matson, *The Broken Image: Man, Science and Society* (New York, 1964); B. Alan Wallace, *Choosing Reality: A Contemplative View of Physics and the Mind* (Boston, 1989).

3. "Common Sense as the Obstacle to the Progress of Modern Physics," *Current Opinion* 68 (April 1920):501–2; José Ortega y Gasset, "Adam en el Paraiso," in *Obras completas* (1910; Madrid, 1946), 1:471.

4. Marshall Berman, "'All that is solid melts into air': Marx, Modernism, and Modernization," *Dissent* 25 (Winter 1978):54–73; Gertrude Stein, *Picasso* (1938; Boston, 1959), 21; Gertrude Stein, "Realism in Novels," 4, unpublished manuscript, Gertrude Stein Papers, Beinecke Library, Yale University. Modernists, said a sympathetic art critic, looked "beneath the shell of facts" to perceive "the real inwardness of life as opposed to that outward reality glorified by the realists." J. Nilson Laurvik, "The Greatest Exhibition of Insurgent Art Ever Held," *Current Opinion* 54 (March 1913):231.

There is a vast literature on modernism. I found the following sources most useful: Daniel Bell, *The Cultural Contradictions of Capitalism* (New York, 1976); Herschel B. Chipp, ed., *Theories of Modern Art* (Berkeley, 1968); Astradur Eysteinsson, *The Concept of Modernism* (Ithaca, 1990); Robert Hughes, *The Shock of the New* (New York, 1981); Donald M. Lowe, *History of Bourgeois Perception* (Chicago, 1982); Daniel J. Singal, "Towards a Definition of American Modernism," *American Quarterly* 39 (Spring 1987):7–26; Wylie Sypher, *Loss of the Self in Modern Literature and Art* (New York, 1962).

5. John Francis Strauss, *Camera Notes* 5 (July 1901):30; Wallace Stevens to Elsie Moll, 17 January 1909, in *Letters of Wallace Stevens*, ed. Holly Stevens (New York, 1966), 122.

6. WJ, "The Meaning of Truth," in *The Works of William James—The Meaning of Truth*, ed. Frederick Burkhardt (Cambridge, Mass., 1975), 40.

7. WJ, *The Varieties of Religious Experience* (1902; New York, 1929), 509; WJ, *A Pluralistic Universe* (1909; Cambridge, Mass., 1977), 222–23; WJ, "The Notion of Consciousness," in *The Works of William James—Essays in Radical Empiricism* (Cambridge, Mass., 1976), 110; WJ, "Percept and Concept," in *The Works of William James—Some Problems in Philosophy* (Cambridge, Mass., 1979), 56.

8. R.W.B. Lewis, *The Jameses: A Family Narrative* (New York, 1991), 444; WJ, *Principles of Psychology* (New York, 1890), 1:239.

9. HJ, *Portrait of a Lady* (1881; New York: Penguin, 1984), 253; HJ, *The Spoils of Poynton* (New York, 1908), 24, 30–31.

10. Robert P. Falk, *The Victorian Mode in American Fiction, 1865–1885* (East Lansing, Mich., 1964), 17; HJ, "The Art of Fiction," *Longman's Magazine* 4 (September 1884):502–21. For an incisive exploration of the impact of the "new" psychology on literature, see Judith Ryan, *The Vanishing Subject: Early Psychology and Literary Modernism* (Chicago, 1991), 75–88.

11. HJ, *The Ambassadors* (1903; Oxford, 1985), 63.

12. Ibid., 201. For a contextual reading of *The Ambassadors*, see Robert Dawidoff, *The Genteel Tradition and the Sacred Rage: High Culture Versus Democracy in Adams, James & Santayana* (Chapel Hill, 1992), 75–141.

13. HJ, "The Lesson of Balzac," AM 96 (August 1905):166–80.

14. HB, *An Introduction to Metaphysics*, trans. T. E. Hulme (London, 1913), 8; HB, "On the Pragmatism of William James," in *The Creative Mind* (New York, 1946), 255; HB, *Time and Free Will*, trans. F. L. Pogson (1910; London, 1950), 132–33.

15. HB, *An Introduction to Metaphysics* (1903; New York, 1949), 24; HB, *Creative Evolution* (London, 1907), 401; Harold A. Larrabee, ed., *Selections from Bergson* (New York, 1949), 49. On the differences between James and Bergson, see Horace Kallen, *William James and Henri Bergson: A Study in Contrasting Theories of Life* (Chicago, 1915).

16. HB, "What Is the Object of Art?" *Camera Work* (August 1912):24; Marcel Proust, *Time Regained*, trans. Stephen Hudson (London, 1951), 239–40, 429, 430. See also Shiv K. Kumar, *Bergson and the Stream of Consciousness Novel* (New York, 1963); Wylie Sypher, *Loss of the Self in Modern Art and Literature* (Westport, Conn., 1962), 60–61.

17. Ralph Barton Perry, *The Thought and Character of William James* (Boston, 1935):2:605–34; Charles Leonard Moore, "The Return of the Gods," *Dial* (16 November 1912):372; Dewey quoted in *Selections from Bergson*, ed. Harold Larrabee (New York, 1949), xix; HB, *Matter and Memory* (London, 1950), 264–65; Alvan F. Sanborn, "Bergson: Creator of a New Philosophy," *Outlook* 103 (15 February 1913):355.

An intense young journalist and progressive reformer named Walter Lippmann said that Bergson "is to thought what Roosevelt is to action: a fountain of energy, brilliant, terrifying, and important." Walter Lippmann, "The Most Dangerous Man in the World," *Everybody's* 27 (July 1912):100. For other contemporary assessments of Bergson, see Hutchins Hapgood, "Science and Intuition," *New York Globe and Commercial Advertiser*, 18 October 1912; "An Impression of Bergson," *New York Globe and Commercial Advertiser*, 19 February 1913. Thomas Quirk provides a revealing overview of Bergson's influence in "Bergson in America," *Prospects* 7 (1987):453–90.

18. "The Birth of a New Art," *Independent* 78 (6 April 1914):8. "The motion picture camera," reported one journalist, had "brought time under the same control as space." See "The Motion Picture Movement," *Independent* 72 (11 January 1912):109.

19. Léger quoted in John Russell, *The Meanings of Modern Art* (New York, 1974), 104; McCameron quoted in "Change in Our Skyline Foretells a New Art," *New York Times*, 5 January 1913.

20. WJ quoted in Ralph Barton Perry, *The Thought and Character of William James*, 2 vols. (Boston, 1936), 2:168; Smith Ely Jelliffe, "Sigmund Freud and Psychiatry: A

Partial Appraisal," *American Journal of Sociology* 45 (November 1939):326–40; Sigmund Freud, "An Autobiographical Study," in *The Complete Psychological Works of Sigmund Freud*, ed. James Starchey (London, 1955–1966), 20:52. The impact of Freud in America is the focus of Nathan G. Hale, *Freud and the Americans: The Beginnings of Psychoanalysis in the United States, 1876–1917* (New York, 1971), and F. H. Matthews, "The Americanization of Sigmund Freud: Adaptations of Psychoanalysis Before 1917," *Journal of American Studies* 1 (April 1969):39–62.

21. Sigmund Freud, *The Interpretation of Dreams*, trans. A. A. Brill (New York, 1931), 463.

22. Glaspell quoted in Justin Kaplan, *Lincoln Steffens: A Biography* (New York, 1974), 191; Walter Pach, *Queer Thing, Painting: Forty Years in the World of Art* (New York, 1938), 45.

23. Hutchins Hapgood, "The Insurgents in Art," *New York Globe*, 24 October 1911.

24. Dorothy Richardson, *Pilgrimage I* (1913–38; London, 1979), 1:9, 4:93.

25. Walter Pach, "The Point of View of the 'Moderns,'" *Century* 87 (April 1914):864. An American art critic announced that such art-for-art's-sake modernism substituted "a new individualism for the old realism," and this "new individualism" drew much of its anarchic vitalism from the insights of the physical and social sciences. J. Nilsen Laurvik, *Is It Art? Post-Impressionism, Futurism, Cubism* (New York, 1913), 19.

26. Townsend Ludington, *Marsden Hartley: The Biography of an American Artist* (Boston, 1992), 114; Hartley to AS, December 1912, Alfred Stieglitz Papers, Beinecke Library, Yale University; Hartley to Rockwell Kent, 22 September 1912, Rockwell Kent Papers, Archives of American Art, Washington, D.C.; Marsden Hartley, "A Painter's Faith," *Seven Arts* 1 (August 1917):506. For a lucid analysis of the relationship between modern science and modern art, see Leonard Shlain, *Art and Physics: Parallel Visions in Space, Time and Light* (New York, 1991).

27. Roxanna Robinson, *Georgia O'Keefe: A Life* (New York, 1989), 58; Bram Dijkstra, "America and Georgia O'Keefe," in *Georgia O'Keefe: The New York Years* (New York, 1991), 109; Wendy Slatkin, ed., *The Voices of Women Artists* (Englewood Cliffs, N.J., 1993), 223. See also Benita Eisler, *O'Keefe and Stieglitz: An American Romance* (New York, 1993), and Charles C. Eldredge, *Georgia O'Keefe: American and Modern* (New Haven, 1993).

28. Laurvik, "Greatest Exhibition of Insurgent Art," 231, 232. The American painter Andrew Dasburg, for example, explained that he eliminated all "recognizable objects" from his paintings in order to "obtain a pure aesthetic emotion." His compelling interest was in the "improbable" and the "invisible" rather than the "real." Dasburg quoted in George Soule, "The Musical Analogy in Painting," *NR* 6 (15 April 1916):284.

29. *John Sloan's New York Scene*, ed. Bruce St. John (New York, 1965), 402. Those interested in exploring the emergence of modernism in American culture should consult the following: Edward Abrahams, *The Lyrical Left: Randolph Bourne, Alfred Stieglitz and the Origins of Cultural Radicalism in America* (Charlottesville, 1986); Robert Crunden, *American Salons: Encounters with European Modernism, 1885–1917* (New York, 1992); William Innes Homer, *Alfred Stieglitz and the American Avant-garde* (Boston, 1977); Henry May, *The End of American Innocence: A Study of the First Years of Our Own Time, 1912–1917* (New York, 1959); Daniel J. Singal, ed., *Modernist Culture in America* (Belmont, Calif., 1991); Steven Watson, *Strange Bedfellows: The First American Avant-garde* (New York, 1990); Arthur Wertheim, *The New York Little Renaissance: Iconoclasm, Modernism, and Nationalism in American Culture, 1908–1917* (New York, 1976).

30. AS, Editor's Statement, *Camera Work* no. 18 (April 1907):37; "Some Remarkable Work . . .," *New York Evening Sun*, 27 April 1912; "Paintings by Young Americans," *Camera Work* no. 30 (April 1910):54. The best study of Stieglitz remains Homer, *Alfred Stieglitz*.

31. "Alfred Stieglitz, Artist and His Search for the Human Soul," *New York Herald*, 8 March 1908, p. 8.

32. AS, *Twice a Year* (Fall, Winter 1940):138, 142.

33. AS, "The Hand Camera—Its Present Importance," *American Annual of Photography for 1897*, 19–26; TD, "The Camera Club of New York," *Ainslee's* 4 (October 1899):329. The striking "tone of reality" in Stieglitz's urban scenes so captivated Dreiser that he wrote three fulsome articles about the photographer.

34. AS, "Our Illustrations," *Camera Work* no. 49/50 (June 1917):36; AS, "The Pictures in This Number," *Camera Work* no. 1 (January 1903):63; Dorothy Norman, *Alfred Stieglitz, Portrait of an American Seer* (New York, 1960), 25. (Note: People have questioned the authenticity of many of the statements Norman attributes to Stieglitz.)

35. Joseph T. Keiley, "The Photo-Secession Exhibition of the Pennsylvania Academy of Fine Arts," *Camera Work* no. 16 (October 1906):51.

36. TD, "A Remarkable Art: The New Pictorial Photography," *Great Round World* 19 (3 May 1902):434. On Stieglitz's transition from realist to modernist, see Neil Leonard, "Alfred Stieglitz and Realism," *Art Quarterly* 29 (1966):277–86.

37. "Two Academies," *New York Evening Sun*, 25 March 1916; AS to W. Orison Underwood, 22 December 1911, Stieglitz Papers.

38. Frederick A. Sweet, *Miss Mary Cassatt* (Norman, Okla., 1966), 96; Monroe quoted in Homer, *Alfred Stieglitz*, 122.

39. Milton Brown, *The Story of the Armory Show* (Greenwich, Conn., 1963), 30.

40. Kuhn quoted in Garnett McCoy, "Reaction and Revolution, 1900–1930," *Art in America* 53 (August 1965):70; William Innes Homer, *Robert Henri and His Circle* (Ithaca, 1969), 166. See also Martin Green, *New York 1913: The Armory Show and the Paterson Strike Pageant* (New York, 1990); Judith K. Zilczer, "The Armory Show and the American Avant-garde: A Reevaluation," *Arts Magazine* 53 (September 1978): 126–30.

41. Kuhn quoted in Barbara Rose, *American Art Since 1900: A Critical History* (New York, 1967), 70.

42. John Weichsel, "Cosmism or Amorphism," *Camera Work* no. 43 (July 1913):71; "History of Modern Art at the International Exhibition Illustrated by Paintings and Sculpture," *New York Times*, 23 February 1913, sec. 5, p. 15.

43. Kenyon Cox, "The 'Modern' Spirit in Art," *HW* (15 March 1913): 10; Frank Jewett Mather, "Newest Tendencies in Art," *Independent* 74 (6 March 1913):504–12; *New York Times*, 16 March 1913; Cox, "Modern Spirit in Art," 10. See also J. Meredith Neil, "The Impact of the Armory Show," *South Atlantic Quarterly* 79 (1980):375–85.

44. James Huneker, "The Field of Art," *Scribner's* 56 (July 1914):136; Christian Brinton, "Evolution Not Revolution in Art," *International Studio* 49 (April 1913):xxvii–xxxv.

45. AS, "The First Great Clinic to Revitalize Art," *New York American*, 26 January 1913.

46. Bennard B. Perlman, *The Immortal Eight: American Painting from Eakins to the Armory Show* (New York, 1962), 213–14.

47. Stuart Davis, "Autobiography" (New York, 1945), n.p.; Glackens quoted in Charles Hirshfield, "'Ash Can' versus 'Modern' Art in America," *Western Humanities*

Review 10 (Autumn 1956):353–73; Lawson quoted in Ira Glackens, *William Glackens and the Ashcan Group: The Emergence of Realism in American Art* (New York, 1957), 184.

48. Walt Kuhn, "The Story of the Armory Show," *Arts Magazine* (Summer 1984):140; Jerome Myers, *An Artist in Manhattan* (New York, 1940), 36; "The Fantastic George Luks," *New York Herald-Tribune*, 10 September 1933, p. 17. Theodore Dreiser's first contact with cubism befuddled him. In *A Traveler at Forty*, he described cubism as a "radical departure from conventional forms in which, if my impressions are correct, the artist passes from any attempt at transcribing the visible scene and becomes wholly geometric, metaphysical, and symbolic." TD, *A Traveler at Forty* (New York, 1920), 66.

49. "Sloan in Luks Job," *Art Digest* 8 (December 1933):24; John Sloan, unpublished notes, p. 146A, John Sloan Trust, Delaware Art Center; Peter Morse, *John Sloan's Prints* (New Haven, 1969), 386; Henry McBride, "What Is Happening in the World of Art," *New York Sun*, 8 November 1914; André Tridon, "America's First Aesthetician," *Forum* 55 (January 1916):124. See also Charles Hirschfield, "Ash Can Versus 'Modern Art.'

50. Mabel Dodge, "Speculations on Post-Impressionism in Prose," *Arts and Decoration* 3 (1913):172–74, and "Speculations," *Camera Work*, special number (June 1913):7; Floyd Dell to Arthur Davison Ficke, 17 December 1915, Floyd Dell Papers, Newberry Library, Chicago.

51. Charles L. Moore, "Modernity in Literature and the Next Movement," *Dial* 54 (16 February 1913):123; Ezra Pound, "The New Sculpture," *Egoist* 1 (16 February 1914):6–8, and "Affirmations/Vorticism," *New Age* 16 (14 January 1915):277.

52. "Collapse of Howells's Realism in the Light of Freudian Romanticism," *Current Opinion* 63 (October 1917):270; Alexander Harvey, *William Dean Howells: A Study of the Artist* (New York, 1917).

53. Ezra Pound, "Editorial," *Little Review* 4 (May 1917):6; Margaret Anderson, "Argument," *Little Review* 4 (July 1917):29.

54. Frost quoted in Robert Lowell, *Collected Prose* (New York, 1987), 9; Wallace Stevens, *Opus Posthumous: Poems, Plays, Prose*, ed. Samuel French Morse (New York, 1982) 169, 214. See also Tom Quirk, "Realism, the 'Real,' and the Poet of Reality: Some Reflections on American Realists and the Poetry of Wallace Stevens," *American Literary Realism* 21 (Winter 1989):34–53.

55. Charles Caffin, "The Art of Edmund Tarbell," *HM* 117 (June 1908):66.

56. Willard Huntington Wright, *Modern Painting* (New York, 1915), 328, 332, 333, 342, 334.

57. Ibid., 46, 7, 31, 72, 252, 23.

58. James Oppenheim, "Editorial," *Seven Arts* 1 (December 1916):152. "The ailment of modern art," noted Frank Jewett Mather in 1911, "is its isolation from the choicer social forces, and a consequent overvaluation of the mere idiosyncrasy of the artist." Frank Jewett Mather, Jr., "The Present State of Art," *Nation* 93 (14 December 1911):588. John Carey explores this theme in *The Intellectuals and the Masses: Pride and Prejudice Among the Literary Intelligentisia, 1880–1939* (Oxford, Eng., 1991).

59. "Editorial," *Seven Arts* 1 (November 1916):52–53; Waldo Frank, "Concerning a Little Theatre," *Seven Arts* 1 (December 1916):164. For a comprehensive study of the outlook of these young intellectuals during the Progressive era, see Casey Blake, *Beloved Community: The Cultural Criticism of Randolph Bourne, Van Wyck Brooks, Waldo Frank, & Lewis Mumford* (Chapel Hill, 1990). Also useful are Abrahams, *The Lyrical Left*; Thomas Bender, *New York Intellect: A History of Intellectual Life in New York*

City (New York, 1987); Crunden, *Ministers of Reform: The Progressives' Achievement in American Civilization, 1889–1920* (Urbana, Ill., 1982); Charles Forcey, *The Crossroads of Liberalism: Croly, Weyl, Lippmann, and the Progressive Era, 1900–1925* (New York, 1961); Wertheim, *New York Little Renaissance.*

60. Walter Lippmann, *A Preface to Politics* (New York, 1913), 306–7, 32.

61. Walter Lippmann to Van Wyck Brooks, 5 February 1914, Walter Lippmann Papers, Beinecke Library, Yale University.

62. James Oppenheim, "Art, Religion and Science," *Seven Arts* 1 (June 1917):234.

63. George Soule, "Irrelevant Art," *NR* 10 (28 April 1917):375, and "Realism as Confession," *NR* 8 (19 August 1916):64; Harold Stearns, "Nibbling at Realism," *NR* 6 (5 February 1916):24.

64. Walter Lippmann, *Drift and Mastery* (New York, 1914), 151; Lippmann, "The Lost Theme," *NR* 6 (8 April 1916):258–60, "The White Passion," *NR* 8 (21 October 1916):293–95, and "A Reply," *Seven Arts* 1 (January 1917):305.

65. Lippmann, *A Preface to Politics*, 200, 224.

66. James Oppenheim, "Editorial," *Seven Arts* 1 (June 1917):200; Walter Lippmann, *Public Opinion* (New York 1923).

67. Sherwood Anderson, *A Story Teller's Story*, ed. Ray Lewis White (Cleveland, 1968), 297; Anderson, "A Note on Realism," *New York Evening Post Literary Review*, 25 October 1924, 1, and "An Apology for Crudity," *The Dial* 63 (8 November 1917):438.

68. Sherwood Anderson, *Winesburg, Ohio* (1919; New York: Penguin, 1976), 163.

69. Ibid., 51.

70. John Dos Passos, "Against American Literature," *NR* 8 (14 October 1916):269–70; John Dos Passos to Arthur McComb, October 1916, in *John Dos Passos's Correspondence with Arthur McComb*, ed. Melvin Landsberg (Niwot, Colo., 1991), 30.

Bourne likewise evoked the memory of Whitman in an article entitled "Transnational America" in which he announced the failure of the "melting-pot" ideal. The outbreak of war in Europe had set off an explosion of volatile ethnic hatreds in the United States. While lamenting the rise of such deeply embedded social tensions, Bourne welcomed the demise of the "melting pot" image, for he did not want to see the distinctive qualities of America's ethnic mosaic "washed out into a tasteless, colorless fluid of uniformity." Far from ending the "great American democratic experiment," the puncturing of the melting-pot myth had opened the way for a new cosmopolitan federation of diverse peoples. "Let us face realistically the America we have arround us," Bourne urged. "Let us work with the forces that are at work. Let us make something of this trans-national spirit instead of outlawing it." A mature America, Bourne declared, accepted differences as a strength rather than a weakness. "All our idealisms must be those of future social goals in which all can participate, the good life of personality lived in the environment of the Beloved Community." Randolph Bourne, "Trans-national America," *AM* 108 (July 1916):86–97.

71. Brooks, "Our Awakeners," *Seven Arts* 2 (June 1917):236; Brooks, "Toward a National Culture," *Seven Arts* 1 (March 1917):535–47; Brooks, "Our Awakeners," 248.

72. Brooks, "America's Coming-of-Age," in Claire Sprague, ed., *Van Wyck Brooks: The Early Years* (New York, 1968), 95, 128, 185.

73. Blake, *Beloved Community*, 294–95.

74. Randolph Bourne, "Twilight of the Idols," *Seven Arts* 2 (October 1917):688–702, and "The War and the Intellectuals," *Seven Arts* 2 (June 1917):133–46.

75. James Oppenheim, "Editorial," *Seven Arts* 2 (July 1917):340.

76. Ellery Sedgwick, "Impressions of the Fifth Year," *AM* 122 (1918):809; Modris

Eksteins, *Rites of Spring: The Great War and the Birth of the Modern Age* (Boston, 1989), 222. See also Michael C. C. Adams, *The Great Adventure: Male Desire and the Coming of World War I* (Bloomington, Ind., 1990); Stanley Cooperman, *World War I and the American Novel* (Baltimore, 1967); John Ellis, *Eye-Deep in Hell: Trench Warfare in World War I* (Baltimore, 1976); Paul Fussell, *The Great War and Modern Memory* (New York, 1975); Eric J. Leed, *No Man's Land: Combat and Identity in World War I* (1979); Robert Wohl, *The Generation of 1914* (Cambridge, Mass., 1979).

77. John Dos Passos to Arthur McComb, September 1917, *Dos Passos's Correspondence*, 71; "War and the Future of World Literature," *Current Opinion* 66 (1919):250.

78. Floyd Dell, "Randolph Bourne," *NR* 12 (4 January 1919):276; Oppenheim, "Editorial," 267–68.

79. Harold Stearns, "America and the Young Intellectual," *Bookman* 53 (March 1921):47; Marsden Hartley, "A Painter's Faith," *Seven Arts* 1 (August 1917):506. "The war and the modern outlook," a writer in *Current Opinion* reported in 1920, "have created, are creating, and must create a new world for us." See "The Safe and Sane Genius of William Dean Howells," *Current Opinion* 69 (July 1920):94.

80. Ernest Hemingway, *A Moveable Feast* (New York, 1964), 29; Robert Herrick, "In General," *Nation* 113 (7 December 1921):658; Van Wyck Brooks, "The Influence of William James," in *Sketches in Criticism* (New York, 1932), 37.

81. "America Turned Realist," *Nation* 108 (4 January 1919):4; St. John Ervine, "Literary Taste in America," *NR* 24 (6 October 1920):147.

82. Waldo Frank, *Our America* (New York, 1919), 9, 96, 156, 157; "The 'Nation' of the Present," *Nation* 101 (8 July 1915):34.

83. H. W. Boynton, "Realism and Recent American Fiction," *Nation* 102 (16 April 1916):380; Nancy Barr Mavity, "A Word About Realism," *Dial* 66 (28 January 1919): 637; Edward Sapir, "Realism in Prose Fiction," *Dial* 63 (22 November 1917):503; Robert Lynd, "Literary Reality," *Living Age* 286 (July 1915):242–44. H. L. Mencken and George Jean Nathan declared in the inaugural issue of their new magazine, *American Mercury*, that their objective was "to attempt a realistic presentation of the whole gaudy, gorgeous American scene." "Editorial," *American Mercury* 1 (January 1924):30.

Epilogue

1. Philip Roth, "Writing American Fiction," *Commentary* 31 (March 1961):223–33.

2. George Steiner, *Real Presences* (Chicago, 1989), 95, 122; Sven Birkerts, "American Fictions," *Wilson Quarterly* 16 (Spring 1992):103.

3. Ronald Sukenick, "The Death of the Novel," in *The Death of the Novel and Other Stories* (New York, 1969), 41. Like his hero Wallace Stevens, Sukenick defines postmodern art as the "invention of reality." Sukenick, "On the New Cultural Conservatism," *Partisan Review* 39 (Summer 1972):450.

4. William H. Gass, *Fiction and the Figures of Life* (New York, 1970), 282; Frank McConnell, "Will Deconstruction Be the Death of Literature?," *Wilson Quarterly* 14 (Winter 1990):108. See also Raymond Federman, "Surfiction—A Position," *Partisan Review* 40 (1973):427–37; Ibn Hassan, *The Dismemberment of Orpheus: Towards a Postmodern Literature* (New York, 1976).

"If the world is absurd," Jerome Klinkowitz has written, "if what passes for reality is distressingly unreal, why spend time representing it?" This "postmodern" insistence that reality is created rather than discovered has permeated all of the arts. In the postmodern scheme of things, as the architectural historian Ada Louise Huxtable reported in 1992, "form follows feeling; desire, not utility, dictates design." Postmodern design has become a function of "dream, invention, wish-fulfillment." Jerome Klinkowitz, *Literary Disruptions: The Making of a Post-Contemporary American Fiction* (Urbana, Ill., 1975), 32; Ada Louise Huxtable, "Inventing American Reality," *New York Review of Books*, 3 December 1992, p. 28.

5. Robert Scholes, *The Fabulators* (New York, 1967), 20, 21; George Levine, "Scientific Realism and Literary Representation," *Raritan* 10 (Spring 1991):18. See also Robert Anchor, "Realism and Ideology: The Question of Order," *History and Theory* 22 (1983):107–19. Critical theory, observes the philosopher Richard Rorty, has become for many practitioners "a means to private perfection rather than to human solidarity." Richard Rorty, *Contingency, Irony, and Solidarity* (Cambridge, 1989), 96.

Recent developments in "critical theory" are summarized in Steven Connor, *Postmodernist Culture: An Introduction to Theories of the Contemporary* (London, 1990); Jonathan Culler, *On Deconstruction: Theory and Criticism After Structuralism* (New York, 1982); Terry Eagleton, *Literary Theory: An Introduction* (Minneapolis, 1983). A feisty defense of realism against the assaults of critical theory is contained in Raymond Tallis, *In Defense of Realism* (London, 1988). Other sympathetic treatments of realism include Gerald Graff, *Literature Against Itself: Literary Ideas and Modern Society* (Chicago, 1979); George Levine, *The Realistic Imagination: English Fiction from "Frankenstein" to "Lady Chatterley"* (Chicago, 1981); and C. Prendergast, *The Order of Mimesis* (Cambridge, 1988). For an acerbic critique of critical theory, see Gertrude Himmelfarb, "Telling It as You Like It: Postmodernist History and the Flight from Fact," *Times Literary Supplement* (16 October 1992):12–15.

6. Bliss Perry, *A Study of Prose Fiction* (New York, 1902), 255; Richard Estes, "Artists in Focus," *American Artist* 50 (October 1986):92. In 1978 art critic Hilton Kramer affirmed that realism remains a major cultural force: "No one with a wide knowledge of contemporary art any longer doubts that Realism is now firmly established as a viable creative enterprise." Hilton Kramer, "The Return of Realism," *New York Times*, 12 March 1978, sec. 2, p. 1. See also "Editorial," *Art in America* (September 1981):5; John Canaday, "Painters Who Put the Real World into Sharp Focus," *Smithsonian* 12 (October 1981):69; "Revival of Realism," *Newsweek* 99 (7 June 1982):64; Nancy Grimes, "Facts of Life," *ARTnews* 87 (December 1988):118; Richard Lack, "Classical Realism: The Other Twentieth Century," *Utne Reader* (July/August 1989):59; Sven Birkerts, "American Fictions," 103.

7. Tom Wolfe, "Stalking the Billion-Footed Beast," *Harper's Magazine* 279 (November 1989): 50; Eudora Welty, *The Eye of the Story* (New York, 1979), 128.

8. Jill Susser, "Phillip Pearlstein: The Figure as Still Life," *American Artist* (February 1973):38; Pat Mainardi, "Philip Pearlstein: Old Master to the New Realists," *ARTnews* 75 (November 1976):72.

9. Carl Van Doren, "American Realism," *NR* 34 (21 March 1923):107; George Santayana, "The Genteel Tradition in American Philosophy," in *Winds of Doctrine* (1913; New York, 1957), 211; Gordon Lloyd Harper, "Interview with Saul Bellow," *Writers at Work: The Paris Review Interviews* (New York, 1967), 180.

10. Cynthia Ozick, "Where Orphans Can Still Become Heiresses," *New York Times Book Review* 92 (8 March 1987):13.

11. Ludwig Lewisohn, "The New Literature in America," *Nation* 112 (23 March 1921):429.

12. Willard Midgette, "Illusionistic Portraiture," *Allan Frumkin Gallery Newsletter*

no. 3 (Spring 1977):5; Irving Howe, "The Treason of the Critics," NR 200 (12 June 1989):30. When people stop being attracted to such empathetic representation of others, literary historian John Seelye has written, "we will know that innocence is indeed dead, and hope, and all." John Seelye, "Wyeth and Hopper," NR 166 (11 March 1972):18. See also Howe, "The Human Factor," NR 200 (8 May 1989):30–34.

13. John Updike, "Howells as Anti-Novelist," *New Yorker* 63 (13 July 1987):88. Howells made a similar statement in 1900: "I do not think it by any means a despicable thing to have hit the fancy of our enormous commonplace average." WDH, "The New Historical Romances," NAR 171 (October 1900):943.

INDEX